Instructor's Manual

Fundamentals of ANALYTICAL CHEMISTRY

7th Edition

Douglas A. Skoog
Stanford University

Donald M. West
San Jose State University

F. James Holler
University of Kentucky

SAUNDERS GOLDEN SUNBURST SERIES

Saunders College Publishing
Harcourt Brace College Publishers

Fort Worth Philadelphia San Diego New York Orlando Austin
San Antonio Toronto Montreal London Sydney Tokyo

Copyright ©1996, 1992, 1988 by Saunders College Publishing

All rights reserved. No part of this publication may be reproduced or transmitted in any form or by any means, electronic or mechanical, including photocopy, recording, or any information storage and retrieval system, without permission in writing from the publisher, except that, until further notice, the contents or parts thereof may be reproduced for instructional purposes by users of FUNDAMENTALS OF ANALYTIC CHEMISTRY, Seventh Edition by Douglas A. Skoog, F. James Holler and Donald M. West.

Printed in the United States of America.

Skoog, West & Holler; Instructor's Manual to accompany
Fundamentals of Analytic Chemistry, 7E.

ISBN 0-03-015688-2

67 017 98765432

PREFACE

This manual is designed as a teaching supplement to *Fundamentals of Analytical Chemistry*, 7th edition. It contains answers to all of the questions appearing at the end of each chapter of the text as well as detailed solutions to all of the problems.

In the solutions to problems we have tried to indicate the uncertainty associated with answers by following the significant figure convention in which the significant figures in a number are all the certain digits plus the first uncertain digit. However, we follow the practice of postponing rounding until the end of a computation in order to avoid accumulating rounding errors. Thus, intermediate results in a computation will often contain several digits that are not significant. In some solutions, we indicate our estimate of the absolute standard deviation of the result. For example, the result $3.82 (\pm 0.02) \times 10^{-3}$ indicates that our estimate of the standard deviation of the computed result is 0.02×10^{-3}.

Preparation of a solutions manual such as this that is entirely free of mistakes and errors, either typographical or computational, is unfortunately impossible. We would appreciate being notified of corrections so that we can incorporate them in later printings of the manual.

Douglas A. Skoog
Donald M. West
F. James Holler
August, 1995

CHAPTER 2

2-1 (a) *Constant errors* are the same magnitude regardless of the sample size. *Proportional errors* are proportional in size to the sample size.

(b) *Random errors* result from uncontrolled variables in an experiment while *systematic errors* are those which can be ascribed to a particular cause and can usually be determined.

(c) The *mean* is the sum of the measurements in a set divided by the number of measurements. The *median* is the central value for a set of data; half of the measurements are larger and half are smaller than the median.

(d) The *absolute error E* and the *relative error* E_r in a measured value x_i are given by the expressions

$$E = x_i - x_t \quad \text{and} \quad E_r = (x_i - x_t)/x_t$$

where x_t is the accepted value.

2-2 (1) Random temperature fluctuations causing random changes in the length of the metal rule; (2) uncertainties arising from having to move and position the rule twice; (3) personal judgment in reading the rule; (4) vibrations in the table and/or rule; (5) uncertainty in locating the rule perpendicular to the edge of the table.

2-3 The three types of systematic error are *instrumental error, method error, and personal error*.

2-4 *Systematic method errors* are detected by (1) analysis of standard samples, (2) independent analysis, or (3) blank determination.

2-5 Constant errors.

2-6 (a) $\dfrac{-0.3 \text{ mg}}{800 \text{ mg}} \times 100\% = -0.038\% = \underline{\underline{-0.04\%}}$

(b) $\dfrac{-0.3 \text{ mg}}{500 \text{ mg}} \times 100\% = \underline{\underline{-0.06\%}}$

(c) $\dfrac{-0.3 \text{ mg}}{100 \text{ mg}} \times 100\% = \underline{\underline{-0.3\%}}$

Chapter 2

(d) $\dfrac{-0.3 \text{ mg}}{25 \text{ mg}} \times 100\% = \underline{\underline{-1.2\%}}$

2-7 (a) $\dfrac{0.3 \text{ mg Au lost}}{\text{mg Au in sample}} \times 100\% = 0.2\%$ error

$\dfrac{0.3 \text{ mg Au} \times 100\%}{0.2\% \text{ error}} = 150 \text{ mg Au in sample}$

$\dfrac{150 \text{ mg Au in sample}}{\text{mg sample}} \times 100\% = 1.2\%$ Au in sample

$\dfrac{150 \text{ mg Au in sample} \times 100\%}{1.2\% \text{ Au in sample}} = 1.25 \times 10^4 \text{ mg sample}$ or $\underline{\underline{13 \text{ g sample}}}$

Proceeding in the same way, we obtain

(b) $\underline{\underline{5 \text{ g}}}$ (c) $\underline{\underline{3 \text{ g}}}$ (d) $\underline{\underline{2 \text{ g}}}$

2-8 (a) $E_r = \dfrac{0.03 \text{ mL}}{50.00 \text{ mL}} \times 100\% = \underline{\underline{0.06\%}}$

Proceeding in the same way, we obtain

(b) $\underline{\underline{0.3\%}}$ (c) $\underline{\underline{0.12\%}}$ (d) $\underline{\underline{0.075\%}}$

2-9 (a) $E_r = \dfrac{-0.4 \text{ mg}}{40 \text{ mg}} \times 100\% = \underline{\underline{-1.0\%}}$

Proceeding in the same way, we obtain

(b) $\underline{\underline{-0.23\%}}$ (c) $\underline{\underline{-0.10\%}}$ (d) $\underline{\underline{-0.067\%}}$

CHAPTER 3

3-1 (a) The *spread* or *range* for a set of replicate data is the numerical difference between the highest and lowest value.

(b) The *coefficient of variation* is the percent relative standard deviation of $(s/\bar{x}) \times 100\%$.

(c) *Significant figures* include all of the digits in a number that are known with certainty plus the first uncertain digit.

(d) A *Gaussian distribution* or *normal distribution* is described by the bell-shaped curve obtained by plotting frequency versus deviation from the mean for measurements that conform to such a distribution.

3-2 (a) The *sample variance*, s^2, is given by the expression

$$s^2 = \frac{\sum_{i=1}^{N}(x_i - \bar{x})^2}{N-1}$$

where $\bar{x}$ is the sample mean.

The *sample standard deviation* is given by

$$s = \sqrt{\frac{\sum_{i=1}^{N}(x_i - \bar{x})^2}{N-1}}$$

(b) Both types of mean are obtained by summing all of the data in a replicate set and dividing by the number of data. For a population mean, the number of data are very large and approach an infinite number. For a sample mean, the number of data are small.

(c) *Accuracy* represents the agreement between an experimentally measured value and the true value. *Precision* describes the agreement among measurements that have been performed in exactly the same way.

Chapter 3

    (d) *Random errors* result from uncontrolled variables in an experiment while *systematic errors* are those which can be ascribed to a particular cause and can usually be determined.

3-3 (a) In statistics, a sample is a small set of replicate measurement. In chemistry, a sample is a portion of a material that is used for analysis.

    (b) The sample standard deviation s is given by

$$s = \sqrt{\frac{\sum_{i=1}^{N}(x_i - \bar{x})^2}{N-1}}$$

The population standard deviation σ is given by

$$\sigma = \sqrt{\frac{\sum_{i=1}^{N}(x_i - \mu)^2}{N}}$$

where $\bar{x}$ is the mean of a small sample of data, μ is the mean of a population of data that approaches infinite, N is the number of data, and x_i is the individual data.

3-4 The standard error of a mean s_m is the standard deviation of the set s divided by the square root of the number of data in the set. That is, $s_m = s/\sqrt{N}$.

3-5 For A:

x_i	x_i^2
2.4	5.76
2.1	4.41
2.1	4.41
2.3	5.29
1.5	2.25
$\Sigma x_i = 10.4$	$\Sigma x_i^2 = 22.12$

(a) $\bar{x}$ = 10.4/5 = 2.08 = <u>2.1</u>

(b) median = 2.1

(c) w = 2.4 - 1.5 = 0.9

(d) $s = \sqrt{\dfrac{\Sigma x_i^2 - (\Sigma x_i)^2/N}{N-1}} = \sqrt{\dfrac{22.12 - (10.4)^2/5}{4}} = 0.3493 = 0.35$

(e) CV = $(s/\bar{x}) \times 100\%$ = $(0.3493/2.08) \times 100\%$ = 17%

Results for sets A - F obtained in a similar way are summarized in the table below.

	A	B	C	D	E	F
N	5	3	4	5	4	5
Σx_i	10.4	209.66	0.3672	12.4	278.13	3.022
Σx_i^2	22.12	14652.45	3.38006×10^{-2}	31.06	19339.2131	1.827534
(a) $\bar{x}$	2.08	69.89	0.0918	2.48	69.53	0.604
(b) median	2.1	69.92	0.0894	2.4	69.635	0.607
(c) w	0.9	0.14	0.0116	0.7	0.44	0.042
(d) s	0.35	0.076	0.0055	0.28	0.22	0.016
(e) CV, %	17	0.11	6.0	11	0.31	2.7

3-6 $E = \bar{x} - x_t \qquad E_r = \dfrac{\bar{x} - x_t}{x_t} \times 1000 \text{ ppt}$

For set A:

(a) E = 2.08 − 2.0 = +0.08

(b) $E_r = \dfrac{2.08 - 2.0}{2.0} \times 1000 \text{ ppt} = \underline{40 \text{ ppt}}$

The remaining sets are treated in the same way.

Chapter 3

Set	E	E_r, ppt
A	+ 0.08	+ 40
B	+ 0.14	+ 2.0
C	− 0.0012	− 13
D	− 0.52	− 173
E	+ 0.48	+ 7.0
F	− 0.031	− 49

3-7 (a) $s_y^2 = (0.03)^2 + (0.001)^2 + (0.001)^2 = 9.02 \times 10^{-4}$

$s_y = \sqrt{9.02 \times 10^{-4}} = \underline{0.030}$

$CV = \dfrac{0.0030}{0.572} \times 100\% = \underline{\underline{5.2\%}}$

$y = \underline{\underline{0.57(\pm 0.03)}}$

(b) $s_y^2 = (0.04)^2 + (0.0001)^2 + (0.08)^2 = 8.00 \times 10^{-3}$

$s_y = \sqrt{8.00 \times 10^{-3}} = \underline{0.089}$

$CV = \dfrac{0.089}{21.263} \times 100\% = \underline{\underline{0.42\%}}$

$y = \underline{\underline{21.3(\pm 0.1)}}$

(c) $\left(\dfrac{s_y}{y}\right)^2 = \left(\dfrac{0.3}{67.1}\right)^2 + \left(\dfrac{0.02 \times 10^{-17}}{1.03 \times 10^{-17}}\right)^2 = 3.970 \times 10^{-4}$

$\dfrac{s_y}{y} = 1.99 \times 10^{-2}$ $CV = 1.99 \times 10^{-2} \times 100\% = \underline{\underline{2.0\%}}$

$$s_y = (1.99 \times 10^{-2})(6.9113 \times 10^{-16}) = 0.137 \times 10^{-16} = \underline{\underline{0.14 \times 10^{-16}}}$$

$$y = \underline{\underline{6.9(\pm 0.1) \times 10^{-16}}}$$

(d)
$$\left(\frac{s_y}{y}\right)^2 = \left(\frac{1}{243}\right)^2 + \left(\frac{2}{760}\right)^2 + \left(\frac{0.006}{1.006}\right)^2$$

$$= (1.693 \times 10^{-5}) + (6.925 \times 10^{-6}) + (3.557 \times 10^{-5})$$

$$= 5.943 \times 10^{-5}$$

$$\frac{s_y}{y} = \sqrt{5.943 \times 10^{-5}} = 7.709 \times 10^{-3}$$

$$CV = (7.709 \times 10^{-3}) \times 100\% = \underline{\underline{0.77\%}}$$

$$s_y = (183578.5)(7.709 \times 10^{-3}) = 1.415 \times 10^3 = \underline{\underline{1.4 \times 10^3}}$$

$$y = \underline{\underline{1.84(\pm 0.01) \times 10^5}}$$

(e) $s_{\text{num}}^2 = (6)^2 + (3)^2 = 45$

$s_{\text{num}} = \sqrt{45} = 6.71$

$s_{\text{denom}}^2 = (1)^2 + (8)^2 = 65$

$s_{\text{denom}} = \sqrt{65} = 8.06$

$$y = \frac{143 - 64}{1249 + 77} = \frac{79}{1326} = 5.9578 \times 10^{-2}$$

$$\left(\frac{s_y}{y}\right)^2 = \left(\frac{6.71}{79}\right)^2 + \left(\frac{8.06}{1326}\right)^2 = 7.2511 \times 10^{-3}$$

$$\frac{s_y}{y} = 8.515 \times 10^{-2} \qquad CV = (8.515 \times 10^{-2}) \times 100\% = \underline{\underline{8.5\%}}$$

Chapter 3

$$s_y = (8.515 \times 10^{-2})(5.9578 \times 10^{-2}) = \underline{\underline{5.1 \times 10^{-3}}}$$

$$y = \underline{\underline{6.0(\pm 0.5) \times 10^{-2}}}$$

(f) $\left(\dfrac{s_y}{y}\right)^2 = \left(\dfrac{0.01}{1.97}\right)^2 + \left(\dfrac{3}{243}\right)^2 = 1.782 \times 10^{-4}$

$$\dfrac{s_y}{y} = 1.335 \times 10^{-2} \qquad CV = (1.335 \times 10^{-2}) \times 100\% = \underline{\underline{1.3\%}}$$

$$s_y = (1.335 \times 10^{-2})(8.106996 \times 10^{-3}) = \underline{\underline{1.1 \times 10^{-4}}}$$

$$y = \underline{\underline{8.1(\pm 0.1) \times 10^{-3}}}$$

3-8 (a) $s_y^2 = (0.02 \times 10^{-7})^2 + (0.2 \times 10^{-8})^2 = 8.0 \times 10^{-18}$

$$s_y = 2.8 \times 10^{-9} = \underline{\underline{0.03 \times 10^{-7}}}$$

$$CV = \dfrac{s_y}{y} \times 100\% = \dfrac{2.8 \times 10^{-9}}{1.374 \times 10^{-7}} \times 100\% = \underline{\underline{2\%}}$$

$$y = \underline{\underline{-1.37(\pm 0.03) \times 10^{-7}}}$$

(b) $s_y^2 = (0.08)^2 + (0.06)^2 + (0.004)^2 = 0.010$

$$s_y = \underline{\underline{0.10}}$$

$$CV = \dfrac{0.10}{0.780} \times 100\% = \underline{\underline{13\%}}$$

$$y = \underline{\underline{0.8 \pm 0.1}}$$

(c) $\left(\dfrac{s_y}{y}\right)^2 = \left(\dfrac{0.0005}{0.0010}\right)^2 + \left(\dfrac{0.020}{18.10}\right)^2 + \left(\dfrac{1}{200}\right)^2 = 0.25$

$$\frac{s_y}{y} = 0.50 \qquad CV = 0.50 \times 100\% = \underline{\underline{50\%}}$$

$$s_y = 0.50\,y = (0.50)(3.62) = \underline{\underline{1.8}}$$

$$y = \underline{\underline{4 \pm 2}}$$

(d) $$\left(\frac{s_y}{y}\right)^2 = \left(\frac{0.03 \times 10^{-14}}{1.73 \times 10^{-14}}\right)^2 + \left(\frac{0.04 \times 10^{-16}}{1.63 \times 10^{-16}}\right)^2 = \left(\frac{0.03}{1.73}\right)^2 + \left(\frac{0.04}{1.63}\right)^2$$

$$= (0.0173)^2 + (0.0245)^2 = 9.03 \times 10^{-4}$$

$$\frac{s_y}{y} = 3.00 \times 10^{-2} \qquad CV = (3.00 \times 10^{-2}) \times 100\% = \underline{\underline{3\%}}$$

$$s_y = (3.00 \times 10^{-2})(106.13497) = 3.189 = \underline{\underline{3}}$$

$$y = \underline{\underline{106 \pm 3}}$$

(e) $$\left(\frac{s_y}{y}\right)^2 = \left(\frac{1}{100}\right)^2 + \left(\frac{1}{2}\right)^2 = 0.25$$

$$\frac{s_y}{y} = 0.5 \qquad CV = 0.5 \times 100\% = \underline{\underline{50\%}}$$

$$s_y = (0.5)(50) = 25 = \underline{\underline{2 \times 10^1}}$$

$$y = \underline{\underline{5(\pm 2) \times 10^1}}$$

(f) Numerator: $$s_n^2 = (0.02 \times 10^{-2})^2 + (0.06 \times 10^{-3})^2 = 4.36 \times 10^{-8}$$

$$s_n = 2.09 \times 10^{-4}$$

$$n = (1.43 \times 10^{-2}) - (4.76 \times 10^{-3}) = 9.54 \times 10^{-3}$$

Chapter 3

$$\frac{s_n}{n} = \frac{2.09 \times 10^{-4}}{9.54 \times 10^{-3}} = 2.19 \times 10^{-2}$$

Denominator: $s_d^2 = (0.7)^2 + (0.08)^2 = 0.496$

$$s_d = 0.704$$

$d = 24.3 + 8.06 = 32.26$

$$\frac{s_d}{d} = \frac{0.704}{32.36} = 0.0218$$

$$\left(\frac{s_y}{y}\right)^2 = \left(\frac{s_n}{n}\right)^2 + \left(\frac{s_d}{d}\right)^2 = (0.0219)^2 + (0.0218)^2 = 9.55 \times 10^{-4}$$

$$\frac{s_y}{y} = 3.09 \times 10^{-2} \qquad CV = (3.09 \times 10^{-2}) \times 100\% = \underline{\underline{3.1\%}}$$

$$s_y = (3.09 \times 10^{-2})(2.948 \times 10^{-4}) = \underline{\underline{9.1 \times 10^{-6}}}$$

$$y = 2.94(\pm 0.09) \times 10^{-4} \qquad \text{or} \qquad \underline{\underline{2.9(\pm 0.1) \times 10^{-4}}}$$

3-9 (a) From Equation (4), Table 3-4

$$s_y = 0.434\, s_a/a$$

$$s_y = 0.434 \times (\pm 0.03 \times 10^{-4})/(2.00 \times 10^{-4}) = \underline{\underline{\pm 0.0065}}$$

$$y = \underline{\underline{-3.699(\pm 0.007)}} = \underline{\underline{-3.70 \pm 0.01}}$$

$$CV = \frac{\pm 0.0065}{3.699} \times 100\% = \underline{\underline{\pm 0.18\%}}$$

(b) $s_y = 0.434 \times (\pm 0.01 \times 10^{37})/(4.42 \times 10^{37}) = \pm 9.82 \times 10^{-4} = \underline{\underline{0.001}}$

$$y = \underline{\underline{37.645(\pm 0.001)}}$$

Chapter 3

$$\text{CV} = \frac{\pm 9.82 \times 10^{-4}}{37.645} \times 100\% = \pm 0.0026\% = \underline{\underline{\pm 0.003\%}}$$

(c) From Equation (5), Table 3-4

$$\frac{s_y}{y} = 2.303 s_a = 2303 \times (\pm 0.003) = \pm 0.00691 = \pm 0.007$$

$$s_y = 15.849 \times (\pm 0.00691) = \underline{\underline{\pm 0.11}}$$

$$y = \underline{\underline{15.8(\pm 0.1)}}$$

$$\text{CV} = \pm 0.007 \times 100\% = \underline{\underline{\pm 0.7\%}}$$

(d) $$\frac{s_y}{y} = 2.303 \times 0.04 = \pm 0.092$$

$$s_y = 3.467 \times 10^{49} \times (\pm 0.092) = \underline{\underline{\pm 0.32 \times 10^{49}}}$$

$$y = \underline{\underline{3.5(\pm 0.3) \times 10^{49}}}$$

$$\text{CV} = (\pm 0.092) \times 100\% = \underline{\underline{\pm 9.2\%}}$$

3-10 From Equation (3), Table 3-4

$$\frac{s_y}{y} = x \frac{s_a}{a}$$

where x is the numerical exponent.

(a) $$\frac{s_y}{y} = 3 \times \frac{(\pm 0.03 \times 10^{-4})}{4.73 \times 10^{-4}} = \pm 0.019$$

$$\text{CV} = \pm 0.019 \times 100\% = \underline{\underline{\pm 1.9\%}}$$

$$s_y = (\pm 0.019) \times 105.82 = \underline{\underline{\pm 2.0}}$$

Chapter 3

$$y = \underline{\underline{105(\pm 2)}}$$

(b) $\dfrac{s_y}{y} = \dfrac{1}{4} \times \dfrac{(\pm 0.002)}{2.145} = \pm 2.33 \times 10^{-4}$

$CV = \pm 2.33 \times 10^{-4} \times 100\% = \underline{\underline{\pm 0.023\%}}$

$s_y = 1.210199 \times (\pm 2.33 \times 10^{-4}) = \underline{\underline{\pm 2.8 \times 10^{-4}}}$

$y = \underline{\underline{1.210(\pm 0.0003)}}$

3-11 For diameter measurement

$$\bar{d} = \dfrac{5.4 + 5.2 + 5.5 + 5.2}{4} = 5.33 \text{ m}$$

$$s_d = \sqrt{\dfrac{(0.070)^2 + (-0.13)^2 + (0.17)^2 + (-0.13)^2}{4-1}} = 0.150 \text{ m}$$

For height measurement

$$\bar{h} = (9.8 + 9.9 + 9.6)/3 = 9.77 \text{ m}$$

$$s_h = \sqrt{\dfrac{0.030)^2 + (0.133)^2 + (-0.170)^2}{2}} = 0.153 \text{ m}$$

$$V_m = \pi \left(\dfrac{d}{2}\right)^2 h = \pi \times \left(\dfrac{5.33}{2}\right)^2 \times 9.77 = 218.0 \text{ m}^3$$

$$V_L = 218.0 \text{ m}^3 \times \dfrac{1 \text{ L}}{10^{-3} \text{ m}^3} = 2.18 \times 10^5 \text{ L}$$

$$\dfrac{s_d}{d} = \dfrac{0.150}{5.33}$$

For $d^2/2$

$$s_{d^2} = \frac{2 \times 0.150}{5.33} \quad \text{[see Equation (3), Table 3-4]}$$

The relative standard deviation of the calculated volume is given by

$$(s_V)_r = \sqrt{(s_{d^2})^2 + (s_h)^2} = \sqrt{\left(\frac{2 \times 0.150}{5.33}\right)^2 + \left(\frac{0.153}{9.77}\right)^2} = 0.0584$$

and the absolute standard deviation for the calculated volume is

$$s_V = 0.0584 \times 2.18 \times 10^5 \, L = \underline{\underline{0.13 \times 10^5 \, L}}$$

$$V_L = \underline{\underline{2.2(\pm 0.1) \times 10^5 \, L}}$$

3-12 (a) $T = (0.273 + 0.276 + 0.274 + 0.268)/4 = 0.2728 = 0.273$

$$c_X = \frac{-\log 0.2728}{2505} = 2.252 \times 10^{-4} \, M = \underline{\underline{2.25 \times 10^{-4} \, M}}$$

(b) To obtain the standard deviation of T, we employ Equation 3-3 and write

$$s_T = \sqrt{\frac{(0.00020)^2 + (0.00332)^2 + (-0.0048)^2 + (0.0012)^2}{4-1}} = \pm 0.0034$$

The standard deviation for $-\log T$ is given by Equation (4) in Table 3-4. Thus

$$s_y = \frac{0.434 \times (\pm 0.0034)}{-\log 0.2728} = \pm 0.00262$$

The relative standard deviation of c_X is given by Equation (2), Table 3-4

$$\frac{s_c}{c_X} = \sqrt{\left(\frac{\pm 0.00262}{-\log 0.2728}\right) + \left(\frac{\pm 12}{2505}\right)^2} = 0.0067$$

and

$$s_c = 0.0067 \times 2.252 \times 10^{-4} = \underline{\underline{\pm 0.015 \times 10^{-4}}}$$

Chapter 3

Thus, c_X should be written

$$c_X = 2.25(\pm 0.02) \times 10^{-4} \text{ M}$$

(c) CV $= 0.00667 \times 100\% = \underline{\underline{0.67\%}}$

3-13 (a) For each set,

$$s = \sqrt{\frac{\sum_{i=1}^{N}(x_i - \bar{x})^2}{N-1}}$$

$$s_1 = \sqrt{\frac{(0.13)^2 + (0.09)^2 + (0.07)^2 + (0.05)^2 + (0.06)^2}{4}}$$

$$= \sqrt{\frac{0.0360}{4}} = 0.0949 = \underline{\underline{0.095}}$$

The results for all of the sets are summarized below.

Sample	$\sum(x_i - \bar{x})^2$	s, % K^+
1	0.0360	0.095
2	0.0289	0.12
3	0.0378	0.11
4	0.0326	0.10
5	0.0435	0.10
Total	0.1788	

(b)
$$s_{\text{pooled}} = \sqrt{\frac{0.036 + 0.0289 + 0.0378 + 0.0326 + 0.0435}{5 + 3 + 4 + 4 + 5 - 5}}$$

$$= \sqrt{\frac{0.1788}{16}} = 0.1057 = \underline{\underline{0.11\% \text{ K}^+}}$$

3-14 (a) $s_1 = \sqrt{\dfrac{(0.050)^2 + (0.10)^2 + (0.08)^2}{2}} = \sqrt{\dfrac{0.0189}{2}} = 0.0972 = \underline{\underline{0.097}}$

The results for all sets are summarized below.

Set	$\sum(x_i - \bar{x})^2$	s
1	0.0189	0.097
2	0.0178	0.077
3	0.0282	0.084
4	0.0242	0.090
5	0.0230	0.107
6	0.0205	0.083
Total	0.1326	

(b) $s_{\text{pooled}} = \sqrt{\dfrac{0.0189 + 0.0178 + 0.0282 + 0.0242 + 0.0230 + 0.0205}{23 - 6}}$

$= \sqrt{\dfrac{0.1326}{17}} = \underline{\underline{0.088}}$

3-15

Sample	$\bar{x}$	$\sum(x_i - \bar{x})^2$	Sample	$\bar{x}$	$\sum(x_i - \bar{x})^2$
1	2.255	4.50×10^{-4}	6	1.045	1.25×10^{-3}
2	8.55	4.50×10^{-2}	7	14.60	8.00×10^{-2}
3	7.55	5.00×10^{-3}	8	21.50	3.20×10^{-1}
4	12.25	2.45×10^{-1}	9	8.60	8.00×10^{-2}
5	4.25	5.00×10^{-3}		$\Sigma\Sigma$	0.7817

Chapter 3

$$s_{pooled} = \sqrt{\frac{0.7817}{18-9}} = \underline{\underline{0.29\% \text{ heroin}}}$$

3-16

Sample	Number Replicated	$\bar{x}$	$\Sigma(x_i - \bar{x})^2$
1	4	13.00	26.00
2	3	37.67	0.67
3	5	26.40	27.20
	12	$\Sigma\Sigma$	53.87

$$s_{pooled} = \sqrt{\frac{53.87}{12-3}} = \underline{\underline{2.4 \text{ ppb NTA}}}$$

CHAPTER 4

4-1 $$\text{CL} = \bar{x} \pm \frac{ts}{\sqrt{N}} \quad \text{(Equation 4-4)}$$

In the solution to Problem 3-5 the following values of $\bar{x}$ and s were obtained for the same six sets of data:

	$\bar{x}$	s	N		$\bar{x}$	s	N
A	2.08	0.35	5	D	2.48	0.28	5
B	69.89	0.076	3	E	69.53	0.22	4
C	0.0918	0.0055	4	F	0.604	0.016	5

For set A: $N = 5$, $s = 0.35$, $\bar{x} = 2.08$, and $t = 2.78$ for 4 degrees of freedom and 95% confidence (Table 4-2). Thus,

$$\text{CL} = 2.08 \pm \frac{(2.78)(0.35)}{\sqrt{5}} = 2.08 \pm 0.44 \text{ or } \underline{\underline{2.1 \pm 0.4}}$$

Similarly for the remaining sets:

Set B:
$$\text{CL} = 69.89 \pm \frac{(4.30)(0.076)}{\sqrt{3}} = 69.89 \pm 0.19 \text{ or } \underline{\underline{69.9 \pm 0.2}}$$

Set C:
$$\text{CL} = 0.0918 \pm \frac{(3.18)(0.0055)}{\sqrt{4}} = 0.0918 \pm 0.0087 \text{ or } \underline{\underline{0.092 \pm 0.009}}$$

Set D:
$$\text{CL} = 2.48 \pm \frac{(2.78)(0.28)}{\sqrt{5}} = 2.48 \pm 0.348 \text{ or } \underline{\underline{2.5 \pm 0.3}}$$

Set E:
$$\text{CL} = 69.53 \pm \frac{(3.18)(0.22)}{\sqrt{4}} = 69.53 \pm 0.35 \text{ or } \underline{\underline{69.5 \pm 0.3}}$$

Set F:
$$\text{CL} = 0.604 \pm \frac{(2.78)(0.016)}{\sqrt{5}} = 0.604 \pm 0.020 \text{ or } \underline{\underline{0.60 \pm 0.02}}$$

The 95% confidence limit is an interval around a sample mean within which the population mean is expected to lie with 95% probability.

Chapter 4

4-2 For $s \to \sigma$, the $CL = \bar{x} \pm z\sigma/\sqrt{N}$ (Equation 4-2). At the 95% level $z = 1.96$, so for set A: $\sigma = 0.20$ and we have

(a) Set A: $$CL = 2.1 \pm \frac{(1.96)(0.20)}{\sqrt{5}} = 2.08 \pm 0.18 \quad \text{or} \quad \underline{\underline{2.1 \pm 0.2}}$$

For the remaining sets:

(b) Set B: $$CL = 69.89 \pm \frac{(1.96)(0.050)}{\sqrt{3}} = \underline{\underline{68.9 \pm 0.06}}$$

(c) Set C: $$CL = 0.0918 \pm \frac{(1.96)(0.0070)}{\sqrt{4}} = 0.0918 \pm 0.0069 \quad \text{or} \quad \underline{\underline{0.092 \pm 0.007}}$$

(d) Set D: $$CL = 2.48 \pm \frac{(1.96)(0.50)}{\sqrt{5}} = \underline{\underline{2.5 \pm 0.4}}$$

(e) Set E: $$CL = 69.53 \pm \frac{(1.96)(0.15)}{\sqrt{4}} = 69.53 \pm 0.15 \quad \text{or} \quad \underline{\underline{69.5 \pm 0.2}}$$

(f) Set F: $$CL = 0.604 \pm \frac{(1.96)(0.015)}{\sqrt{5}} = \underline{\underline{0.60 \pm 0.01}}$$

4-3 $$Q_{exp} = \frac{|x_q - x_n|}{w}$$

For set A: $$Q_{expt} = \frac{|1.5 - 2.1|}{2.4 - 1.5} = 0.67$$

At 95% confidence, $Q_{crit} = 0.71$ (see Table 4-4)

Thus the outlier *cannot be rejected* by the Q test.

Set	Q_{expt}	Q_{crit}	Result
A	0.67	0.710	retain
B	0.86	0.970	retain
C	0.84	0.829	reject
D	0.43	0.710	retain
E	0.95	0.829	reject
F	0.33	0.710	retain

4-4 $\text{CL} = \bar{x} \pm z\sigma/\sqrt{N}$ (Equation 4 – 2)

(a) $80\% \text{ CL} = 18.5 \pm \dfrac{(1.29)(2.4)}{\sqrt{1}} = 18.5 \pm 3.1 \, \mu\text{g/mL}$ or $\underline{\underline{18 \pm 3 \, \mu\text{g/mL}}}$

$95\% \text{ CL} = 18.5 \pm \dfrac{(1.96)(2.4)}{\sqrt{1}} = 18.5 \pm 4.7 \, \mu\text{g/mL}$ or $\underline{\underline{18 \pm 5 \, \mu\text{g/mL}}}$

(b) $80\% \text{ CL} = 18.5 \pm \dfrac{(1.29)(2.4)}{\sqrt{2}} = 18.5 \pm 2.2 \, \mu\text{g/mL}$ or $\underline{\underline{18 \pm 2 \, \mu\text{g/mL}}}$

$95\% \text{ CL} = 18.5 \pm \dfrac{(1.96)(2.4)}{\sqrt{2}} = 18.5 \pm 3.3 \, \mu\text{g/mL}$ or $\underline{\underline{18 \pm 3 \, \mu\text{g/mL}}}$

(c) $80\% \text{ CL} = 18.5 \pm \dfrac{(1.29)(2.4)}{\sqrt{4}} = 18.5 \pm 1.5 \, \mu\text{g/mL}$ or $\underline{\underline{18 \pm 2 \, \mu\text{g/mL}}}$

$95\% \text{ CL} = 18.5 \pm \dfrac{(1.96)(2.4)}{\sqrt{4}} = 18.5 \pm 2.4 \, \mu\text{g/mL}$ or $\underline{\underline{18 \pm 2 \, \mu\text{g/mL}}}$

4-5 (a) $90\% \text{ CL} = 8.53 \pm \dfrac{(1.64)(0.32)}{\sqrt{1}} = 8.53 \pm 0.52 \, \mu\text{g/mL}$ or $\underline{\underline{8.5 \pm 0.5 \, \mu\text{g/mL}}}$

$99\% \text{ CL} = 8.53 \pm \dfrac{(2.58)(0.32)}{\sqrt{1}} = 8.53 \pm 0.83 \, \mu\text{g/mL}$ or $\underline{\underline{8.5 \pm 0.8 \, \mu\text{g/mL}}}$

(b) $90\% \text{ CL} = 8.53 \pm \dfrac{(1.64)(0.32)}{\sqrt{4}} = 8.53 \pm 0.26 \, \mu\text{g/mL}$ or $\underline{\underline{8.5 \pm 0.3 \, \mu\text{g/mL}}}$

Chapter 4

$$99\% \text{ CL} = 8.53 \pm \frac{(2.58)(0.32)}{\sqrt{4}} = 8.53 \pm 0.41 \text{ µg/mL} \quad \text{or} \quad \underline{\underline{8.5 \pm 0.4 \text{ µg/mL}}}$$

(c) $$90\% \text{ CL} = 8.53 \pm \frac{(1.64)(0.32)}{\sqrt{16}} = 8.53 \pm 0.13 \text{ µg/mL} \quad \text{or} \quad \underline{\underline{8.5 \pm 0.1 \text{ µg/mL}}}$$

$$99\% \text{ CL} = 8.53 \pm \frac{(2.58)(0.32)}{\sqrt{16}} = 8.53 \pm 0.21 \text{ µg/mL} \quad \text{or} \quad \underline{\underline{8.5 \pm 0.2 \text{ µg/mL}}}$$

4-6 $z\sigma/\sqrt{N} = $ µg/mL $\sqrt{N} = z\sigma/(1.5 \text{ µg/mL})$ (Equation 4 – 2)

95% $\sqrt{N} = (1.96)(2.4)/1.5 = 3.14$

$N = (3.14)^2 = 9.83 = \underline{\underline{10 \text{ measurements}}}$

99% $\sqrt{N} = (2.58)(2.4)/1.5 = 4.13$

$N = (4.13)^2 = 17.0 = \underline{\underline{17 \text{ measurements}}}$

4-7 $\sqrt{N} = z\sigma/(0.2 \text{ µg/mL})$ (Equation 4 – 2)

95% $\sqrt{N} = (1.96)(0.32)/0.2 = 3.14$

$N = (3.14)^2 = 9.83 = \underline{\underline{10 \text{ measurements}}}$

99% $\sqrt{N} = (2.58)(0.32)/0.2 = 4.13$

$N = (4.13)^2 = 17.0 = \underline{\underline{17 \text{ measurements}}}$

4-8 (a)

x_i, meq/L	x_i^2
3.15	9.9225
3.25	10.5625
3.26	10.6276

$\sum x_i = 9.66$ $\sum x_i^2 = 31.1126$ $\bar{x} = 9.66/3 = 3.22$

$$s = \sqrt{\frac{\sum x_i^2 - (\sum x_i)^2/N}{N-1}} = \sqrt{\frac{31.1126 - (9.66)^2/3}{2}} = \sqrt{0.00370} = 0.061 \text{ mmol/L}$$

Substituting into Equation 4-4 gives

Chapter 4

$$95\% \text{ CL} = \bar{x} \pm ts/\sqrt{N} = 3.22 \pm (4.30)(0.061)/\sqrt{3} = \underline{\underline{3.22 \pm 0.15 \text{ mmol/L}}}$$

(b) $95\% \text{ CL} = \bar{x} \pm z\sigma/\sqrt{N} = 3.22 \pm (1.96)(0.056)/\sqrt{3}$

$$= 3.22 \pm 0.063 \text{ mmol/L} \quad \text{or} \quad \underline{\underline{3.22 \pm 0.06 \text{ mmol/L}}}$$

4-9 (a)

x_i, %	x_i^2
7.47	55.8009
6.98	48.7204
7.27	52.8529

$\sum x_i = 21.72 \qquad \sum x_i^2 = 157.3742 \qquad \bar{x} = \sum x_i/N = 21.72/3 = 7.24$

$$s = \sqrt{\frac{157.3742 - (21.72)^2/3}{2}} = \sqrt{\frac{0.1214}{2}} = 0.246 = 0.25\%$$

$$90\% \text{ CL} = \bar{x} \pm ts/\sqrt{N} = 7.24 \pm (2.92)(0.25)/\sqrt{3} = \underline{\underline{7.24 \pm 0.42\%}}$$

(b) $90\% \text{ CL} = \bar{x} \pm z\sigma/\sqrt{N} = 7.24 \pm (1.64)(0.28)/\sqrt{3} = \underline{\underline{7.24 \pm 0.27\%}}$

4-10 (a) For 99% confidence, $z\sigma/\sqrt{N} = (2.58)(0.040)/\sqrt{N} = 0.03$

$$\sqrt{N} = (2.58)(0.040)/0.03 = 3.44$$

$$N = 11.8 = \underline{\underline{12 \text{ measurements}}}$$

(b) For 95% confidence, $\sqrt{N} = (1.96)(0.040)/0.03 = 2.61$

$$N = 6.83 = \underline{\underline{7 \text{ measurements}}}$$

(c) For 90% confidence, $\sqrt{N} = (1.64)(0.040)/0.02 = 3.28$

$$N = 10.75 = \underline{\underline{11 \text{ measurements}}}$$

4-11 (a) $(\bar{x} - \mu)_{\text{actual}} = 30.26 - 30.15 = 0.11$

$$(\bar{x} - \mu)_{\text{exp}} = zs_{\text{pooled}}/\sqrt{N} = (1.96)(0.094)/\sqrt{4} = 0.092 < 0.11$$

<u>Systematic error is indicated at 95% confidence.</u>

Chapter 4

(b) $(\bar{x} - \mu)_{exp} = ts_{exp}/\sqrt{N} = (3.18)(0.085/\sqrt{4}) = 0.135 > 0.11$

No systematic error is demonstrated.

4-12 Carbon

$(\bar{x} - \mu)_{actual} = 68.5\% - 68.8\% = -0.3\%$

$(\bar{x} - \mu)_{exp} = z\sigma/\sqrt{N} = (1.96)(0.004 \times 68.8\%)/\sqrt{2} = 0.38\%$

No systematic error is demonstrated.

Hydrogen

$(\bar{x} - \mu)_{actual} = 4.882\% - 4.953\% = 0.071\%$

$(\bar{x} - \mu)_{exp} = (1.96)(0.006 \times 4.953\%)/\sqrt{2} = 0.041\%$

Systematic error is suggested.

4-13 $(\bar{x}_1 - \bar{x}_2)_{actual}$ is compared with $(\bar{x}_1 - \bar{x}_2)_{calcu} = \pm\sigma\sqrt{\frac{N_1+N_2}{N_1 N_2}}$

where N_1 and $N_2 = 3$ and $z_{99\%} = 2.58$.

For As, $(\bar{x}_1 - \bar{x}_2)_{actual} = 129 - 119 = 10$ ppm

$(\bar{x}_1 - \bar{x}_2)_{calc} = \pm(2.58)(9.5)\sqrt{\frac{3+3}{3 \times 3}} = \pm 20$ ppm

Results from similar calculations for the elements are listed below.

Chapter 4

	Concentration, ppm		Difference, ppm			Difference at
Element	Clothes	Window	Actual	Calc	σ	99% CL
As	129	119	+ 10	± 20	9.5	no
Co	0.53	0.60	− 0.07	± 0.053	0.025	yes
La	3.92	3.52	+ 0.40	± 0.42	0.20	no
Sb	2.75	2.71	+ 0.04	± 0.53	0.25	no
Th	0.61	0.73	− 0.12	± 0.090	0.043	yes

4-14

$$s_{pooled} = \sqrt{\frac{\sum(x_{i(top)} - \overline{x}_{(top)})^2 + \sum(x_{i(bot)} - \overline{x}_{(bot)})^2}{N_1 + N_2 - 2}}$$

$$= \sqrt{\frac{0.0037 + 0.00926}{4 + 3 - 2}} = \sqrt{\frac{0.1296}{5}} = 0.51$$

(a) $(\overline{x}_1 - \overline{x}_2)_{actual} = 26.355 - 26.303 = 0.052$

$$(\overline{x}_1 - \overline{x}_2)_{calc} = \pm t s_{pooled} \sqrt{\frac{N_1 + N_2}{N_1 N_2}} = \pm(2.57)(0.051)\sqrt{\frac{4+3}{4 \times 3}} = 0.10$$

Nonhomogeneity is not demonstrated.

(b)
$$(\overline{x}_1 - \overline{x}_2)_{calc} = \pm z\sigma\sqrt{\frac{N_1 + N_2}{N_1 N_2}} = \pm(1.96)(0.03)\sqrt{\frac{7}{12}} = 0.045$$

Nonhomogeneity is indicated.

Chapter 4

4-15

$$\Delta x_{min} = \bar{x}_1 - \bar{x}_b > ts_b \sqrt{\frac{N_1 + N_b}{N_1 N_b}}$$

x_b, mg	s_b^2
0.1	0.01
-0.2	0.04
0.3	0.09
0.2	0.04
0.0	0.00
-0.1	0.01

$\sum x_b = 0.3 \quad \sum x_b^2 = 0.19$

$$s_b = \sqrt{\frac{\sum x_b^2 - (\sum x_b)^2/N}{N-1}} = \sqrt{\frac{0.19 - (0.3)^2/6}{5}} = 0.19$$

(a) For 2 analyses at 99% CL, DF = 6 + 2 − 2 = 6

$$ts_b \sqrt{\frac{N_1 + N_b}{N_1 N_b}} = (3.71)(0.19)\sqrt{\frac{2+6}{2 \times 6}} = \underline{\underline{0.58 \text{ mg}}}$$

(b) For 4 analyses at 99% Cl, DF = 6 + 4 − 2 = 8

$$ts_b \sqrt{\frac{N_1 + N_b}{N_1 N_b}} = (3.36)(0.19)\sqrt{\frac{4+6}{4 \times 6}} = \underline{\underline{0.41 \text{ mg}}}$$

(c) For 6 analyses at 99% CL, DF = 6 + 6 − 2 = 10

$$ts_b \sqrt{\frac{N_1 + N_b}{N_1 N_b}} = (3.17)(0.19)\sqrt{\frac{6+6}{6 \times 6}} = \underline{\underline{0.35 \text{ mg}}}$$

4-16 (a)

$$F = \frac{s_B^2}{s_A^2} = \frac{(0.12)^2}{(0.10)^2} = 1.44$$

$F_{12,6} = 4.00 \quad \text{(Table 4−3)}$

No difference demonstrated.

(b) $$F = \frac{s_A^2}{s_B^2} = \frac{(0.07)^2}{(0.04)^2} = 3.06$$

$$F_{12,20} = 2.28$$

Difference is indicated at 90% CL.

(c) $$F = \frac{s_B^2}{s_A^2} = \frac{(0.07)^2}{(0.05)^2} = 1.96$$

$$F_{6,20} = 2.60$$

No difference demonstrated.

(d) $$F = \frac{s_B^2}{s_A^2} = \frac{(0.035)^2}{(0.020)^2} = 3.06$$

Difference is indicated at the 90% CL.

4-17 (a) $$Q_{expt} = \frac{|x_q - x_n|}{w} = \frac{|41.27 - 41.61|}{41.84 - 41.27} = 0.60$$

At 95% $Q_{crit} = 0.829$ (Table 4-4) > 0.60

Therefore the outlier is retained.

(b) $$Q_{exp} = \frac{|7.388 - 7.295|}{|7.388 - 7.284|} = 0.89$$

$$Q_{crit} = 0.829$$

Therefore the outlier is rejected.

4-18 (a) $$Q_{expt} = \frac{|85.10 - 84.70|}{|85.10 - 84.62|} = 0.833$$

$Q_{crit} = 0.970 > 0.83$ Therefore retain.

Chapter 4

(b) $Q_{expt} = \dfrac{|85.10 - 84.70|}{|85.10 - 84.62|} = 0.833$

$Q_{crit} = 0.829 < 0.833$ Therefore reject.

4-19 (b) See Section 4E-2.

mg SO_4^{2-}/L, x_i	Reading, y_i	x_i^2	y_i^2	$x_i y_i$
0.00	0.06	0.00	0.0036	0.00
5.00	1.48	25.00	2.1904	7.40
10.00	2.28	100.00	5.1984	22.80
15.00	3.98	225.00	15.8404	59.70
20.00	4.61	400.00	21.1521	92.20
$\sum x_i = 50.00$	$\sum y_i = 12.41$	$\sum x_i^2 = 750.00$	$\sum y_i^2 = 44.4849$	$\sum x_i y_i = 182.10$

$\bar{x} = 50.00/5 = 10.00$ $\bar{y} = 12.41/5 = 2.482$

$S_{xx} = \sum x_i^2 - (\sum x_i)^2/N = 750.00 - (50.00)^2/5 = 250.00$ (Equation 4 – 10)

$S_{yy} = \sum y_i^2 - (\sum y_i)^2/N = 44.4849 - (12.41)^2/5 = 13.68328$ (Equation 4 – 11)

$S_{xy} = \sum x_i y_i - (\sum x_i \sum y_i)/N = 182.10 - 50.00 \times 12.41/5 = 58.00$

                                                                                                                       (Equation 4 – 12)

and

$m = S_{xy}/S_{xx} = 58.00/250.00 = 0.2320$ (Equation 4 – 13)

$b = \bar{y} - m\bar{x} = 2.482 - 0.232 \times 10 = 0.162$ (Equation 4 – 14)

Letting $C_x = $ concn of $SO_4^{2-} = y_i$

$R = $ meter reading $= x_i$

$\underline{\underline{R = 0.232\, C_x + 0.162}}$

(d) $3.67 = 0.162 + 0.232\, C_x$

$$C_x = (3.67 - 0.162)/0.232 = \underline{\underline{15.1 \text{ mg SO}_4^{2-}/\text{L}}}$$

In order to calculate the standard deviation for C_x, we must first calculate s_r by means of Equation 4-15

$$s_r = \sqrt{\frac{S_{yy} - m^2 S_{xx}}{N-2}} = \sqrt{\frac{13.68328 - (0.232)^2 \times 25.00}{5-2}} = 0.275$$

Then

$$s_c = \frac{s_r}{m}\sqrt{\frac{1}{M} + \frac{1}{N} + \frac{(\bar{y}_c - \bar{y})^2}{m^2 S_{xx}}} \quad \text{(Equation 4-18)}$$

where $\bar{y}_c$ is the mean for M measurement of y_c. For $\bar{y}_c = 3.67$ for 1 measurement ($M = 1$),

$$s_c = \frac{0.275}{0.232}\sqrt{\frac{1}{1} + \frac{1}{5} + \frac{(3.67 - 2.482)^2}{(0.232)^2 \times 250.0}}$$

$$= \underline{\underline{1.4 \text{ mg SO}_4^{2-}/\text{L}}}$$

$$\text{CV} = \frac{1.4}{15.1} \times 100\% = \underline{\underline{9.3\%}}$$

(e) For the mean of 6 measurements

$$s_c = \frac{0.275}{0.232}\sqrt{\frac{1}{6} + \frac{1}{5} + \frac{(3.07 - 2.482)^2}{(0.232)^2 \times 250.0}} = \underline{\underline{0.81 \text{ mg SO}_4^{2-}/\text{L}}}$$

$$\text{CV} = \frac{0.81}{15.1} \times 100\% = \underline{\underline{5.4\%}}$$

4-20 Proceeding as in Solution 4-19, we obtain

(b) $\underline{\underline{E = -29.74 \text{ pCa} + 92.86}}$

Chapter 4

(c) $\underline{\text{pCa} = 2.44}$

$s_c = \underline{0.080}$

$(s_c)_r = \text{CV} = \underline{3.3\%}$

(d) For $M = 2$, $s_c = \underline{0.061}$ and $\text{CV} = \underline{2.5\%}$

For $M = 8$, $s_c = \underline{0.043}$ and $\text{CV} = \underline{1.8\%}$

4-21 Proceeding as in Solution 4-19, we obtain:

(a) Letting A_x = peak area and c_{MVK} = mmol/L MVK,

$$\underline{\underline{A_x = 5.57\, c_{\text{MVK}} + 0.902}}$$

(c) $c_x = (6.3 - 0.90)/5.57 = 0.969 = \underline{\underline{0.97 \text{ mmol MVK/L}}}$

(d) For $M = 1$, $s_c = 0.086 = \underline{\underline{0.09 \text{ mmol MVK/L}}}$

$(s_c)_r = \text{CV} = \underline{\underline{8.8\%}}$

For $M = 4$, $s_c = 0.058 = \underline{\underline{0.06 \text{ mmol MVK/L}}}$

$\text{CV} = \underline{\underline{6.0\%}}$

(e) $c_{\text{MVK}} = \underline{\underline{4.78 \text{ mmol } MVK/L}}$

For $M = 1$, $s_c = 0.084 = \underline{\underline{0.08 \text{ mmol MVK/L}}}$

$\text{CV} = \underline{\underline{1.8\%}}$

For $M = 4$, $s_c = 0.056 = \underline{\underline{0.06 \text{ mmol MVK/L}}}$

$\text{CV} = \underline{\underline{1.2\%}}$

CHAPTER 5

5-1 (a) *Mass, m*, is an invariant measure of the amount of matter in an object. *Weight, w,* is the force of attraction between an object and earth.

(b) An *empirical formula* gives the simplest whole-number ratio of the number of atoms in a chemical compound. A *chemical formula* expresses the actual number of various atoms in a compound.

(c) The individual particles of a *colloid* are smaller than about 10^{-5} mm in diameter, while those of a *crystalline precipitate* are larger. As a consequence, crystalline precipitates settle out of solution relatively rapidly, whereas colloidal particles do not unless they can be caused to agglomerate.

(d) A *specific reagent* reacts with a single analyte under a given set of conditions. Dimethylglyoxime (page 95) is a prime example of a specific precipitating reagent. A *selective reagent* reacts with several analytes. Most of the reagents listed in Table 5-4 are selective.

(e) *Precipitation* is the process by which a solid phase forms and is carried out of solution when the solubility product of a species is exceeded. *Coprecipitation* is the process in which a normally soluble species is carried out of solution during the formation of a precipitate.

(f) *Coagulation* is the process in which colloidal particles coalesce to form larger aggregates. *Peptization* is the reverse process. Coagulation is brought about by heating and by adding an electrolyte. Washing a coagulated colloid with water may remove sufficient electrolyte to permit re-establishment of the repulsive forces that favor return to the colloidal state.

(g) *Occlusion* is a type of coprecipitation in which an impurity is entrapped in a pocket formed by a rapidly growing crystal. *Mixed-crystal formation* is a type of coprecipitation in which a foreign ion is incorporated into a growing crystal in a lattice position that is ordinarily occupied by one of the ions of the precipitate.

(h) *Nucleation* is the process by which a small number of ions, atoms, or molecules come together to form a particle of the solid phase. This process competes with *particle growth* in which precipitation occurs by deposition on a nucleus or particle

Chapter 5

already in the solution. If nucleation is more favorable, a large number of small particles is formed. A small number of large particles results if particle growth is favored.

5-2 (a) The dalton, which is synonymous with the atomic mass unit, is a relative mass unit that is equal to 1/12 of the mass of one neutral ^{12}C atom.

(b) The mole is the amount of a chemical species that contains 6.022×10^{23} particles (atoms, molecules, ions, electrons, ion pairs, or subatomic particles).

(c) Stoichiometry is the mass relationship among reacting chemical species.

(d) A *gravimetric method* is a quantitative analytical method that is based upon determining the mass of the analyte or the mass of a pure compound containing the analyte.

(e) *Digestion* is a process for improving the purity and filterability of a precipitate by heating the solid in contact with the solution from which it is formed (the *mother liquor*).

(f) *Adsorption* results when ions of electrolyte are attracted to the surface of precipitate particles by ions of the opposite charge in the solid. This results in an adsorbed layer of ions on the surface of the particles.

(g) In *reprecipitation*, a precipitate is filtered, washed, redissolved, and then reformed from the new solution. Because the concentration of contaminant is lower in this new solution than in the original, the second precipitate contains less coprecipitated impurity.

(h) *Precipitation from homogeneous solution* is a technique in which the precipitating reagent is synthesized or generated in the reaction mixture. The rate of generation of the reagent is arranged to be relatively slow so that particle growth is favored over nucleation.

(i) The *electric double layer* consists of lattice ions adsorbed on the surface of a solid (the primary adsorption layer) and a volume of solution surrounding the particle (the counter-ion layer) in which an excess of ions of opposite charge exists.

(j) The *specific surface* expresses the surface area of a unit mass of a solid that is exposed to its surroundings. Specific surface area commonly has the units of cm^2/g.

Chapter 5

(k) *Relative supersaturation* is given by the expression

$$\text{relative supersaturation} = \frac{Q - S}{S}$$

where Q is the concentration of a solute in a solution at any instant and S is its equilibrium solubility ($Q > S$).

5-3 A *chelating agent* is an organic compound that contains two or more electron-donor groups located in such a configuration that five- or six-membered rings are formed when the donor groups complex a cation.

5-4 *Relative supersaturation* can be regulated through control of reagent concentration, temperature, and the rate at which reagents are combined.

5-5 (a) positive charge (b) adsorbed Ag^+ (c) NO_3^-

5-6 The AgCl produced in a gravimetric silver determination will have a primary adsorption layer in which chloride ion predominates. Use of an acidic wash liquid will cause hydrogen ions to exist in the counter-ion layer and HCl to be adsorbed. Hydrogen chloride will volatilize at the drying temperature for the solid. The same will not hold true for AgCl from a gravimetric chloride determination; here silver ions will predominate in the primary adsorption layer and a non-volatile silver salt will be adsorbed.

5-7 *Peptization* is the process in which a coagulated colloid returns to its original dispersed state as a consequence of a decrease in the electrolyte concentration of the solution in contact with the precipitate. Peptization of a coagulated colloid can be avoided by washing with an electrolyte solution rather than with pure water.

5-8 Chloroplatinic acid H_2PtCl_6 forms the precipitate K_2PtCl_6 with K^+ but does not form analogous precipitates with Li^+ or Na^+. Thus K^+ can be separated by this means.

5-9 (a) Generate hydroxide ions from urea.

(b) Generate dimethylglyoxime from biacetyl and hydroxylamine.

(c) Generate hydrogen sulfide from thioacetamide.

(d) Generate hydrogen sulfide from thioacetamide.

5-10 The atomic mass of iron is 55.85 daltons, or 55.85 amu. Its molar mass is 55.85 grams.

Chapter 5

5-11 $\mathcal{M} = 9.109 \times 10^{-31} \dfrac{\text{kg}}{\text{electron}} \times 1000 \dfrac{\text{g}}{\text{kg}} \times 6.022 \times 10^{23} \text{ electrons}$

$= \underline{\underline{5.485 \times 10^{-4} \text{ g/mol}}}$

5-12 (a) $6.84 \text{ g B}_2\text{O}_3 \times \dfrac{1 \text{ mol B}_2\text{O}_3}{69.62 \text{ g B}_2\text{O}_3} = \underline{\underline{0.0982 \text{ mol B}_2\text{O}_3}}$

(b) $296 \text{ mg} \times \dfrac{1 \text{ g}}{1000 \text{ mg}} \times \dfrac{1 \text{ mol}}{381.37 \text{ g}} = \underline{\underline{7.76 \times 10^{-4} \text{ mol Na}_2\text{B}_4\text{O}_7 \cdot 10\text{H}_2\text{O}}}$

(c) $8.75 \text{ g Mn}_3\text{O}_4 \times \dfrac{1 \text{ mol Mn}_3\text{O}_4}{228.81 \text{ g Mn}_3\text{O}_4} = \underline{\underline{0.0382 \text{ mol Mn}_3\text{O}_4}}$

(d) $67.4 \text{ mg CaC}_2\text{O}_4 \times \dfrac{1 \text{ g}}{1000 \text{ mg}} \times \dfrac{1 \text{ mol CaC}_2\text{O}_4}{128.10 \text{ g CaC}_2\text{O}_4} = \underline{\underline{5.26 \times 10^{-4} \text{ mol CaC}_2\text{O}_4}}$

5-13 (a) $79.8 \text{ mg H}_2 \times \dfrac{10^{-3} \text{ g H}_2}{\text{mg H}_2} \times \dfrac{1 \text{ mmol H}_2}{2.016 \times 10^{-3} \text{ g H}_2} = \underline{\underline{39.6 \text{ mmol H}_2}}$

(b) $8.43 \text{ g SO}_2 \times \dfrac{1 \text{ mmol SO}_2}{0.06406 \text{ g SO}_2} = \underline{\underline{131.6 \text{ mmol SO}_2}}$

(c) $64.4 \text{ g Na}_2\text{CO}_3 \times \dfrac{1 \text{ mmol Na}_2\text{CO}_3}{0.1060 \text{ g Na}_2\text{CO}_3} = \underline{\underline{608 \text{ mmol Na}_2\text{CO}_3}}$

(d) $411 \text{ mg KMnO}_4 \times \dfrac{10^{-3} \text{ g KMnO}_4}{\text{mg KMnO}_4} \times \dfrac{1 \text{ mmol KMnO}_4}{0.1580 \text{ g KMnO}_4} = \underline{\underline{2.60 \text{ mmol KMnO}_4}}$

5-14 (a) $0.666 \text{ mol HNO}_3 \times \dfrac{63.01 \text{ g HNO}_3}{\text{mol HNO}_3} \times \dfrac{1000 \text{ mg}}{\text{g}} = \underline{\underline{4.20 \times 10^4 \text{ mg HNO}_3}}$

(b) $300 \text{ mmol MgO} \times \dfrac{40.30 \text{ mg MgO}}{\text{mmol MgO}} = \underline{\underline{1.21 \times 10^4 \text{ mg MgO}}}$

(c) $19.0 \text{ mol NH}_4\text{NO}_3 \times \dfrac{80.04 \text{ g NH}_4\text{NO}_3}{1 \text{ mol NH}_4\text{NO}_3} \times \dfrac{1000 \text{ mg}}{\text{g}} = \underline{\underline{1.52 \times 10^6 \text{ mg NH}_4\text{NO}_3}}$

Chapter 5

(d) $5.32 \text{ mol} \times \dfrac{548.22 \text{ g}}{\text{mol}} \times \dfrac{1000 \text{ mg}}{\text{g}} = \underline{\underline{2.92 \times 10^6 \text{ mg}}}$

5-15 (a) $32.1 \text{ mmol H}_2\text{O}_2 \times \dfrac{0.03402 \text{ g H}_2\text{O}_2}{\text{mmol H}_2\text{O}_2} = \underline{\underline{1.09 \text{ g H}_2\text{O}_2}}$

(b) $0.466 \text{ mol NH}_4\text{VO}_3 \times \dfrac{117.0 \text{ g NH}_4\text{VO}_3}{\text{mol NH}_4\text{VO}_3} = \underline{\underline{54.5 \text{ g NH}_4\text{VO}_3}}$

(c) $5.38 \text{ mol MgNH}_4\text{PO}_4 \times \dfrac{137.3 \text{ g MgNH}_4\text{PO}_4}{\text{mol MgNH}_4\text{PO}_4} = \underline{\underline{739 \text{ g MgNH}_4\text{PO}_4}}$

(d) $26.7 \text{ mmol KH(IO}_3)_2 \times \dfrac{0.3899 \text{ g KH(IO}_3)_2}{\text{mmol KH(IO}_3)_2} = \underline{\underline{10.4 \text{ g KH(IO}_3)_2}}$

5-16 (a) $\dfrac{1 \text{ mol CO}_2}{\text{mol BaCO}_3}$ (f) $\dfrac{3 \times \text{mol UO}_2}{\text{mol U}_3\text{O}_8}$

(b) $\dfrac{2 \times \text{mol Mg}}{\text{mol Mg}_2\text{P}_2\text{O}_7}$ (g) $\dfrac{1 \text{ mol C}_8\text{H}_6\text{O}_3\text{Cl}_2}{2 \times \text{mol AgCl}}$

(c) $\dfrac{1 \text{ mol K}_2\text{O}}{2 \times \text{mol (C}_6\text{H}_5)_4\text{BK}}$ (h) $\dfrac{3 \times \text{mol CoSiF}_6 \cdot 6\text{H}_2\text{O}}{\text{mol Co}_3\text{O}_4}$

(d) $\dfrac{2 \times \text{mol Bi}}{\text{mol Bi}_2\text{O}_3}$ (i) $\dfrac{1 \text{ mol CoSiF}_6 \cdot 6\text{H}_2\text{O}}{6 \times \text{mol H}_2\text{O}}$

(e) $\dfrac{1 \text{ mol H}_2\text{S}}{\text{mol CdSO}_4}$ (j) $\dfrac{1 \text{ mol CoSiF}_6 \cdot 6\text{H}_2\text{O}}{6 \times \text{mol PbClF}}$

5-17 $\% \text{ A} = \dfrac{\text{mass ppt} \times \text{stoichiometric factor} \times 100\%}{\text{mass sample}}$

Rearranging

$\text{mass ppt} = \dfrac{\% \text{ A} \times \text{mass sample}}{\text{stoichiometric factor}} \times 100\%$

Thus, for a given mass of sample, the mass precipitate becomes larger as the stoichiometric factor becomes smaller.

Chapter 5

5-18 (a) $1.200 \text{ g AgNO}_3 \times \dfrac{1 \text{ mol AgNO}_3}{169.873 \text{ g AgNO}_3} \times \dfrac{1 \text{ mol Ag}_2\text{CrO}_4}{2 \text{ mol AgNO}_3} \times \dfrac{331.730 \text{ g Ag}_2\text{CrO}_4}{\text{mol Ag}_2\text{CrO}_4} =$

$\underline{\underline{1.172 \text{ g Ag}_2\text{CrO}_4}}$

(b) $1.200 \text{ g K}_2\text{CrO}_4 \times \dfrac{1 \text{ mol K}_2\text{CrO}_4}{194.190 \text{ g K}_2\text{CrO}_4} \times \dfrac{1 \text{ mol Ag}_2\text{CrO}_4}{\text{mol K}_2\text{CrO}_4} \times \dfrac{331.730 \text{ g Ag}_2\text{CrO}_4}{\text{mol Ag}_2\text{CrO}_4} =$

$\underline{\underline{2.050 \text{ g Ag}_2\text{CrO}_4}}$

5-19 (a) $0.750 \text{ g Ba(OH)}_2 \times \dfrac{1 \text{ mol Ba(OH)}_2}{171.342 \text{ g Ba(OH)}_2} \times \dfrac{1 \text{ mol Mg(OH)}_2}{\text{mol Ba(OH)}_2} \times \dfrac{58.320 \text{ g Mg(OH)}_2}{\text{mol Mg(OH)}_2} =$

$\underline{\underline{0.2553 \text{ g Mg(OH)}_2}}$

(b) $0.750 \text{ g MgCl}_2 \times \dfrac{1 \text{ mol MgCl}_2}{95.210 \text{ g MgCl}_2} \times \dfrac{1 \text{ mol Mg(OH)}_2}{\text{mol MgCl}_2} \times \dfrac{58.320 \text{ g Mg(OH)}_2}{\text{mol Mg(OH)}_2} =$

$\underline{\underline{0.4594 \text{ g Mg(OH)}_2}}$

5-20 (a) $0.400 \text{ g sple} \times \dfrac{41.3 \text{ g KCl}}{100 \text{ g sple}} \times \dfrac{1 \text{ mol KCl}}{74.551 \text{ g KCl}} \times \dfrac{1 \text{ mol AgCl}}{\text{mol KCl}} \times \dfrac{143.321 \text{ g AgCl}}{\text{mol AgCl}} = \underline{\underline{0.318 \text{ g AgCl}}}$

(b) $0.400 \text{ g sple} \times \dfrac{41.3 \text{ g MgCl}_2}{100 \text{ g sple}} \times \dfrac{1 \text{ mol MgCl}_2}{95.210 \text{ g MgCl}_2} \times \dfrac{2 \text{ mol AgCl}}{\text{mol MgCl}_2} \times \dfrac{143.321 \text{ g AgCl}}{\text{mol AgCl}} =$

$\underline{\underline{0.497 \text{ g AgCl}}}$

(c) $0.400 \text{ g sample} \times \dfrac{41.3 \text{ g FeCl}_3}{100 \text{ g sample}} \times \dfrac{1 \text{ mol FeCl}_3}{162.205 \text{ g FeCl}_3} \times \dfrac{3 \text{ mol AgCl}}{\text{mol FeCl}_3} \times \dfrac{143.321 \text{ g AgCl}}{\text{mol AgCl}} =$

$\underline{\underline{0.438 \text{ g AgCl}}}$

5-21 $\mathcal{M}_{\text{BaCl}_2 \cdot 2\text{H}_2\text{O}} = 244.26 \text{ g/mol}$

$\mathcal{M}_{\text{NaIO}_3} = 197.89 \text{ g/mol}$

$\mathcal{M}_{\text{Ba(IO}_3)_2} = 487.13 \text{ g/mol}$

Initial amounts:

Chapter 5

$$0.200 \text{ g BaCl}_2 \cdot \text{H}_2\text{O} \times \frac{\text{mol BaCl}_2 \cdot 2\text{H}_2\text{O}}{244.26 \text{ g}} \times \frac{1 \text{ mol Ba}^{2+}}{\text{mol BaCl}_2 \cdot 2\text{H}_2\text{O}} =$$

$$8.188 \times 10^{-4} \text{ mol Ba}^{2+}$$

$$0.300 \text{ g NaIO}_3 \times \frac{\text{mol NaIO}_3}{197.89 \text{ g}} \times \frac{1 \text{ mol IO}_3^-}{\text{mol NaIO}_3} = 1.5160 \times 10^{-3} \text{ mol IO}_3^-$$

Complete reaction of Ba^{2+} would require $(2 \times 8.188 \times 10^{-4}) = 1.6375 \times 10^{-3}$ mol of iodate ion. Since $1.5160 \times 10^{-3} < 1.6376 \times 10^{-3}$, barium ion is present in excess.

(a) no. mol $Ba(IO_3)_2$ = no. mol $IO_3^-/2$ = $1.5160 \times 10^{-3} / 2 = 7.580 \times 10^{-4}$

$$\text{mass Ba(IO}_3)_2 = 7.580 \times 10^{-4} \text{ mol} \times \frac{487.14 \text{ g Ba(IO}_3)_2}{\text{mol}} = \underline{\underline{0.369 \text{ g Ba(IO}_3)_2}}$$

(b) no. mol $BaCl_2 \cdot 2H_2O$ remaining $= (8.188 - 7.580) \times 10^{-4}$ mol

$$= 6.08 \times 10^{-5} \text{ mol}$$

$$\text{mass BaCl}_2 \cdot 2\text{H}_2\text{O} = 6.08 \times 10^{-5} \text{ mol BaCl}_2 \cdot 2\text{H}_2\text{O} \times \frac{244.26 \text{ g BaCl}_2 \cdot 2\text{H}_2\text{O}}{\text{mol BaCl}_2 \cdot 2\text{H}_2\text{O}}$$

$$= \underline{\underline{0.0149 \text{ g BaCl}_2 \cdot 2\text{H}_2\text{O}}}$$

5-22 $\mathcal{M}_{CO_2} = 44.010$ g/mol

$\mathcal{M}_{MgCO_3} = 84.31$ g/mol

$\mathcal{M}_{K_2CO_3} = 138.21$ /mol

no. mol CO_2 = no. mol $MgCO_3$ + no. mol K_2CO_3 =

$$1.204 \text{ g sample} \times \frac{36.0 \text{ g MgCO}_3}{100 \text{ g sample}} \times \frac{1 \text{ mol MgCO}_3}{84.31 \text{ g MgCO}_3} +$$

$$1.204 \text{ g sample} \times \frac{44.0 \text{ g K}_2\text{CO}_3}{100 \text{ g sample}} \times \frac{1 \text{ mol K}_2\text{CO}_3}{138.21 \text{ g K}_2\text{CO}_3}$$

no. mol $CO_2 = (5.141 \times 10^{-3}) + (3.833 \times 10^{-3}) = 8.974 \times 10^{-3}$ mol

Chapter 5

$$\text{mass CO}_2 = 8.974 \times 10^{-3} \text{ mol} \times 44.010 \text{ g/mol} = \underline{\underline{0.395 \text{ g CO}_2}}$$

5-23 $\mathcal{M}_{AgNO_3} = 169.87$ g/mol

$\mathcal{M}_{K_2CrO_4} = 194.19$ g/mol

$\mathcal{M}_{Ag_2CrO_4} = 331.73$ g/mol

$$\text{no. mol AgNO}_3 = 0.500 \text{ g AgNO}_3 \times \frac{1 \text{ mol AgNO}_3}{169.87 \text{ g AgNO}_3} = 2.943 \times 10^{-3} \text{ mol AgNO}_3$$

$$\text{no. mol K}_2\text{CrO}_4 = 0.300 \text{ g K}_2\text{CrO}_4 \times \frac{1 \text{ mol K}_2\text{CrO}_4}{194.19 \text{ g K}_2\text{CrO}_4} = 1.545 \times 10^{-3} \text{ mol}$$

$$\text{no. mol K}_2\text{CrO}_4 \text{ required} = \text{no. mol AgNO}_3 \times \frac{1 \text{ mol K}_2\text{CrO}_4}{2 \text{ mol AgNO}_3} = 1.472 \times 10^{-3} \text{ mol}$$

Thus K_2CrO_4 is in excess.

no. mol K_2CrO_4 remaining = $(1.545 - 1.472) \times 10^{-3}$ mol = 7.3×10^{-5} mol

(a) no. mol Ag_2CrO_4 = no. mol K_2CrO_4 required = 1.472×10^{-3} mol

$$\text{mass Ag}_2\text{CrO}_4 = 1.472 \times 10^{-3} \text{ mol} \times \frac{331.73 \text{ g Ag}_2\text{CrO}_4}{\text{mol Ag}_2\text{CrO}_4} = \underline{\underline{0.488 \text{ g}}}$$

(b) $$\text{mass K}_2\text{CrO}_4 = 7.3 \times 10^{-5} \text{ mol K}_2\text{CrO}_4 \times \frac{194.19 \text{ g K}_2\text{CrO}_4}{\text{mol K}_2\text{CrO}_4} = \underline{\underline{0.0142 \text{ g}}}$$

5-24 $\mathcal{M}_{AgCl} = 143.32$ g/mol $\mathcal{M}_{KCl} = 74.55$ g/mol

$$\frac{0.7332 \text{ g AgCl} \times \frac{1 \text{ mol AgCl}}{143.32 \text{ g AgCl}} \times \frac{1 \text{ mol KCl}}{1 \text{ mol AgCl}} \times \frac{74.55 \text{ g KCl}}{\text{mol KCl}}}{0.4000 \text{ g sample}} \times 100\% = \underline{\underline{95.35\% \text{ KCl}}}$$

5-25 $\mathcal{M}_{Al_2O_3} = 101.96$ g/mol $\mathcal{M}_{NH_4Al(SO_4)_2} = 237.15$ g/mol

Chapter 5

(a) $0.1798 \text{ g Al}_2\text{O}_3 \times \dfrac{1 \text{ mol Al}_2\text{O}_3}{101.96 \text{ g Al}_2\text{O}_3} \times \dfrac{2 \text{ mol NH}_4\text{Al(SO}_4)_2}{\text{mol Al}_2\text{O}_3} =$

$3.5269 \times 10^{-3} \text{ mol NH}_4\text{Al(SO}_4)_2$

$\dfrac{3.5269 \times 10^{-3} \text{ mol NH}_4\text{Al(SO}_4)_3 \times \frac{237.15 \text{ g NH}_4\text{Al(SO}_4)_2}{\text{mol NH}_4\text{Al(SO}_4)_2}}{1.200 \text{ g sample}} \times 100\% = \underline{\underline{69.70\% \text{ NH}_4\text{Al(SO}_4)_3}}$

(b) $\dfrac{0.1798 \text{ g Al}_2\text{O}_3}{1.200 \text{ g sample}} \times 100\% = \underline{\underline{14.98\% \text{ Al}_2\text{O}_3}}$

(c) no. mol Al = no. mol $\text{NH}_4\text{Al(SO}_4)_3 = 3.5269 \times 10^{-3}$

$\dfrac{3.5269 \times 10^{-3} \text{ mol Al} \times 26.981 \text{ g Al/mol}}{1.200 \text{ g sample}} \times 100\% = \underline{\underline{7.930\% \text{ Al}}}$

5-26 $\mathcal{M}_{\text{BaCO}_3} = 197.34 \text{ g/mol}$

$\dfrac{0.5613 \text{ g BaCO}_3 \times \frac{1 \text{ mol BaCO}_3}{197.34 \text{ g BaCO}_3} \times \frac{1 \text{ mol C}}{\text{mol BaCO}_3} \times \frac{12.011 \text{ g C}}{\text{mol C}}}{0.1799 \text{ g sample}} \times 100\% = \underline{\underline{18.99\% \text{ C}}}$

5-27 $\mathcal{M}_{\text{CO}_2} = 44.010 \text{ g/mol}$

$\dfrac{0.1881 \text{ g CO}_2 \times \frac{1 \text{ mol CO}_2}{44.040 \text{ g CO}_2} \times \frac{1 \text{ mol Mg}}{\text{mol CO}_2} \times \frac{24.305 \text{ g Mg}}{\text{mol Mg}}}{0.7406 \text{ g sample}} \times 100\% = \underline{\underline{14.02\% \text{ Mg}}}$

5-28 $\mathcal{M}_{\text{CdSO}_4} = 208.47 \text{ g/mol} \qquad \mathcal{M}_{\text{H}_2\text{S}} = 34.08 \text{ g/mol}$

$\dfrac{0.108 \text{ g CdSO}_4 \times \frac{1 \text{ mol CdSO}_4}{208.47 \text{ g CdSO}_4} \times \frac{1 \text{ mol H}_2\text{S}}{1 \text{ mol CdSO}_4} \times \frac{34.08 \text{ g H}_2\text{S}}{\text{mol H}_2\text{S}}}{50.0 \text{ g sample}} \times 100\% = \underline{\underline{0.0353\% \text{ H}_2\text{S}}}$

5-29 $\mathcal{M}_{\text{NH}_3} = 17.0306 \text{ g/mol} \qquad \mathcal{M}_{\text{Pt}} = 195.08 \text{ g/mol}$

$\dfrac{0.5881 \text{ g Pt}}{0.2213 \text{ g sample}} \times \dfrac{1 \text{ mol Pt}}{195.08 \text{ g}} \times \dfrac{2 \text{ mol NH}_3}{1 \text{ mol Pt}} \times \dfrac{17.0306 \text{ g NH}_3}{\text{mol NH}_3} \times 100\% = \underline{\underline{46.40\% \text{ NH}_3}}$

5-30 $\mathcal{M}_{\text{BaSO}_4} = 233.39 \text{ g/mol} \qquad \mathcal{M}_{\text{C}_{21}\text{H}_{29}\text{NS}_2} = 359.6 \text{ g/mol}$

Chapter 5

$$\frac{0.3343 \text{ g BaSO}_4}{8 \text{ tablets}} \times \frac{1 \text{ mol BaSO}_4}{233.39 \text{ g BaSO}_4} \times \frac{1 \text{ mol C}_{21}\text{H}_{29}\text{NS}_2}{2 \text{ mol BaSO}_4} \times \frac{359.6 \text{ g C}_{21}\text{H}_{29}\text{NH}_2}{\text{mol}} =$$

$$0.03219 \text{ g C}_{21}\text{H}_{29}\text{NH}_2/\text{tablet}$$

5-31 $\mathcal{M}_{\text{PbMoO}_4} = 367.14$ g/mol $\mathcal{M}_{\text{P}_2\text{O}_5} = 141.94$ g/mol

$$0.2752 \text{ g PbMoO}_4 \times \frac{1 \text{ mol PbMoO}_4}{367.14 \text{ g PbMoO}_4} \times \frac{1 \text{ mol P}}{12 \text{ mol PbMoO}_4} \times \frac{1 \text{ mol P}_2\text{O}_5}{2 \text{ mol P}} =$$

$$3.1233 \times 10^{-5} \text{ mol P}_2\text{O}_5$$

$$\frac{3.1233 \times 10^{-5} \text{ mol P}_2\text{O}_5 \times \frac{141.94 \text{ g P}_2\text{O}_5}{\text{mol P}_2\text{O}_5}}{0.2374 \text{ g sample}} \times 100\% = \underline{\underline{1.867 \% \text{ P}_2\text{O}_5}}$$

5-32 $\mathcal{M}_{\text{MnO}_2} = 86.94$ g/mol $\mathcal{M}_{\text{AlCl}_3} = 133.34$ g/mol

$$(0.6447 - 0.3521) \text{ g MnO}_2 \times \frac{1 \text{ mol MnO}_2}{86.94 \text{ g MnO}_2} \times \frac{2 \text{ mol Cl}}{1 \text{ mol MnO}_2} \times \frac{1 \text{ mol AlCl}_3}{3 \text{ mol Cl}} =$$

$$2.24369 \times 10^{-3} \text{ mol AlCl}_3$$

$$\frac{2.24369 \times 10^{-3} \text{ mol AlCl}_3 \times \frac{133.34 \text{ g AlCl}_3}{\text{mol AlCl}_3}}{1.1402 \text{ g sample}} \times 100\% = \underline{\underline{26.24 \% \text{ AlCl}_3}}$$

5-33 $\mathcal{M}_{\text{Sn}} = 118.71$ g/mol $\mathcal{M}_{\text{C}_6\text{H}_5\text{NO}_2} = 123.11$ g/mol

$$(1.044 - 0.338) \text{ g Sn} \times \frac{1 \text{ mol Sn}}{118.71 \text{ g Sn}} \times \frac{2 \text{ mol C}_6\text{H}_5\text{NO}_2}{3 \text{ mol Sn}} \times \frac{123.11 \text{ g C}_6\text{H}_5\text{NO}_2}{\text{mol C}_6\text{H}_5\text{NO}_2} =$$

$$0.48811 \text{ g C}_6\text{H}_5\text{NO}_2$$

$$\frac{0.48811 \text{ g C}_6\text{H}_5\text{NO}_2}{0.5078 \text{ g sample}} \times 100\% = \underline{\underline{96.12 \% \text{ C}_6\text{H}_5\text{NO}_2}}$$

5-34 Let S_w = mass of sample in grams

$\mathcal{M}_{\text{BaSO}_4} = 233.39$ g/mol $\mathcal{M}_{\text{SO}_4^{2-}} = 96.064$ g/mol

$$0.300 \text{ g BaSO}_4 \times \frac{1 \text{ mol BaSO}_4}{233.39 \text{ g BaSO}_4} \times \frac{1 \text{ mol SO}_4^{2-}}{\text{mol BaSO}_4} = 1.2854 \times 10^{-3} \text{ mol SO}_4^{2-}$$

$$\frac{1.2854 \times 10^{-3} \text{ mol SO}_4^{2-} \times \frac{96.064 \text{ g SO}_4^{2-}}{\text{mol}} \text{SO}_4^{2-}}{S_w \text{ g sample}} \times 100\% = 20\% \text{ SO}_4^{2-}$$

$$S_w = \frac{0.12348 \text{ g SO}_4^{2-} \times 100\%}{20\% \text{ SO}_4^{2-}} = \underline{\underline{0.617 \text{ g sample}}}$$

We selected the smaller of the two percentages because it yields the smaller amount of BaSO$_4$. This quantity of sample yields at a level of 55% SO$_4^{2-}$:

$$0.617 \text{ g sample} \times \frac{55 \text{ g SO}_4^{2-}}{100 \text{ g sample}} \times \frac{1 \text{ mol SO}_4^{2-}}{96.064 \text{ g SO}_4^{2-}} \times \frac{1 \text{ mol BaSO}_4}{1 \text{ mol SO}_4^{2-}} \times \frac{233.39 \text{ g BaSO}_4}{\text{mol BaSO}_4} =$$

$$\underline{\underline{0.824 \text{ g BaSO}_4}}$$

5-35 (a) $\mathcal{M}_{AgCl} = 143.32 \text{ g/mol} \qquad \mathcal{M}_{ZrCl_4} = 233.03 \text{ g/mol}$

Let S_w = mass of sample in grams

$$\frac{0.400 \text{ g AgCl} \times \frac{1 \text{ mol AgCl}}{143.32 \text{ g}} \times \frac{1 \text{ mol ZrCl}_4}{4 \text{ mol AgCl}} \times \frac{233.03 \text{ g ZrCl}_4}{\text{mol ZrCl}_4}}{S_w \text{ g sample}} \times 100\% = 68\%$$

$$\frac{0.400 \text{ AgCl} \times 0.40649 \text{ g ZrCl}_4/\text{g AgCl}}{S_w \text{ g sample}} \times 100\% = 68\%$$

$$S_w = 0.1626 \times 100/68 = \underline{\underline{0.239 \text{ g sample}}}$$

(b) $0.239 \text{ g sample} \times \frac{84 \text{ g ZrCl}_4}{100 \text{ g sample}} \times \frac{1 \text{ mol ZrCl}_4}{233.03 \text{ g ZrCl}_4} \times \frac{4 \text{ mol AgCl}}{1 \text{ mol ZrCl}_4} \times \frac{143.32 \text{ g AgCl}}{\text{mol AgCl}} =$

$$\underline{\underline{0.494 \text{ g AgCl}}}$$

(c) The second equation in part (a) can be written as

$$\% \text{ ZrCl}_4 = \frac{\text{g AgCl} \times 0.40649 \text{ g ZrCl}_4/\text{g AgCl}}{S_w \text{ g sample}} \times 100\%$$

Chapter 5

which arranges to

$$\frac{\% \text{ ZrCl}_4}{\text{mass AgCl}} = \frac{40.649}{S_w} = 100 \qquad S_w = \underline{\underline{0.4065 \text{ g}}}$$

5-36 Let S_w = mass of sample in grams

$$\mathcal{M}_{\text{Ni(DMG)}_2} = 288.92 \text{ g/mol} \qquad \mathcal{M}_{\text{Ni}} = 58.693 \text{ g/mol}$$

$$0.175 \text{ g Ni(DMG)}_2 \times \frac{1 \text{ mol Ni(DMG)}_2}{288.92 \text{ g Ni(DMG)}_2} \times \frac{1 \text{ mol Ni}}{1 \text{ mol Ni(DMG)}_2} = 6.057 \times 10^{-4} \text{ mol Ni}$$

$$\frac{6.057 \times 10^{-4} \text{ mol Ni} \times \frac{58.693 \text{ g Ni}}{1 \text{ mol Ni}}}{S_w \text{ g sample}} \times 100\% = 35\% \text{ Ni}$$

$$S_w = \frac{0.035549 \text{ g Ni} \times 100\%}{35\% \text{ Ni}} = \underline{\underline{0.102 \text{ g sample}}}$$

We chose the larger of the two percentages because this percentage corresponds to the maximum quantity of precipitate.

5-37 $\mathcal{M}_{\text{AgCl}} = 143.32 \text{ g/mol} \qquad \mathcal{M}_{\text{AgI}} = 234.77 \text{ g/mol}$

$$\text{mass AgCl} + \text{mass AgI} = 0.4430 \text{ g} \qquad (1)$$

$$\text{mass AgCl} + \text{mass AgI} \times \frac{1 \text{ mol AgI}}{234.77 \text{ g}} \times \frac{1 \text{ mol AgCl}}{1 \text{ mol AgI}} \times \frac{143.32 \text{ g}}{\text{mol AgCl}} = 0.3181 \text{ g}$$

or

$$\text{mass AgCl} + \text{mass AgI} \left(\frac{143.32}{234.77}\right) = 0.3181 \text{ g} \qquad (2)$$

By subtracting equation (2) from equation (1), we have

$$\text{mass AgI} \left(1 - \frac{143.32}{234.77}\right) = 0.1249 \text{ g}$$

$$\text{mass AgI} = \frac{0.1249 \text{ g}}{0.3895} = 0.3206 \text{ g}$$

40

Thus, mass AgCl = (0.4430 − 0.3206) = 0.1224 g

$$\frac{0.1224 \text{ g AgCl}}{0.6407 \text{ g sample}} \times \frac{1 \text{ mol AgCl}}{143.32 \text{ g}} \times \frac{1 \text{ mol Cl}^-}{\text{mol AgCl}} \times \frac{35.453 \text{ g}}{\text{mol Cl}^-} \times 100\% = \underline{\underline{4.72\% \text{ Cl}^-}}$$

$$\frac{0.3206 \text{ g AgI}}{0.6407 \text{ g sample}} \times \frac{1 \text{ mol AgI}}{234.77 \text{ g}} \times \frac{1 \text{ mol I}^-}{\text{mol AgI}} \times \frac{126.904 \text{ g}}{\text{mol I}^-} \times 100\% = \underline{\underline{27.05\% \text{ I}^-}}$$

5-38 $\mathcal{M}_{Mg_2P_2O_7}$ = 222.55 g/mol

$\mathcal{M}_{NaCl}$ = 58.44 g/mol

$\mathcal{M}_{MgCl_2 \cdot 6H_2O}$ = 203.32 g/mol

$$0.1796 \text{ g Mg}_2\text{P}_2\text{O}_7 \times \frac{1 \text{ mol Mg}_2\text{P}_2\text{O}_7}{222.55 \text{ g Mg}_2\text{P}_2\text{O}_7} \times \frac{2 \text{ mol MgCl}_2 \cdot 6\text{H}_2\text{O}}{1 \text{ mol Mg}_2\text{P}_2\text{O}_7} =$$

$$1.61402 \times 10^{-3} \text{ mol MgCl}_2 \cdot 6\text{H}_2\text{O}$$

$$0.5923 \text{ g AgCl} \times \frac{1 \text{ mol AgCl}}{143.32 \text{ g AgCl}} = 4.1327 \times 10^{-3} \text{ mol AgCl}$$

$$1.61402 \times 10^{-3} \text{ mol MgCl}_2 \cdot 6\text{H}_2\text{O} \times \frac{2 \text{ mol AgCl}}{\text{mol MgCl}_2 \cdot 6\text{H}_2\text{O}} = 3.2280 \times 10^{-3} \text{ mol AgCl}$$

$$(4.1327 \times 10^{-3} - 3.2280 \times 10^{-3}) \text{ mol AgCl} \times \frac{1 \text{ mol NaCl}}{\text{mol AgCl}} = 0.9047 \times 10^{-3} \text{ mol NaCl}$$

$$\frac{1.61402 \times 10^{-3} \text{ mol MgCl}_2 \cdot 6\text{H}_2\text{O} \times \frac{203.31 \text{ g MgCl}_2 \cdot 6\text{H}_2\text{O}}{\text{mol MgCl}_2 \cdot 6\text{H}_2\text{O}} \times \frac{500 \text{ mL}}{50.00 \text{ mL}}}{6.881 \text{ g sample}} \times 100\% =$$

$$\underline{\underline{47.69\% \text{ MgCl}_2 \cdot 6\text{H}_2\text{O}}}$$

$$\frac{0.9047 \times 10^{-3} \text{ mol NaCl} \times \frac{58.44 \text{ g NaCl}}{\text{mol NaCl}} \times \frac{500 \text{ mL}}{50.00 \text{ mL}}}{6.881 \text{ g sample}} \times 100\% = \underline{\underline{7.684\% \text{ NaCl}}}$$

Chapter 5

5-39 $\mathcal{M}_{Ba(IO_3)_2}$ = 487.13 g/mol $\qquad \mathcal{M}_{KI}$ = 166.00 g/mol

$$0.0720 \text{ g Ba(IO}_3)_2 \times \frac{1 \text{ mol Ba(IO}_3)_2}{487.13 \text{ g Ba(IO}_3)_2} \times \frac{2 \text{ mol KI}}{\text{mol Ba(IO}_3)_2} = 2.9561 \times 10^{-4} \text{ mol KI}$$

$$\frac{2.9561 \times 10^{-4} \text{ mol KI} \times \frac{166.00 \text{ g KI}}{\text{mol KI}}}{2.72 \text{ g sample}} \times 100\% = \underline{\underline{1.80\% \text{ KI}}}$$

5-40 $\mathcal{M}_{AgIO_3}$ = 282.77 g/mol $\qquad \mathcal{M}_{Cu(IO_3)_2}$ = 413.35 g/mol

$\mathcal{M}_{Ag}$ = 107.87 g/mol $\qquad \mathcal{M}_{Cu}$ = 63.546 g/mol

(a) g Ag + g Cu = 0.2175 $\qquad\qquad\qquad$ (1)

g AgIO$_3$ + g Cu(IO$_3$)$_2$ = 0.7391 $\qquad\qquad$ (2)

$$\text{g AgIO}_3 = \text{g Ag} \times \frac{1 \text{ mol Ag}}{107.87 \text{ g Ag}} \times \frac{1 \text{ mol AgIO}_3}{\text{mol Ag}} \times \frac{282.77 \text{ g AgIO}_3}{\text{mol AgIO}_3} = 2.6214 \text{ g Ag}$$

$$\text{g Cu(IO}_3)_2 = \text{g Cu} \times \frac{1 \text{ mol Cu}}{63.546 \text{ g Cu}} \times \frac{1 \text{ mol Cu(IO}_3)_2}{\text{mol Cu}} \times \frac{413.35 \text{ g Cu(IO}_3)_2}{\text{mol Cu(IO}_3)_2} = 6.50533 \text{ g Cu}$$

Substituting these two equations into equation (2) gives

$\qquad$ 0.7391 = 2.6214 g Ag + 6.5033 g Cu $\qquad\qquad$ (3)

But from equation (1)

$\qquad$ g Cu = 0.2175 - g Ag

and substituting into equation (3) gives

$\qquad$ 0.7391 = 2.6214 g Ag + 6.5033 (0.2175 - g Ag)

Solving for g Ag gives

$\qquad$ g Ag = 0.1740

$\qquad$ g Cu = 0.2175 - 0.1740 = 0.0435

% Ag = 0.1740 × 100/0.2175 = 80.00%

% Cu = 0.0435 × 100/0.2175 = 20.00%

Proceeding in the same way, we obtain

(b) 72.00% Ag and 28.00% Cu

(c) 90.00% Ag and 10.00% Cu

(d) 60.00% Ag and 40.00% Cu

(e) 50.00% Ag and 50.00% Cu

CHAPTER 6

6-1
$$\text{amount A} = \frac{\text{mass A (g)}}{\text{mass A(g)/millimole A}} = \text{millimole A}$$

$$\text{amount A} = \text{volume (mL)} \times c_A \text{ (mmol A/mL)} = \text{millimole A}$$

6-2 (a) The *millimole* is the amount of an elementary species, such as an atom, an ion, a molecule, or an electron. A millimole contains

$$6.02 \times 10^{23} \frac{\text{particles}}{\text{mole}} \times 10^{-3} \frac{\text{mole}}{\text{millimole}} = 6.02 \times 10^{20} \frac{\text{particles}}{\text{millimole}}$$

(b) A *titration* involves measuring the quantity of a reagent of known concentration required to react with a measured quantity of sample of unknown concentration. The concentration of the sample is then determined from the quantities of reagent and sample, the concentration of the reagent, and the stoichiometry of the reaction.

(c) The *stoichiometric factor* is the molar ratio of two species that appear in a balanced chemical equation.

(d) The *molar mass* of a substance is the mass in grams of one mole of the substance.

6-3 (a) The *equivalence point* in a titration is that point at which sufficient titrant has been added so that stoichiometrically equivalent amounts of analyte and titrant are present. The *end point* in a titration is the point at which an observable physical change signals the equivalence point.

(b) *Density* is given by the expression $d = m/V$, where m is the mass of a sample of a substance and V is its volume. Density usually has units of g/mL for liquids or for solids. *Specific gravity* is the ratio of the mass of a sample of a substance to the mass of an equal volume of water at 4°C. Specific gravity has no units.

(c) A *primary standard* is a highly purified substance that serves as the basis for a titrimetric method. It is used either (1) to prepare a standard solution directly by mass or (2) to standardize a solution to be used in a titration.

A *secondary standard* is a material or solution whose concentration is determined from the stoichiometry of its reaction with a primary standard material. Secondary standards are employed when a reagent is not available in primary standard quality. For example, solid sodium hydroxide is hygroscopic and cannot be used to prepare a standard solution directly. A secondary standard solution of the reagent is readily prepared, however, by standardizing a solution of sodium hydroxide against a primary standard reagent such as potassium hydrogen phthalate.

6-4 For a dilute aqueous solution $1 \text{ L} = 1000 \text{ mL} = 1000 \text{ g}$, so

$$\frac{\text{mg}}{\text{L}} = \frac{10^{-3} \text{ g solute}}{1000 \text{ g soln}} = \frac{1 \text{ g solute}}{1,000,000 \text{ g soln}} = 1 \text{ ppm}$$

6-5 (a) $\dfrac{1 \text{ mol H}_2\text{NNH}_2}{2 \text{ mol I}_2}$ (c) $\dfrac{1 \text{ mol Na}_2\text{B}_4\text{O}_7 \cdot 10\text{H}_2\text{O}}{2 \text{ mol H}^+}$

(b) $\dfrac{5 \text{ mol H}_2\text{O}_2}{2 \text{ mol MnO}_4^-}$ (d) $\dfrac{2 \text{ mol S}}{3 \text{ mol KIO}_3}$

6-6 (a) $2.00 \text{ L} \times 2.76 \times 10^{-3} \dfrac{\text{mol}}{\text{L}} \times \dfrac{1000 \text{ mmol}}{\text{mol}} = \underline{\underline{5.52 \text{ mmol}}}$

(b) $750 \text{ mL} \times \dfrac{0.0416 \text{ mmol}}{\text{mL}} = \underline{\underline{31.2 \text{ mmol}}}$

(c) $\dfrac{4.20 \text{ g CuSO}_4}{10^6 \text{ g soln}} \times \dfrac{1.00 \text{ g soln}}{\text{mL soln}} \times \dfrac{1 \text{ mmol CuSO}_4}{0.1596/\text{g CuSO}_4} \times 250 \text{ mL soln} = \underline{\underline{6.58 \times 10^{-3} \text{ mmol}}}$

(d) $3.50 \text{ L} \times 0.276 \dfrac{\text{mol}}{\text{L}} \times \dfrac{1000 \text{ mmol}}{\text{mol}} = \underline{\underline{966 \text{ mmol}}}$

6-7 (a) $4.25 \text{ mL} \times \dfrac{0.0917 \text{ mmol KH}_2\text{PO}_4}{\text{mL}} = \underline{\underline{0.390 \text{ mmol}}}$

(b) $0.1020 \text{ L} \times \dfrac{0.0643 \text{ mol HgCl}_2}{\text{L}} \times \dfrac{10^3 \text{ mmol}}{\text{mol}} = \underline{\underline{6.56 \text{ mmol}}}$

(c) $\dfrac{49.0 \text{ g Mg(NO}_3)_2}{10^6 \text{ g soln}} \times \dfrac{1000 \text{ g soln}}{\text{L soln}} \times \dfrac{1 \text{ mol Mg(NO}_3)_2}{148.31 \text{ g Mg(NO}_3)_2} \times 2.81 \text{ L} \times \dfrac{10^3 \text{ mmol}}{\text{mol}} = \underline{\underline{0.928 \text{ mmol}}}$

(d) $79.8 \text{ mL} \times \dfrac{0.1379 \text{ mmol}}{\text{mL}} = \underline{\underline{11.00 \text{ mmol}}}$

6-8 (a) $26.0 \text{ mL} \times \dfrac{0.150 \text{ mmol sucrose}}{\text{mL}} \times \dfrac{0.342 \text{ g}}{\text{mmol sucrose}} \times \dfrac{1000 \text{ mg}}{\text{g}} = \underline{\underline{1.33 \times 10^3 \text{ mg}}}$

Chapter 6

(b) $2.92 \text{ L} \times \dfrac{5.23 \times 10^{-3} \text{ mol H}_2\text{O}_2}{\text{L}} \times \dfrac{34.02 \text{ g}}{\text{mol H}_2\text{O}_2} \times \dfrac{1000 \text{ mg}}{\text{g}} = \underline{\underline{520 \text{ mg}}}$

(c) $\dfrac{6.38 \text{ mg Pb(NO}_3)_2}{\text{L}} \times \dfrac{1 \text{ L}}{1000 \text{ mL}} \times 737 \text{ mL} = \underline{\underline{4.70 \text{ mg}}}$ $\left(1 \text{ ppm} = \dfrac{1 \text{ mg}}{\text{L}}\right)$

(d) $6.75 \text{ mL} \times \dfrac{0.0619 \text{ mmol}}{\text{mL}} \times \dfrac{101.10 \text{ mg KNO}_3}{\text{mmol}} = \underline{\underline{42.2 \text{ mg}}}$

6-9 (a) $450 \text{ mL} \times \dfrac{0.164 \text{ mol H}_2\text{O}_2}{\text{L}} \times \dfrac{34.02 \text{ g}}{\text{mol H}_2\text{O}_2} \times \dfrac{1 \text{ L}}{1000 \text{ mL}} = \underline{\underline{2.51 \text{ g}}}$

(b) $27.0 \text{ mL} \times \dfrac{8.75 \times 10^{-5} \text{ mol}}{\text{L}} \times \dfrac{122.1 \text{ g}}{\text{mol}} \times \dfrac{1 \text{ L}}{1000 \text{ mL}} = \underline{\underline{2.88 \times 10^{-3} \text{ g}}}$

(c) $3.50 \text{ L} \times \dfrac{21.7 \text{ mg SnCl}_2}{\text{L}} \times \dfrac{\text{g}}{1000 \text{ mg}} = \underline{\underline{7.60 \times 10^{-2} \text{ g}}}$ $\left(1 \text{ ppm} = \dfrac{1 \text{ mg}}{\text{L}}\right)$

(d) $21.7 \text{ mL} \times \dfrac{0.0125 \text{ mol}}{\text{L}} \times \dfrac{167.0 \text{ g}}{\text{mol}} \times \dfrac{1 \text{ L}}{1000 \text{ mL}} = \underline{\underline{4.53 \times 10^{-2} \text{ g}}}$

6-10 $\dfrac{50.0 \text{ g NaOH}}{100 \text{ g soln}} \times \dfrac{1.52 \text{ g soln}}{\text{mL soln}} \times \dfrac{1 \text{ mmol NaOH}}{0.0400 \text{ g NaOH}} = 19.0 \dfrac{\text{mmol NaOH}}{\text{mL soln}} = \underline{\underline{19.0 \text{ M}}}$

6-11 $\dfrac{20.0 \text{ g KCl}}{100 \text{ g soln}} \times \dfrac{1.13 \text{ g soln}}{\text{mL soln}} \times \dfrac{1 \text{ mmol KCl}}{0.07455 \text{ g KCl}} = 3.03 \dfrac{\text{mmol KCl}}{\text{mL soln}} = \underline{\underline{3.03 \text{ M}}}$

6-12 Proceeding as in Solutions 6-10 and 6-11, we obtain

(a) 6.161 M   (b) 1.738 M   (c) 4.669 M   (d) 0.4639 M

6-13 (a) $500 \text{ mL soln} \times \dfrac{16.0 \text{ g EtOH}}{100 \text{ mL soln}} = 80.0 \text{ g EtOH}$

Dilute 80 g ethanol to 500 mL with water.

(b) $500 \text{ mL soln} \times \dfrac{16.0 \text{ mL EtOH}}{100 \text{ mL soln}} = 80.0 \text{ mL EtOH}$

Dilute 80.0 mL ethanol to 500 mL with water.

Chapter 6

(c) $\quad 500 \text{ g soln} \times \dfrac{16.0 \text{ g EtOH}}{100 \text{ g soln}} = 80.0 \text{ g EtOH}$

<u>Dilute 80.0 g ethanol with 420 g water.</u>

6-14 Proceeding as in Solution 6-13, we obtain:

(a) dilute 50.0 g acetone to 250 mL with water.

(b) dilute 50.0 mL acetone to 250 mL with water.

(c) dilute 50.0 g acetone with 200 g water.

6-15 $\quad c_{HClO_4} = \dfrac{1.66 \times 10^3 \text{ g reagent}}{\text{L reagent}} \times \dfrac{70 \text{ g HClO}_4}{100 \text{ g reagent}} \times \dfrac{1 \text{ mol HClO}_4}{100.5 \text{ g HClO}_4} = 11.56 \dfrac{\text{mol HClO}_4}{\text{L reagent}}$

Amount $HClO_4$ required $= 2.00 \text{ L} \times \dfrac{0.150 \text{ mol HClO}_4}{\text{L}} = 0.300 \text{ mol HClO}_4$

Vol concd reagent $= 0.300 \text{ mol HClO}_4 \times \dfrac{1 \text{ L reagent}}{11.56 \text{ mol HClO}_4}$

$= 0.0260 \text{ L} \quad \text{or} \quad 26.0 \text{ mL}$

<u>Dilute 26 mL of the concentrated reagent to 2.0 L.</u>

6-16 Proceeding as in Solution 6-15 we find that the concentration of the concentrated reagent is 14.3 M.

Amount NH_3 required $= 800 \text{ mL} \times \dfrac{0.400 \text{ mmol}}{\text{mL}} = 320 \text{ mmol}$

Vol concd reagent $= 320 \text{ mmol} \times \dfrac{1 \text{ mL}}{14.3 \text{ mmol}} = 22.4 \text{ mL}$

Therefore <u>dilute about 22 mL of the concentrated reagent to about 800 mL</u>.

6-17 (a) $\quad 500 \text{ mL} \times \dfrac{1 \text{ L}}{1000 \text{ mL}} \times \dfrac{0.0750 \text{ mol}}{1 \text{ L}} \times \dfrac{169.87 \text{ g}}{1 \text{ mol}} = \underline{\underline{6.37 \text{ g}}}$

<u>Dissolve 6.37 g of AgNO$_3$ in water and dilute to 500 mL.</u>

(b) $V_{concd} c_{concd} = V_{dil} c_{dil}$

47

Chapter 6

$$V_{concd} = \frac{V_{dil}c_{dil}}{c_{concd}} = \frac{1.00 \text{ L} \times 0.315 \text{ M}}{6.00 \text{ M}} = 0.0525 \text{ L} = \underline{\underline{52.5 \text{ mL}}}$$

Dilute 52.5 mL of 6.00 M HCl to 1.00 L.

(c) $600 \text{ mL} \times \dfrac{1 \text{ L}}{1000 \text{ mL}} \times \dfrac{0.0825 \text{ mol K}^+}{\text{L}} \times \dfrac{1 \text{ mol K}_4\text{Fe(CN)}_6}{4 \text{ mol K}^+} \times \dfrac{368.35 \text{ g}}{1 \text{ mol K}_2\text{Fe(CN)}_6} = \underline{\underline{4.56 \text{ g}}}$

Dissolve 4.56 g K$_4$Fe(CN)$_6$ in water and dilute to 600 mL.

(d) $400 \text{ mL} \times \dfrac{3.00 \text{ g}}{100 \text{ mL}} \times \dfrac{1 \text{ mol}}{208.23 \text{ g}} \times \dfrac{1 \text{ L}}{0.400 \text{ mol}} = 0.144 \text{ L} = \underline{\underline{144 \text{ mL}}}$

Dilute 144 mL of 0.400 M BaCl$_2$ to a volume of 400 mL.

(e) $c_{HClO_4} = \dfrac{1.60 \times 10^3 \text{ g reagent}}{\text{L reagent}} \times \dfrac{60 \text{ g HClO}_4}{100 \text{ g reagent}} \times \dfrac{1 \text{ mol HClO}_4}{100.5 \text{ g HClO}_4} = 9.55 \dfrac{\text{mol HClO}_4}{\text{L reagent}}$

Amount HClO$_4$ required $= 2.00 \text{ L} \times 0.120 \dfrac{\text{mol HClO}_4}{\text{L}} = 0.240 \text{ mol HClO}_4$

Vol concd reagent $= 0.240 \text{ mol HClO}_4 \times \dfrac{1 \text{ L reagent}}{9.55 \text{ mol HClO}_4} \times \dfrac{10^3 \text{ mL}}{\text{L}} = \underline{\underline{25.1 \text{ mL}}}$

Dilute 25 mL of the commercial reagent to a volume of 2.0 L.

(f) $9.00 \times 10^3 \text{ mL soln} \times \dfrac{1.00 \text{ g soln}}{\text{mL soln}} \times \dfrac{60 \text{ g Na}^+}{10^6 \text{ g soln}} \times \dfrac{1 \text{ mol Na}^+}{22.99 \text{ g soln}} \times \dfrac{1 \text{ mol Na}_2\text{SO}_4}{2 \text{ mol Na}^+} \times$

$\dfrac{142.0 \text{ g Na}_2\text{SO}_4}{\text{mol Na}_2\text{SO}_4} = \underline{\underline{1.67 \text{ g Na}_2\text{SO}_4}}$

Dissolve 1.67 g Na$_2$SO$_4$ in water and dilute to 9.00 L.

6-18 (a) $5.00 \text{ L} \times \dfrac{0.150 \text{ mol}}{\text{L}} \times \dfrac{158.03 \text{ g}}{\text{mol}} = \underline{\underline{118.5 \text{ g}}}$

Dissolve 118.5 g of KMnO$_4$ in water and make up to 5.00 L.

(b) $V_{concd}c_{concd} = V_{dil}c_{dil}$

$$V_{concd} = \frac{V_{dil}c_{dil}}{c_{concd}} = \frac{4.00 \text{ L} \times 0.175 \text{ M}}{8.00 \text{ M}}$$

$$= 0.0875 \text{ L} = \underline{\underline{87.5 \text{ mL}}}$$

Dilute 87.5 mL of the concentrated reagent to 4.00 L.

(c) $400 \text{ mL} \times \dfrac{0.0500 \text{ mol I}^-}{\text{L}} \times \dfrac{1 \text{ L}}{1000 \text{ mL}} \times \dfrac{1 \text{ mol MgI}_2}{2 \text{ mol I}^-} \times \dfrac{278.11 \text{ g MgI}_2}{\text{mol MgI}_2} = \underline{\underline{2.78 \text{ g MgI}_2}}$

Dissolve 2.78 g MgI$_2$ in water and dilute to 400 mL.

(d) $200 \text{ mL} \times \dfrac{1.00 \text{ g CuSO}_4}{100 \text{ mL}} \times \dfrac{1 \text{ mol CuSO}_4}{159.61 \text{ g CuSO}_4} \times \dfrac{1 \text{ L}}{0.218 \text{ mol}} \times \dfrac{1000 \text{ mL}}{1 \text{ L}} = \underline{\underline{57.5 \text{ mL}}}$

Dilute 57.5 mL of the 0.218 M solution to a volume of 200 mL.

(e) $c_{NaOH} = \dfrac{1.525 \times 10^3 \text{ g reagent}}{\text{L reagent}} \times \dfrac{50 \text{ g NaOH}}{100 \text{ g reagent}} \times \dfrac{1 \text{ mol NaOH}}{40.00 \text{ g NaOH}} = 19.06 \dfrac{\text{mol NaOH}}{\text{L}}$

Amount NaOH required $= 1.50 \text{ L} \times \dfrac{0.215 \text{ mol NaOH}}{\text{L}} = 0.3225 \text{ mol NaOH}$

Vol concd reagent $= 0.3225 \text{ mol NaOH} \times \dfrac{1 \text{ L reagent}}{19.06 \text{ mol}} \times \dfrac{10^3 \text{ mL}}{\text{L}} = \underline{\underline{16.9 \text{ mL}}}$

Dilute 16.8 mL of the concentrated reagent to 1.50 L.

(f) $1.50 \text{ L soln} \times \dfrac{1000 \text{ g soln}}{\text{L soln}} \times \dfrac{12.0 \text{ g K}^+}{10^6 \text{ g soln}} \times \dfrac{1 \text{ mol K}^+}{39.10 \text{ g K}^+} \times \dfrac{1 \text{ mol K}_4\text{Fe(CN)}_6}{4 \text{ mol K}^+} \times$

$\dfrac{368.3 \text{ g K}_4\text{Fe(CN)}_6}{\text{mol K}_5\text{Fe(CN)}_6} = \underline{\underline{0.0424 \text{ g K}_4\text{Fe(CN)}_6}}$

Dissolve 0.0424 g K$_4$Fe(CN)$_6$ in water and dilute to 1.50 L.

6-19 $\mathcal{M}_{HgO} = 216.59 \text{ g/mol}$

$$\dfrac{0.3745 \text{ g HgO} \times \dfrac{1 \text{ mmol HgO}}{0.21659 \text{ g HgO}} \times \dfrac{2 \text{ mmol OH}^-}{1 \text{ mmol HgO}} \times \dfrac{1 \text{ mmol HClO}_4}{1 \text{ mmol OH}^-}}{37.79 \text{ mL soln}} = \underline{\underline{9.151 \times 10^{-2} \text{ M}}}$$

Chapter 6

6-20 $m\mathcal{M}_{Na_2CO_3}$ = 0.10599 g/mol

$$\frac{0.3367 \text{ g Na}_2\text{CO}_3 \times \frac{1 \text{ mmol Na}_2\text{CO}_3}{0.10599 \text{ g Na}_2\text{CO}_3} \times \frac{2 \text{ mmol H}^+}{1 \text{ mmol Na}_2\text{CO}_3} \times \frac{1 \text{ mmol H}_2\text{SO}_4}{2 \text{ mmol H}^+}}{28.66 \text{ mL soln}} = \underline{\underline{0.1108 \text{ M H}_2\text{SO}_4}}$$

6-21 $\mathcal{M}_{Na_2SO_4}$ = 142.04 g/mol

$$\frac{0.3396 \text{ g sample} \times \frac{96.4 \text{ g Na}_2\text{SO}_4}{100 \text{ g sample}} \times \frac{1 \text{ mmol Na}_2\text{SO}_4}{0.14204 \text{ g Na}_2\text{SO}_4} \times \frac{1 \text{ mmol BaCl}_2}{1 \text{ mmol Na}_2\text{SO}_4}}{37.70 \text{ mL}} = \underline{\underline{0.06114 \text{ M BaCl}_2}}$$

6-22 $\dfrac{V_{HClO_4}}{V_{NaOH}} = \dfrac{27.43 \text{ mL HClO}_4}{25.00 \text{ mL NaOH}} = 1.0972 \dfrac{\text{mL HClO}_4}{\text{mL NaOH}}$

The volume of $HClO_4$ needed to titrate 0.4793 g of Na_2CO_3 is

$$40.00 \text{ mL HClO}_4 - 8.70 \text{ mL NaOH} \times \frac{1.0972 \text{ mL HClO}_4}{\text{mL NaOH}} = 30.45 \text{ mL}$$

Thus

$$\frac{0.4793 \text{ g Na}_2\text{CO}_3}{30.45 \text{ mL HClO}_4} \times \frac{1 \text{ mmol Na}_2\text{CO}_3}{0.10599 \text{ g Na}_2\text{CO}_3} \times \frac{2 \text{ mmol HClO}_4}{\text{mmol Na}_2\text{CO}_3} = \underline{\underline{0.2970 \text{ M HClO}_4}}$$

and

$$c_{NaOH} = c_{HClO_4} \times \frac{V_{HClO_4}}{V_{NaOH}}$$

$$= \frac{0.2970 \text{ mmol HClO}_4}{\text{mL HClO}_4} \times \frac{1.0972 \text{ mL HClO}_4}{\text{mL NaOH}} \times \frac{1 \text{ mmol NaOH}}{\text{mmol HClO}_4} = \underline{\underline{0.3259 \text{ M}}}$$

6-23 $\dfrac{50.00 \text{ mL Na}_2\text{C}_2\text{O}_4 \times 0.05251 \times \frac{1 \text{ mmol Na}_2\text{C}_2\text{O}_4}{\text{mL Na}_2\text{C}_2\text{O}_4} \times \frac{2 \text{ mmol KMnO}_4}{5 \text{ mmol Na}_2\text{C}_2\text{O}_4}}{38.71 \text{ mL KMnO}_4} = \underline{\underline{0.02713 \text{ M}}}$

6-24 $\dfrac{0.1238 \text{ g KIO}_3 \times \frac{1 \text{ mmol KIO}_3}{0.21400 \text{ g KIO}_3} \times \frac{6 \text{ mmol Na}_2\text{S}_2\text{O}_3}{\text{mmol KIO}_3}}{41.27 \text{ mL Na}_2\text{S}_2\text{O}_3} = \underline{\underline{0.08411 \text{ M}}}$

6-25 total mmol NaOH = 25.00 mL × 0.00923 mmol/mL = 0.23075

mmol HCl added = 13.33 mL × 0.01007 mmol/mL = 0.13423

mmol NaOH consumed by analyte = 0.23075 − 0.13423 = 0.09652 mmol NaOH

$$\frac{0.0965 \text{ mmol NaOH} \times \frac{1 \text{ mmol H}_2\text{SO}_4}{2 \text{ mmol NaOH}} \times \frac{1 \text{ mmol S}}{1 \text{ mmol H}_2\text{SO}_4} \times \frac{0.03207 \text{ g S}}{\text{mmol S}}}{4.476 \text{ g sample}} \times 10^6 \text{ ppm} = \underline{\underline{345.8 \text{ ppm S}}}$$

6-26 total no. mmol Fe^{2+} = $25.00 \text{ mL} \times \frac{0.002107 \text{ mmol Cr}_2\text{O}_7^{2-}}{\text{mL}} \times \frac{6 \text{ mmol Fe}^{2+}}{1 \text{ mmol Cr}_2\text{O}_7^{2-}} = 0.3160$

= no. mmol analyte Fe^{2+} + no. mmol Fe^{2+} back titrated

no. mmol analyte Fe^{2+} = 0.3160 mmol − 7.47 mL × 0.00979 mmol/mL

= 0.2429 mmol Fe^{2+}

$$\frac{0.2429 \text{ mmol Fe}}{100 \text{ mL soln}} \times \frac{0.055847 \text{ g Fe}}{\text{mmol Fe}} \times \frac{1 \text{ mL soln}}{1 \text{ g soln}} \times 10^6 \text{ ppm} = \underline{\underline{135.7 \text{ ppm Fe}}}$$

6-27 $\mathcal{M}_{\text{As}_2\text{O}_3}$ = 197.84 g/mol

amount Ag$^+$ required for Ag$_3$AsO$_4$ = total mmol Ag$^+$ − no. mmol KSCN

= 40.00 mL × 0.07891 M − 11.27 mL × 0.1000 M

= 2.0294 mmol Ag$^+$

$$\frac{2.0294 \text{ mmol Ag}^+ \times \frac{1 \text{ mmol As}}{3 \text{ mmol Ag}} \times \frac{1 \text{ mmol As}_2\text{O}_3}{2 \text{ mmol As}} \times \frac{0.19784 \text{ g As}_2\text{O}_3}{\text{mmol As}_2\text{O}_3}}{1.223 \text{ g sample}} \times 100\% = \underline{\underline{5.471\% \text{ As}_2\text{O}_3}}$$

6-28
$$\frac{37.31 \text{ mL Hg}^{2+} \times 0.009372 \frac{\text{mmol Hg}^{2+}}{\text{mL Hg}^{2+}} \times \frac{4 \text{ mmol analyte}}{\text{mmol Hg}^{2+}} \times 0.07612 \frac{\text{g analyte}}{\text{mmol analyte}}}{1.455 \text{ g sample}} \times 100\% =$$

$$\underline{\underline{7.317\% \text{ (NH}_2)_2\text{CS}}}$$

6-29 total amount KOH = $40.00 \text{ mL KOH} \times 0.04672 \frac{\text{mmol KOH}}{\text{mL KOH}} = 1.8688 \text{ mmol KOH}$

amount KOH reacting with H$_2$SO$_4$ = $3.41 \text{ mL H}_2\text{SO}_4 \times 0.05042 \frac{\text{mmol H}_2\text{SO}_4}{\text{mL H}_2\text{SO}_4} \times \frac{2 \text{ mmol KOH}}{\text{mmol H}_2\text{SO}_4}$

= 0.34386 mmol KOH

Chapter 6

$$\text{mass EtOAc} = (1.8688 - 0.34386) \text{ mmol KOH} \times \frac{1 \text{ mmol EtOAc}}{\text{mmol KOH}} \times 0.08811 \frac{\text{g EtOAc}}{\text{mmol EtOAc}}$$

$$= 0.13436 \text{ g EtOAc}$$

$$\frac{\text{mass EtOAc}}{100 \text{ mL sample}} = \frac{0.13436 \text{ g EtOAc}}{10.0 \text{ mL sample} \times 20 \text{ mL}/100 \text{ mL}} \times \frac{100 \text{ mL sample}}{100 \text{ mL sample}}$$

$$= \underline{\underline{6.718 \text{ g EtOAc}/100 \text{ mL sample}}}$$

6-30 (a) $$\frac{0.1016 \text{ g } C_6H_5COOH \times \frac{1 \text{ mmol } C_6H_5COOH}{0.12212 \text{ g } C_6H_5COOH} \times \frac{1 \text{ mmol Ba(OH)}_2}{2 \text{ mmol } C_6H_5COOH}}{44.42 \text{ mL}} = \underline{\underline{9.36 \times 10^{-3} \text{ M Ba(OH)}_2}}$$

(b) $$\left(\frac{\sigma_M}{M}\right)^2 = \left(\frac{\sigma_W}{W}\right)^2 + \left(\frac{\sigma_V}{V}\right)^2 = \left(\frac{0.2 \text{ mg}}{101.6 \text{ mg}}\right)^2 + \left(\frac{0.03 \text{ mL}}{44.42 \text{ mL}}\right)^2 = 4.33 \times 10^{-6}$$

$$\frac{\sigma_M}{M} = \sqrt{4.33 \times 10^{-6}} = 2.08 \times 10^{-3} = 2.1 \times 10^{-3}$$

$$\sigma_M = 2.08 \times 10^{-3} \times 9.36 \times 10^{-3} = \underline{\underline{1.9 \times 10^{-5} \text{ M}}}$$

(c) Assuming that the systematic error in volume is zero

$$\frac{\Delta M}{M} = \frac{\Delta W}{W} = \frac{-0.3 \text{ mg}}{101.6 \text{ mg}} = -2.95 \times 10^{-3} \text{ or } \underline{\underline{-3 \text{ ppt}}}$$

6-31 $$43.17 \text{ mL Ba(OH)}_2 \times \frac{0.1475 \text{ mmol Ba(OH)}_2}{\text{mL Ba(OH)}_2} \times \frac{2 \text{ mmol HOAc}}{1 \text{ mmol Ba(OH)}_2} = 12.735 \text{ mmol HOAc}$$

$$\frac{(43.17 \times 0.1475 \times 2) \text{ mmol HOAc} \times 0.06005 \frac{\text{g HOAc}}{\text{mmol HOAc}}}{50.00 \text{ mL HOAc}} \times 100\% = \underline{\underline{1.529\% \text{ (w/v) HOAc}}}$$

Similar calculations for samples 2 through 4 yield the following data.

Chapter 6

Sample	% HOAc (w/v)	x_i^2
1	1.5295	2.33937025
2	1.5274	2.33295076
3	1.5213	2.31435369
4	1.5352	2.35683904

$\sum x_i = 6.1134 \quad \sum x_i^2 = 9.34351374$

(a) $\bar{x} = \dfrac{\sum x_i}{4} = \dfrac{6.1134}{4} = \underline{\underline{1.528}}$

(b) $s = \sqrt{\dfrac{\sum x_i^2 - (\sum x_i)^2/4}{3}} = \sqrt{\dfrac{9.34351374 - (6.1134)^2/4}{3}} = \underline{\underline{0.00574\% \text{ HOAc}}}$

(c) $CI_{90\%} = \bar{x} \pm \dfrac{ts}{\sqrt{4}} = 1.528 \pm \dfrac{2.35 \times 0.00574}{\sqrt{4}} = \underline{\underline{1.528(\pm 0.007)\% \text{ HOAc}}}$

The values of 1.535 and 1.521 can be considered for rejection.

Applying the Q test we find

$Q_{expt} = \dfrac{1.535 - 1.529}{1.535 - 1.521} = 0.43 \qquad Q_{expt} = \dfrac{1.527 - 1.521}{1.535 - 1.521} = 0.43$

Both are less than $Q_{crit} = 0.76$, so <u>neither should be rejected</u>.

(d) $\dfrac{\Delta \% \text{HOAc}}{\% \text{HOAc}} = \dfrac{\Delta V}{V}$

For sample 1, $\dfrac{\Delta \% \text{HOAc}}{\% \text{HOAc}} = -\dfrac{0.05 \text{ mL}}{50.00 \text{ mL}} = -0.001$

Sample	Relative Error
1	- 0.0010
2	- 0.0010
3	- 0.0020
4	- 0.0010

$\sum x = -0.0050$

Mean relative systematic error = $- 0.0050/4 = - 0.00125$

Chapter 6

For the mean % HOAc, $\Delta\%\text{HOAc} = -0.00125 \times 1.528 = \underline{\underline{-0.002\% \text{ HOAc}}}$

6-32 (a) $\dfrac{0.3147 \text{ g}}{31.672 \text{ g soln}} \times \dfrac{1 \text{ mmol Na}_2\text{C}_2\text{O}_4}{0.13400 \text{ g}} \times \dfrac{2 \text{ mmol KMnO}_4}{5 \text{ mmol Na}_2\text{C}_2\text{O}_4} =$

$$\underline{\underline{\dfrac{0.02966 \text{ mmol KMnO}_4}{\text{g soln}}}} = c_{w(\text{KMnO}_4)}$$

(b) $26.753 \text{ g soln} \times \dfrac{0.02966 \text{ mmol}}{\text{g soln}} \times \dfrac{5 \text{ mmol Fe}}{1 \text{ mmol KMnO}_4} \times \dfrac{1 \text{ mmol Fe}_2\text{O}_3}{2 \text{ mmol Fe}} \times \dfrac{0.15969 \text{ g}}{\text{mmol Fe}_2\text{O}_3} \times 100\%$

$$= \underline{\underline{31.68\% \text{ Fe}_2\text{O}_3}}$$

6-33 (a) $\dfrac{0.1752 \text{ g AgNO}_3}{502.3 \text{ g AgNO}_3 \text{ soln}} \times \dfrac{1 \text{ mmol AgNO}_3}{0.16987 \text{ g AgNO}_3} = \underline{\underline{\dfrac{2.0533 \times 10^{-3} \text{ mmol AgNO}_3}{\text{g AgNO}_3 \text{ soln}}}} = c_{w(\text{AgNO}_3)}$

(b) $\dfrac{23.765 \text{ g AgNO}_3 \text{ soln}}{25.171 \text{ g KSCN soln}} \times \dfrac{2.0533 \times 10^{-3} \text{ mmol AgNO}_3}{\text{g AgNO}_3 \text{ soln}} \times \dfrac{1 \text{ mmol KSCN}}{\text{mmol AgNO}_3} = \underline{\underline{1.9386 \times 10^{-3}}}$

$$= c_{w(\text{KSCN})}$$

(c) $20.102 \times 2.0533 \times 10^{-3} - 7.543 \times 1.9386 \times 10^{-6} = 2.665 \times 10^{-2}$ mmol AgNO$_3$

2.665×10^{-2} mmol AgNO$_3 \times \dfrac{1 \text{ mmol BaCl}_2 \cdot 2\text{H}_2\text{O}}{2 \text{ mmol AgNO}_3} = 1.3326 \times 10^{-2}$ mmol BaCl$_2 \cdot$ 2H$_2$O

$\dfrac{1.3326 \times 10^{-2} \text{ mmol BaCl}_2 \cdot 2\text{H}_2\text{O}}{0.7120 \text{ g sample}} \times \dfrac{0.24428 \text{ g BaCl}_2 \cdot 2\text{H}_2\text{O}}{\text{mmol BaCl}_2 \cdot 2\text{H}_2\text{O}} \times 100\% = \underline{\underline{0.4572\% \text{ BaCl}_2 \cdot 2\text{H}_2\text{O}}}$

6-34 (a) $c = \dfrac{10.12 \text{ g} \times \frac{1 \text{ mol}}{277.85 \text{ g}}}{2.000 \text{ L}} = \underline{\underline{1.821 \times 10^{-2} \text{ M}}}$

(b) $[\text{Mg}^{2+}] = \underline{\underline{1.821 \times 10^{-2} \text{ M}}}$

(c) $[\text{Cl}^-] = 2[\text{Mg}^{2+}] + [\text{K}^+] = 3(1.821 \times 10^{-2}) = \underline{\underline{5.463 \times 10^{-2} \text{ M}}}$

(d) $\dfrac{10.12 \text{ g}}{2.00 \text{ L}} \times \dfrac{1 \text{ L}}{1000 \text{ mL}} \times 100\% = \underline{\underline{0.506\% \text{ (w/v)}}}$

54

Chapter 6

(e) $\dfrac{1.821 \times 10^{-2} \text{ mmol KCl} \cdot \text{MgCl}_2}{\text{mL}} \times \dfrac{3 \text{ mmol Cl}^-}{\text{mmol KCl} \cdot \text{MgCl}_2} \times 25.0 \text{ mL} = \underline{\underline{1.366 \text{ mmol Cl}^-}}$

(f) $1.821 \times 10^{-2} \dfrac{\text{mmol salt}}{\text{mL}} \times \dfrac{1 \text{ mmol K}^+}{\text{mmol salt}} \times \dfrac{39.10 \text{ mg}}{\text{mmol K}^+} \times \dfrac{1000 \text{ mL}}{\text{L}} = \dfrac{712 \text{ mg K}^+}{\text{L}} = \underline{\underline{712 \text{ ppm K}^+}}$

6-35 (a) $c_{K_3Fe(CN)_6} = \dfrac{367 \text{ mg} \times \frac{1 \text{ g}}{1000 \text{ mg}} \times \frac{1 \text{ mol}}{329.2 \text{ g}}}{750 \text{ mL} \times \frac{1 \text{ L}}{1000 \text{ mL}}} = \underline{\underline{1.49 \times 10^{-3} \text{ M}}}$

(b) $[K^+] = 3 c_{K_3Fe(CN)_6} = 3[Fe(CN)_6^{3-}] = \underline{\underline{4.46 \times 10^{-3} \text{ M}}}$

(c) $[Fe(CN)_6^{3-}] = c_{K_3Fe(CN)_6} = \underline{\underline{1.49 \times 10^{-3} \text{ M}}}$

(d) $\dfrac{367 \text{ mg K}_3Fe(CN)_6}{750 \text{ mL}} \times \dfrac{1 \text{ g}}{1000 \text{ mg}} \times 100\% = \underline{\underline{4.89 \times 10^{-2} \% \text{ (m/v)}}}$

(e) $\dfrac{1.49 \times 10^{-3} \text{ mmol K}_3Fe(CN)_6}{\text{mL}} \times \dfrac{3 \text{ mmol K}^+}{\text{mmol K}_3Fe(CN)_6} \times 50 \text{ mL} = \underline{\underline{2.24 \times 10^{-1} \text{ mmol K}^+}}$

(f) $\mathcal{M} Fe(CN)_6^{3-} = 212.0 \text{ g/mol}$

$1.49 \times 10^{-3} \dfrac{\text{mol}}{\text{L}} \times \dfrac{212.0 \text{ g}}{\text{mol}} \times \dfrac{1 \text{ L}}{1000 \text{ g}} \times 10^6 \text{ ppm} = \underline{\underline{316 \text{ ppm}}}$

CHAPTER 7

7-1 (a) A weak electrolyte is a substance that ionizes only partially in a solvent.

(b) A Brønsted-Lowry acid is a substance that is a proton donor.

(c) The conjugate base of a Brønstead-Lowry acid is the species formed when the acid has donated a proton.

(d) Neutralization is a process in which an acid reacts with a base form a conjugate base and a conjugate acid.

(e) An amphiprotic solute is one that can act either as an acid or as a base when dissolved in a solvent.

(f) A zwitterion is a product of an internal acid/base reaction that bears both a positive and a negative charge. A common source of zwitterions is the amino acids such as glycine. When dissolved in water it forms a zwitterion having the formula $NH_3^+CH_2COO^-$.

(g) Autoprotolysis is self-ionization of a solvent to give a conjugate acid and a conjugate base.

(h) A strong acid is an acid that dissociates completely in a solvent.

(i) The Le Châtelier principle states that the position of equilibrium in a system always shifts in a direction that tends to relieve an applied stress to the system.

(j) The common ion effect arises when one of the participating ions in an equilibrium is introduced into the system. The effect is to cause a shift in the equilibrium in a direction that consumes part of the added ion.

7-2 (a) An amphiprotic solvent is a solvent that acts as a base with acidic solutes and as an acid with basic solutes.

(b) A differentiating solvent is one in which acids (or bases) dissociate incompletely and to different extents.

(c) A leveling solvent is one in which a series of acids (or bases) all dissociate completely.

(d) The mass action effect arises when one of the participants in an equilibrium is introduced into the system, which causes the equilibrium to shift in a direction such that part of the added species is consumed.

Chapter 7

7-3 For an aqueous equilibrium in which water is a reactant or a product, the concentration of water is normally so much larger than the concentrations of the reactants and products that its concentration can be assumed to be constant and independent of the position of the equilibrium. Thus, its concentration is assumed to be constant and is included in the equilibrium constant. For a solid reactant or product, it is the concentration of that reactant in the solid phase that would influence the position of equilibrium. However, the concentration of a species in the solid phase is constant. Thus, as long as some solid exists as a second phase, its effect on the equilibrium is constant, and its concentration is included in the equilibrium constant.

7-4

	Acid	Conjugate Base
(a)	HCN	CN^-
(b)	H_2O	OH^-
(c)	NH_4^+	NH_3
(d)	HCO_3^-	CO_3^{2-}
(e)	$H_2PO_4^-$	HPO_4^{2-}

7-5

	Base	Conjugate Acid
(a)	H_2O	H_3O^+
(b)	$HONH_2$	$HONH_3^+$
(c)	H_2O	H_3O^+
(d)	HCO_3^-	H_2CO_3
(e)	PO_4^{3-}	HPO_4^{2-}

7-6 (a) $2H_2O \rightleftarrows H_3O^+ + OH^-$

(b) $2CH_3COOH \rightleftarrows CH_3COOH_2^+ + CH_3COO^-$

(c) $2CH_3NH_2 \rightleftarrows CH_3NH_3^+ + CH_3NH^-$

57

Chapter 7

(d) $2CH_3CH_2OH \rightleftarrows CH_3CH_2OH_2^+ + CH_3CH_2O^-$

7-7 (a) $K_{sp} = [Ag^+][I^-] = 8.3 \times 10^{-17}$

(b) $K_{sp} = [Pb^{2+}][Cl^-]^2 = 1.7 \times 10^{-5}$

(c) $K_{sp} = [Ag^+]^2[CrO_4^{2-}] = 1.2 \times 10^{-12}$

(d) $K_{sp} = [Ba^{2+}][CrO_4^{2-}] = 2.1 \times 10^{-10}$

7-8 (a) $C_2H_5NH_2 + H_2O \rightleftarrows C_2H_5NH_3^+ + OH^-$

$$K_b = \frac{K_w}{K_a} = \frac{1.00 \times 10^{-14}}{2.31 \times 10^{-11}} = \frac{[C_2H_5NH_3^+][OH^-]}{[C_2H_5NH_2]} = \underline{\underline{4.33 \times 10^{-4}}}$$

(b) $HCN + H_2O \rightleftarrows H_3O^+ + CN^-$

$$K_a = \frac{[H_3O^+][CN^-]}{[HCN]} = \underline{\underline{6.2 \times 10^{-10}}}$$

(c) $C_5H_5NH^+ + H_2O \rightleftarrows C_5H_5N + H_3O^+$

$$K_a = \frac{[H_3O^+][C_5H_5N]}{[C_5H_5NH^+]} = \underline{\underline{5.90 \times 10^{-6}}}$$

(d) $CN^- + H_2O \rightleftarrows HCN + OH^-$

$$K_b = \frac{K_w}{K_a} = \frac{1.00 \times 10^{-14}}{6.2 \times 10^{-10}} = \underline{\underline{1.6 \times 10^{-5}}} = \frac{[OH^-][HCN]}{[CN^-]}$$

(e) $H_3AsO_4 + 3H_2O \rightleftarrows 3H_3O^+ + AsO_4^{3-}$

$$\beta_3 = K_1K_2K_3 = 5.8 \times 10^{-3} \times 1.1 \times 10^{-7} \times 3.2 \times 10^{-12}$$

$$= \underline{\underline{2.0 \times 10^{-21}}} = \frac{[H_3O^+]^3[AsO_4^{3-}]}{[H_3AsO_4]}$$

(f) $CO_3^{2-} + 2H_2O \rightleftarrows 2OH^- + H_2CO_3$

Chapter 7

$$K_b = \frac{K_w}{K_1 K_2} = \frac{1.00 \times 10^{-14}}{4.45 \times 10^{-7} \times 4.69 \times 10^{-11}}$$

$$= \underline{\underline{4.79 \times 10^2}} = \frac{[OH^-]^2[H_2CO_3]}{[CO_3^{2-}]}$$

7-9 (a) $AgIO_3(s) \rightleftarrows Ag^+ + IO_3^-$ $\qquad K_{sp} = [Ag^+][IO_3^-]$

(b) $Ag_2SO_3(s) \rightleftarrows 2Ag^+ + SO_3^{2-}$ $\qquad K_{sp} = [Ag^+]^3[SO_3^{2-}]$

(c) $Ag_3AsO_4(s) \rightleftarrows 3Ag^+ + AsO_4^{3-}$ $\qquad K_{sp} = [Ag^+]^3[AsO_4^{3-}]$

(d) $PbClF(s) \rightleftarrows Pb^{2+} + Cl^- + F^-$ $\qquad K_{sp} = [Pb^{2+}][Cl^-][F^-]$

(e) $CuI(s) \rightleftarrows Cu^+ + I^-$ $\qquad K_{sp} = [Cu^+][I^-]$

(f) $PbI_2(s) \rightleftarrows Pb^{2+} + 2I^-$ $\qquad K_{sp} = [Pb^{2+}][I^-]^2$

(g) $BiI_3(s) \rightleftarrows Bi^{3+} + 3I^-$ $\qquad K_{sp} = [Bi^{3+}][I^-]^3$

(h) $MgNH_4PO_4(s) \rightleftarrows Mg^{2+} + NH_4^+ + PO_4^{3-}$ $\qquad K_{sp} = [Mg^{2+}][NH_4^+][PO_4^{3-}]$

7-10 (a) $K_{sp} = S^2$ $\qquad$ (e) $K_{sp} = S^2$

(b) $K_{sp} = 4S^3$ $\qquad$ (f) $K_{sp} = 4S^3$

(c) $K_{sp} = 27S^4$ $\qquad$ (g) $K_{sp} = 27S^4$

(d) $K_{sp} = S^3$ $\qquad$ (h) $K_{sp} = S^3$

7-11 (a) $AgSeCN \rightleftarrows Ag^+ + SeCN^-$ $\quad$ Let S = solubility = $[Ag^+]$ = $[SeCN^-]$

$$K_{sp} = [Ag^+][SeCN^-] = S^2 = (2.0 \times 10^{-8})^2 = \underline{\underline{4.0 \times 10^{-16}}}$$

(b) $RaSO_4 \rightleftarrows Ra^{2+} + SO_4^{2-}$

$$K_{sp} = [Ra^{2+}][SO_4^{2-}] = S^2 = (6.6 \times 10^{-6})^2 = \underline{\underline{4.4 \times 10^{-11}}}$$

Chapter 7

(c) $K_{sp} = [Ba^{2+}][BrO_3^-]^2 = 9.2 \times 10^{-3}(2 \times 9.2 \times 10^{-3})^2 = \underline{\underline{3.1 \times 10^{-6}}}$

(d) $K_{sp} = [Pb^{2+}][F^-]^2 = 1.9 \times 10^{-3}(2 \times 1.9 \times 10^{-3})^2 = \underline{\underline{2.7 \times 10^{-8}}}$

(e) $K_{sp} = [Ce^{3+}][IO_3^-]^3 = 1.9 \times 10^{-3}(3 \times 1.9 \times 10^{-3})^3 = \underline{\underline{3.5 \times 10^{-10}}}$

(f) $K_{sp} = [Bi^{3+}][I^-]^3 = 1.3 \times 10^{-5}(3 \times 1.3 \times 10^{-5})^3 = \underline{\underline{7.7 \times 10^{-19}}}$

7-12 (a) $S = [SeCN^-] = \dfrac{K_{sp}}{[Ag^+]} = \dfrac{4.00 \times 10^{-16}}{0.050} = \underline{\underline{8.0 \times 10^{-15} \text{ M}}}$

(b) $S = [SO_4^{2-}] = K_{sp} = \dfrac{4.4 \times 10^{-11}}{0.050} = \underline{\underline{8.8 \times 10^{-10} \text{ M}}}$

(c) $S = \dfrac{1}{2}[BrO_3^-]$

$[BrO_3^-] = \left(\dfrac{K_{sp}}{0.050}\right)^{1/2} = \left(\dfrac{3.1 \times 10^{-6}}{0.050}\right)^{1/2} = 7.87 \times 10^{-2}$

$S = 7.87 \times 10^{-2}/2 = \underline{\underline{3.9 \times 10^{-3} \text{ M}}}$

(d) $S = \dfrac{1}{2}[F^-]$

$[F^-] = (K_{sp}/0.0500)^{1/2} = (2.7 \times 10^{-8}/0.0500)^{1/2} = 7.35 \times 10^{-4}$

$S = 7.35 \times 10^{-4}/2 = \underline{\underline{3.7 \times 10^{-4} \text{ M}}}$

(e) $S = \dfrac{1}{3}[IO_3^-]$

$[IO_3^-] = \left(\dfrac{3.5 \times 10^{-10}}{0.050}\right)^{1/3} = 1.9 \times 10^{-3}$

$S = 1.9 \times 10^{-3}/3 = \underline{\underline{6.4 \times 10^{-4} \text{ M}}}$

Chapter 7

(f) Proceeding in the same way, we obtain solubility = 8.3×10^{-7} M.

7-13 (a) $S = [Ag^+] = K_{sp}/[SeCN^-] = 4.0 \times 10^{-16}/0.050 = \underline{\underline{8.0 \times 10^{-15} \text{ M}}}$

(b) $S = [Ra^{2+}] = 4.4 \times 10^{-11}/0.050 = \underline{\underline{8.8 \times 10^{-10} \text{ M}}}$

(c) $S = [Ba^{2+}] = 3.1 \times 10^{-6}/(0.050)^2 = \underline{\underline{1.2 \times 10^{-3} \text{ M}}}$

(d) $S = [Pb^{2+}] = 2.7 \times 10^{-8}/(0.050)^2 = \underline{\underline{1.1 \times 10^{-5} \text{ M}}}$

(e) $S = [Ce^{3+}] = 3.5 \times 10^{-10}/(0.050)^3 = \underline{\underline{2.8 \times 10^{-6} \text{ M}}}$

(f) $S = [Bi^{3+}] = 7.7 \times 10^{-19}/(0.050)^3 = \underline{\underline{6.2 \times 10^{-15} \text{ M}}}$

7-14 (a) $Tl_2CrO_4 \rightleftarrows 2Tl^+ + CrO_4^{2-}$ $\quad K_{sp} = [Tl^+]^2[CrO_4^{2-}]$

$[CrO_4^{2-}] = K_{sp}/[Tl^+]^2 = 9.8 \times 10^{-13}/(2.12 \times 10^{-3})^2 = \underline{\underline{2.18 \times 10^{-7} \text{ M}}}$

(b) $[CrO_4^{2-}] = 9.8 \times 10^{-13}/(1.00 \times 10^{-6})^2 = \underline{\underline{0.98 \text{ M}}}$

7-15 (a) $[Fe^{3+}][OH^-]^3 = 2 \times 10^{-39}$

$[Fe^{3+}] = 2 \times 1.00 \times 10^{-3} = 2 \times 10^{-3}$

$[OH^-] = \left(\dfrac{2 \times 10^{-39}}{2 \times 10^{-3}}\right)^{1/3} = \underline{\underline{1 \times 10^{-12} \text{ M}}}$

(b) $[OH^-] = \left(\dfrac{2 \times 10^{-39}}{1.0 \times 10^{-9}}\right)^{1/3} = \underline{\underline{1 \times 10^{-10} \text{ M}}}$

7-16 (a) $\dfrac{50.0 \text{ mL} \times 0.0500 \text{ M Ce}^{3+}}{50.00 \text{ mL} + 50.0 \text{ mL}} = \underline{\underline{0.0250 \text{ M Ce}^{3+}}}$

(b) Since the Ce^{3+} is in excess, assume that essentially all of the IO_3^- reacts so that

61

Chapter 7

$$\text{no. mmol Ce}^{3+} = \text{initial no. mmol Ce}^{3+} - \frac{1}{3}\text{initial no. mmol IO}_3^-$$

$$= 50.0 \text{ mL} \times 0.0500 \text{ M} - \frac{50.0 \text{ mL} \times 0.0500 \text{ M}}{3} = 1.67 \text{ mmol}$$

$$c_{Ce^{3+}} = \frac{1.67 \text{ mmol}}{100.0 \text{ mL}} = 1.67 \times 10^{-2} \text{ M}$$

$[Ce^{3+}] = 1.67 \times 10^{-2} + S \approx 1.67 \times 10^{-2}$ (where S = solubility)

In order to check this assumption, we write

$[IO_3^-] = 3S$

$$K_{sp} = [Ce^{3+}][IO_3^-]^3 = 1.67 \times 10^{-2}(3S)^3 = 3.2 \times 10^{-10}$$

$$S = \left(\frac{3.2 \times 10^{-10}}{27 \times 1.67 \times 10^{-2}}\right)^{1/3} = 8.9 \times 10^{-4}$$

which is $\ll 1.67 \times 10^{-2}$

and $[Ce^{3+}] = 1.67 \times 10^{-2} = \underline{\underline{1.7 \times 10^{-2} \text{ M}}}$

(c) no. mmol $IO_3^- = 50.0 \text{ mL} \times 0.150 \text{ M} = 7.50 \text{ mmol}$

no. mmol $Ce^{3+} = 50.0 \text{ mL} \times 0.050 \text{ M} = 2.50 \text{ mmol}$

Since these quantities represent the stoichiometric ration, the solubility is calculated from the K_{sp}. Here,

$[Ce^{3+}] = S$ and $[IO_3^-] = 3S$

$$K_{sp} = [Ce^{3+}][IO_3^-]^3 = S \cdot (3S)^3 = 27 S^4 = 3.2 \times 10^{-10}$$

$S^4 = 1.18 \times 10^{-11}$

$[Ce^{3+}] = S = (1.18 \times 10^{-11})^{1/4} = \underline{\underline{1.9 \times 10^{-3} \text{ M}}}$

(d) no. mmol $IO_3^- = 50.0 \text{ mL} \times 0.300 \text{ M} = 15.0 \text{ mmol}$

no. mmol Ce^{3+} = 2.50 mmol

Let S = solubility and

$$[IO_3^-] = \frac{15.0 \text{ mmol} - 3(2.50 \text{ mmol})}{100 \text{ mL}} + 3S = 0.0750 \text{ M} + 3S$$

$$[Ce^{3+}] = S$$

$$K_{sp} = S(0.0750 + 3S)^3 = 3.2 \times 10^{-10}$$

Assume $3S \ll 0.0750$,

$$S = [Ce^{3+}] \, 3.2 \times 10^{-10}/(0.0750)^3 = \underline{\underline{7.6 \times 10^{-7} \text{ M}}}$$

7-17 no. mmol K^+ = 50.0 mL × 0.400 M = 20.0 mmol

(a) no. mmol $PtCl_6^{2-}$ = 50.0 mL × 0.100 M = 5.00 mmol

no. mmol excess K^+ = 20.0 mmol − 2(5.00 mmol) = 10.0 mmol

$$[K^+] = \frac{10.0 \text{ mmol}}{50.0 \text{ mL}} + 2S = 0.200 \text{ M} + 2S \approx \underline{\underline{0.200 \text{ M}}}$$

(b) no. mmol $PtCl_6^{2-}$ = 50.0 mL × 0.200 M = 10 mmol

Since this is exactly the stoichiometric amount relative to K^+, the solubility and thus $[K^+]$ are determined from the K_{sp}, and

$$S = [PtCl_6^{2-}] = \frac{1}{2}[K^+]$$

$$K_{sp} = [K^+]^2[PtCl_6^{2-}] = (2S)^2(S) = 4S^3 = 1.1 \times 10^{-5}$$

$$S = 1.4 \times 10^{-2}; \quad [K^+] = 2S = \underline{\underline{2.8 \times 10^{-2} \text{ M}}}$$

(c) no. mmol $PtCl_6^{2-}$ = 50.0 mL × 0.400 M = 20.0 mmol

no. mmol of $PtCl_6^{2-}$ in excess = 20.0 mmol − $\frac{1}{2}$(20.0 mmol) = 10.0 mmol

Chapter 7

$$[PtCl_6^{2-}] = \frac{10.0 \text{ mmol}}{100 \text{ mL}} + S = 0.100 + S$$

$$[K^+] = 2S$$

$$(2S)^2(0.100 \text{ M} + S) = 1.1 \times 10^{-5}$$

Assume $0.100 \gg S$,

$$(2S)^2 = \frac{1.1 \times 10^{-5}}{0.100} = 1.1 \times 10^{-4}$$

$$S = 5.2 \times 10^{-3}$$

$$[K^+] = 2S = \underline{\underline{1.0 \times 10^{-2}}}$$

The assumption is valid to 5%.

7-18 (a) $S_{TlI} = [Tl^+] = [I^-] = \sqrt{K_{sp}} = \sqrt{6.5 \times 10^{-8}} = \underline{\underline{2.5 \times 10^{-4} \text{ M}}}$

$S_{AgI} = [Ag^+] = [I^-] = \sqrt{8.3 \times 10^{-17}} = \underline{\underline{9.1 \times 10^{-9} \text{ M}}}$

$K_{sp} = [Pb^+][I^-]^2 = S(2S)^2 = 4S^3$

$S_{PbI_2} = (K_{sp}/4)^{1/3} = (7.1 \times 10^{-9}/4)^{1/3} = \underline{\underline{1.2 \times 10^{-3} \text{ M}}}$

$K_{sp} = [Bi^{3+}][I^-]^3 = S(3S)^3 = 27S^4$

$S_{BiI_3} = (K_{sp}/27)^{1/4} = (8.1 \times 10^{-19}/27)^{1/4} = \underline{\underline{1.3 \times 10^{-5} \text{ M}}}$

$\underline{PbI_2 > TlI > BiI_3 > AgI}$

(b) $S_{TlI} = [Tl^+] = K_{sp}/[I^-] = 6.5 \times 10^{-8}/0.10 = \underline{\underline{6.5 \times 10^{-7} \text{ M}}}$

$S_{AgI} = [Ag^+] = K_{sp}/[I^-] = 8.3 \times 10^{-17}/0.10 = \underline{\underline{8.3 \times 10^{-16} \text{ M}}}$

$S_{PbI_2} = [Pb^{2+}] = K_{sp}/[I^-]^2 = 7.1 \times 10^{-9}/(0.10)^2 = \underline{\underline{7.1 \times 10^{-7} \text{ M}}}$

$$S_{BiI_3} = [Bi^{3+}] = K_{sp}/[I^-]^3 = 8.1 \times 10^{-19}/(0.10)^3 = \underline{\underline{8.1 \times 10^{-16} \text{ M}}}$$

$\underline{\underline{PbI_2 > TlI > AgI > BiI_3}}$

(c) $S_{TlI} = [I^-] = K_{sp}/[Tl^+] = 6.5 \times 10^{-8}/0.010 = \underline{\underline{6.5 \times 10^{-6} \text{ M}}}$

$S_{AgI} = [I^-] = K_{sp}/[Ag^+] = 8.3 \times 10^{-17}/0.010 = \underline{\underline{8.3 \times 10^{-15} \text{ M}}}$

$S_{PbI_2} = [I^-]/2 \qquad [I^-] = \sqrt{K_{sp}/[Pb^{2+}]}$

$S_{PbI_2} = \frac{1}{2}\sqrt{K_{sp}/[Pb^{2+}]} = \frac{1}{2}\sqrt{7.1 \times 10^{-9}/0.010} = \underline{\underline{4.2 \times 10^{-4} \text{ M}}}$

$S_{BiI_3} = [I^-]/3 \qquad [I^-] = (K_{sp}/[Bi^{3+}])^{1/3}$

$S_{BiI_3} = \frac{1}{3}(K_{sp}/[Bi^{3+}])^{1/3} = \frac{1}{3}(8.1 \times 10^{-19}/0.010)^{1/3} = \underline{\underline{1.4 \times 10^{-6} \text{ M}}}$

$\underline{\underline{PbI_2 > TlI > BiI_3 > AgI}}$

7-19 (a) For BiOOH, $S = [BiO^+] = [OH^-]$ and $S = \sqrt{4.0 \times 10^{10}} = \underline{\underline{2.0 \times 10^{-5} \text{ M}}}$

For Be(OH)$_2$, $[Be^{2+}] = S$ and $[OH^-] = 2S$. Thus,

$S(2S)^2 = 7.0 \times 10^{-22}$ and $S = \underline{\underline{5.6 \times 10^{-8} \text{ M}}}$

For Tm(OH)$_3$, $[Tm^{3+}] = S$ and $[OH^-] = 3S$. Therefore,

$S(3S)^3 = 3.0 \times 10^{-24} = 27 S^4$ and $S = \underline{\underline{5.8 \times 10^{-7} \text{ M}}}$

For Hf(OH)$_4$, $S = [Hf^{4+}]$ and $[OH^-] = 4S$. Thus,

$S(4S)^4 = 256 S^5 = 4.0 \times 10^{-26}$ and $S = \underline{\underline{2.7 \times 10^{-6} \text{ M}}}$

$\underline{\underline{Be(OH)_2}}$ has the lowest solubility in H$_2$O.

(b) For BiOOH, $S = [BiO^+] = 4.0 \times 10^{-10}/0.100 = \underline{\underline{4.0 \times 10^{-9} \text{ M}}}$

Chapter 7

For Be(OH)$_2$, $S = [Be^{2+}] = 7.0 \times 10^{-22}/(0.100)^2 = \underline{\underline{7.0 \times 10^{-20}\text{ M}}}$

For Tm(OH)$_3$, $S = [Tm^{3+}] = 3.0 \times 10^{-26}/(0.100)^3 = \underline{\underline{3.0 \times 10^{-21}\text{ M}}}$

For Hf(OH)$_4$, $S = [Hf^{4+}] = 4.0 \times 10^{-26}/(0.100)^4 = \underline{\underline{4.0 \times 10^{-22}\text{ M}}}$

$\underline{\underline{\text{Hf(OH)}_2}}$ has the lowest solubility in 0.1 M NaOH.

7-20 At 100°C, $K_w = 49 \times 10^{-14}$

$[H_3O^+] = \sqrt{49 \times 10^{-14}} = \underline{\underline{7.0 \times 10^{-7}\text{ M}}}$

7-21 (a) $\dfrac{[H_3O^+][OCl^-]}{[HOCl]} = 3.0 \times 10^{-8} = K_a$

$[H_3O^+] = [OCl^-]$ and $[HOCl] = 0.0200 - [H_3O^+]$

Assume $[H_3O^+] \ll 0.0200$. Then

$[H_3O^+]^2/0.0200 = 3.0 \times 10^{-8}$

$[H_3O^+] = \sqrt{0.0200 \times 3.0 \times 10^{-8}} = \underline{\underline{2.4 \times 10^{-5}\text{ M}}}$

The assumption appears valid and

$[OH^-] = 1.00 \times 10^{-14}/(2.4 \times 10^{-5}) = \underline{\underline{4.1 \times 10^{-10}\text{ M}}}$

(b) Proceeding as in part (a), we obtain

$[H_3O^+] = \underline{\underline{1.04 \times 10^{-3}\text{ M}}}$ and $[OH^-] = \underline{\underline{9.7 \times 10^{-12}\text{ M}}}$

(c) $CH_3NH_2 + H_2O \rightleftarrows CH_3NH_3^+ + OH^-$ $K_b = \dfrac{1.00 \times 10^{-14}}{K_a} = \dfrac{1.00 \times 10^{-14}}{2.3 \times 10^{-11}}$

$= 4.35 \times 10^{-4}$

$[CH_3NH_3^+] = [OH^-]$

$[CH_3NH_2] = 0.200 - [OH^-] \approx 0.200$

$$\frac{[OH^-]^2}{0.200} = 4.35 \times 10^{-4}$$

$$[OH^-] = \sqrt{4.35 \times 10^{-4} \times 0.200} = \underline{\underline{9.3 \times 10^{-3}}}$$

The assumption leads to an error of less than 5% and

$$[H_3O^+] = 1.00 \times 10^{-14}/9.3 \times 10^{-3} = \underline{\underline{1.1 \times 10^{-12} \text{ M}}}$$

(d) Proceeding as in part (c), we obtain

$$[OH^-] = \underline{\underline{2.52 \times 10^{-3}}} \quad \text{and} \quad [H_3O^+] = \underline{\underline{3.97 \times 10^{-12} \text{ M}}}$$

(e) $OCl^- + H_2O \rightleftharpoons HOCl + OH^-$

$$\frac{[HOCl][OH^-]}{[OCl^-]} = K_b = \frac{K_w}{K_a} = \frac{1.00 \times 10^{-14}}{3.0 \times 10^{-8}} = 3.33 \times 10^{-7}$$

$[OH^-] = [HOCl]$ and $[OCl^-] = 0.120 - [OH^-] \approx 0.120$

$[OH^-]^2/0.120 = 3.33 \times 10^{-7}$

$$[OH^-] = \sqrt{0.120 \times 3.33 \times 10^{-7}} = \underline{\underline{2.0 \times 10^{-4} \text{ M}}}$$

The assumption is valid, and

$$[H_3O^+] = 1.00 \times 10^{-14}/(2.00 \times 10^{-4}) = \underline{\underline{5.0 \times 10^{-11} \text{ M}}}$$

(f) Proceeding as in part (e), we obtain

$$[H_3O^+] = \underline{\underline{1.43 \times 10^{-9} \text{ M}}} \quad \text{and} \quad [OH^-] = \underline{\underline{7.01 \times 10^{-6} \text{ M}}}$$

(g) $HONH_3^+ + H_2O \rightleftharpoons HONH_2 + H_3O^+ \quad K_a = 1.10 \times 10^{-6}$

$[H_3O^+] = [HONH_2]$

$[HONH_3^+] = 0.100 - [H_3O^+] \approx 0.100$

$[H_3O^+]^2 = 0.100 \times 1.10 \times 10^{-6}$

Chapter 7

$$[H_3O^+] = \underline{\underline{3.32 \times 10^{-4}\ M}} \quad \text{and} \quad [OH^-] = 1.00 \times 10^{-14}/3.32 \times 10^{-4} = \underline{\underline{3.02 \times 10^{-11}\ M}}$$

(h) $HOC_2H_4NH_3^+ + H_2O \rightleftarrows HC_2H_4NH_2 + H_3O^+ \qquad K_a = 3.18 \times 10^{-10}$

Proceeding as in part (g), we obtain

$$[H_3O^+] = \underline{\underline{3.99 \times 10^{-6}\ M}} \quad \text{and} \quad [OH^-] = \underline{\underline{2.51 \times 10^{-9}\ M}}$$

7-22 (a) $HA + H_2O \rightleftarrows H_3O^+$

$$K_a = \frac{[H_3O^+][A^-]}{[HA]} = \frac{[H_3O^+]^2}{c_{HA} - [H_3O^+]} = 1.36 \times 10^{-3}$$

Because K_a is relatively large, we must solve the quadratic equation. We do this by successive approximations:

$$[H_3O^+] = \sqrt{K_a(c_{HA} - [H_3O^+])}$$

$$[H_3O^+]_1 = \sqrt{1.36 \times 10^{-3}\,(0.100 - 0)} = 1.17 \times 10^{-2}$$

$$[H_3O^+]_2 = \sqrt{1.36 \times 10^{-3}\,(0.100 - 1.17 \times 10^{-2})} = 1.096 \times 10^{-2}$$

$$= \underline{\underline{1.10 \times 10^{-2}\ M}} \qquad \text{(good to 3 significant figures)}$$

(b) $ClCH_2COO^- + H_2O \rightleftarrows ClCH_2COOH + OH^- \qquad K_b = \dfrac{1.00 \times 10^{-14}}{1.36 \times 10^{-3}} = 7.35 \times 10^{-12}$

$[OH^-] = ClCH_2COOH \quad \text{and} \quad [ClCH_2COO^-] = 0.100 - [OH^-] \approx 0.100$

$$\frac{[OH^-]^2}{0.100} = 7.35 \times 10^{-12}$$

$$[OH^-] = \sqrt{0.100 \times 7.35 \times 10^{-12}} = 8.58 \times 10^{-7}$$

$$[H_3O^+] = 1.00 \times 10^{-14}/8.58 \times 10^{-7} = \underline{\underline{1.17 \times 10^{-8}\ M}}$$

(c) $CH_3NH_2 + H_2O \rightarrow CH_3NH_3^+ + OH^- \qquad K_b = \dfrac{1.00 \times 10^{-14}}{2.3 \times 10^{-11}} = 4.35 \times 10^{-4}$

$$[OH^-] = [CH_3NH_3^+] \quad \text{and} \quad [CH_3NH_2] = 0.0100 - [OH^-] \approx 0.0100$$

$$\frac{[OH^-]^2}{0.0100} = 4.35 \times 10^{-4}$$

$$[OH^-] = \sqrt{4.35 \times 10^{-6}} = \underline{\underline{2.1 \times 10^{-3} \text{ M}}}$$

We see, however, that the approximation is not very good. Thus, we write

$$\frac{[OH^-]^2}{0.0100 - [OH^-]} = 4.38 \times 10^{-4}$$

$$[OH^-]^2 + 4.38 \times 10^{-4}[OH^-] - 4.38 \times 10^{-6} = 0$$

Solving the quadratic gives $[OH^-] = 1.88 \times 10^{-3}$

$$[H_3O^+] = 1.00 \times 10^{-14}/(1.88 \times 10^{-3}) = \underline{\underline{5.3 \times 10^{-12} \text{ M}}}$$

(d) $CH_3NH_3^+ + H_2O \rightleftarrows CH_3NH_2 + H_3O^+ \quad K_a = 2.3 \times 10^{-11}$

$$[H_3O^+] = [CH_3NH_2] \quad \text{and} \quad [CH_3NH_3^+] = 0.0100 - [H_3O^+] \approx 0.0100$$

$$\frac{[H_3O^+]^2}{0.0100} = 2.3 \times 10^{-11}$$

$$[H_3O^+] = \sqrt{0.0100 \times 2.3 \times 10^{-11}} = \underline{\underline{4.8 \times 10^{-7} \text{ M}}}$$

(e) $C_6H_5NH_3^+ + H_2O \rightleftarrows C_6H_5NH_2 + H_3O^+ \quad K_a = 2.51 \times 10^{-5}$

Proceeding as in part (c), we obtain after solving the quadratic expressions, $[H_3O^+] = \underline{\underline{1.46 \times 10^{-4}}}$. (Using the simplifying assumption, we find $[H_3O^+] = 1.58 \times 10^{-4}$.)

(f) $HIO_3 + H_2O \rightleftarrows H_3O^+ + IO_3^- \quad K_a = 1.7 \times 10^{-1}$

$$[H_3O^+] = [IO_3^-] \quad \text{and} \quad [HIO_3] = 0.200 - [H_3O^+]$$

Because K_a is quite large, we cannot simplify. Thus

Chapter 7

$$\frac{[H_3O^+]^2}{0.200 - [H_3O^+]} = 0.17$$

$[H_3O^+]^2 + 0.17\,[H_3O^+] - 0.034 = 0$

$[H_3O^+]$ = <u>0.12 M</u>

Making the simplifying assumption leads to $[H_3O^+]$ = 0.18 M.

CHAPTER 8

8-1 (a) *Activity*, a, is the effective concentration of a species A in solution. The *activity coefficient*, γ_A, is the factor needed to convert a molar concentration to activity:

$$a_A = \gamma_A [A]$$

(b) The *thermodynamic equilibrium constant* refers to an ideal system within which each species is unaffected by any others. A *concentration equilibrium constant* takes account of the influence excerted by solute species upon one another. A thermodynamic constant is based upon activities of reactants and products; a concentration constant is based upon molar concentrations of reactants and products.

8-2 For μ less than 0.1, activity coefficients
- depend upon ionic strength μ
- approach 1.0 as $\mu \to 0$
- are larger for species with multiple charges

8-3 (a) Reaction:

$$MgCl_2 + 2NaOH \to Mg(OH)_2(s) + 2NaCl$$

The addition of NaOH has the effect of replacing a divalent ion (Mg^{2+}) with a chemically equivalent quantity of a univalent ion (Na^+); μ should decrease.

(b) Reaction:

$$HCl + NaOH \to NaCl + H_2O$$

Addition of NaOH will convert HCl to an equivalent amount of NaCl. Because all of the ions involved are singly charged, μ should be unchanged.

(c) Reaction:

$$NaOH + HOAc \to NaOAc + H_2O$$

Addition of NaOH has the effect of replacing a slightly ionized species (HOAc) with a chemically equivalent quantity of water and Na^+; μ should increase.

Chapter 8

8-4 (a) Addition of $FeCl_3$ to a solution of HCl has the effect of increasing the population of charged species; μ should increase.

(b) Reaction:

$$FeCl_3 + 3NaOH \rightarrow Fe(OH)_3(s) + 3NaCl$$

Addition of $FeCl_3$ has the effect of replacing a univalent ion (OH^-) with another univalent ion (Cl^-); μ should be essentially unaffected.

(c) Reaction:

$$FeCl_3 + 3AgNO_3 \rightarrow Fe(NO_3)_3 + 3AgCl(s)$$

Addition of $FeCl_3$ has the effect of replacing a univalent ion (Ag^+) with a chemically equivalent quantity of a trivalent ion (Fe^{3+}) and a univalent ion (Cl^-) with another univalent ion (NO_3^-); μ should increase, owing to the former replacement.

8-5 For a given ionic strength, activity coefficients for ions with multiple charge show greater departures from ideality.

8-6 The species NH_3 is uncharged; therefore its activity coefficient is unity.

8-7 (a) $\mu = \frac{1}{2}[0.040(2)^2 + 0.040(2)^2] = \underline{\underline{0.16}}$

(b) $\mu = \frac{1}{2}[2 \times 0.20(1)^2 + 0.20(2)^2] = \underline{\underline{0.60}}$

(c) $\mu = \frac{1}{2}[0.10(3)^2 + 0.3(1)^2 + 0.20(2)^2 + 0.4(1)^2] = \underline{\underline{1.2}}$

(d) $\mu = \frac{1}{2}[0.060 \times (3)^2 + 3 \times 0.06(1)^2 + 0.030(2)^2 + 0.060(1)^2] = \underline{\underline{0.45}}$

8-8

$$-\log \gamma_X = \frac{0.51\, Z_X^2 \sqrt{\mu}}{1 + 0.33\, \alpha_X \sqrt{\mu}}$$

Chapter 8

	Z	μ	√μ	α	-log γ	1/γ	γ
(a)	3	0.075	0.274	9	0.693	4.93	0.20
(b)	2	0.012	0.110	4.5	0.192	1.56	0.64
(c)	4	0.080	0.283	11	1.139	13.77	0.073
(d)	4	0.060	0.245	11	1.058	11.43	0.088

8-9 (a) $\mu = 0.050$ $\quad \gamma_{Fe^{3+}} = 0.24$

$\mu = 0.10$ $\quad \gamma_{Fe^{3+}} = 0.18$

$\gamma_{int} = 0.18 + \dfrac{0.025}{0.050}(0.24 - 0.18) = \underline{\underline{0.21}}$

(b) $\mu = 0.010$ $\quad \gamma_{Pb^{2+}} = 0.665$

$\mu = 0.050$ $\quad \gamma_{Pb^{2+}} = 0.46$

$\gamma_{int} = 0.46 + \dfrac{0.038}{0.040}(0.665 - 0.46) = \underline{\underline{0.65}}$

(c) $\mu = 0.050$ $\quad \gamma_{Ce^{4+}} = 0.10$

$\mu = 0.10$ $\quad \gamma_{Ce^{4+}} = 0.065$

$\gamma_{int} = 0.065 + \dfrac{0.020}{0.050}(0.10 - 0.065) = \underline{\underline{0.079}}$

(d) $\mu = 0.050$ $\quad \gamma_{Sn^{4+}} = 0.10$

$\mu = 0.100$ $\quad \gamma_{Sn^{4+}} = 0.065$

$\gamma_{int} = 0.065 + \dfrac{0.040}{0.050}(0.10 - 0.065) = \underline{\underline{0.093}}$

8-10 (a) $\mu = 0.050$ $\quad \gamma_{Ag^+} = 0.80$ $\quad \gamma_{SCN^-} = 0.81$

$K'_{sp} = \dfrac{K_{sp}}{\gamma_{Ag^+} \cdot \gamma_{SCN^-}} = \dfrac{1.1 \times 10^{-12}}{(0.80)(0.81)} = \underline{\underline{1.7 \times 10^{-12}}}$

Chapter 8

(b) $\gamma_{Pb^{2+}} = 0.46 \quad \gamma_{I^-} = 0.80$

$$K'_{sp} = \frac{K_{sp}}{\gamma_{Pb^{2+}} \cdot \gamma_{I^-}^2} = \frac{7.9 \times 10^{-9}}{0.46 (0.80)^2} = \underline{2.7 \times 10^{-8}}$$

(c) $\gamma_{La^{3+}} = 0.24 \quad \gamma_{IO_3^-} = 0.82$

$$K'_{sp} = \frac{K_{sp}}{\gamma_{La^{3+}} \cdot \gamma_{IO_3^-}^3} = \frac{1.0 \times 10^{-11}}{0.24 (0.82)^3} = \underline{7.6 \times 10^{-11}}$$

(d) $\gamma_{Mg^{2+}} = 0.52 \quad \gamma_{NH_4^+} = 0.80 \quad \gamma_{PO_4^{3-}} = 0.16$

$$K'_{sp} = \frac{K_{sp}}{\gamma_{Mg^{2+}} \cdot \gamma_{NH_4^+} \cdot \gamma_{PO_4^{3-}}} = \frac{3 \times 10^{-13}}{(0.52)(0.80)(0.16)} = \underline{4.5 \times 10^{-12}}$$

8-11 $Zn(OH)_2(s) \rightleftharpoons Zn^{2+} + 2OH^- \quad K_{sp} = 3.0 \times 10^{-16}$

(a) $\mu = \frac{1}{2}[0.0100(1)^2 + 0.0100(1)^2] = 0.0100$

In Table 8-1, we find at $\mu = 0.0100$

$\gamma_{Zn^{2+}} = 0.675$ and $\gamma_{OH^-} = 0.900$

$$K_{sp} = a_{Zn^{2+}} \cdot a_{OH^-}^2 = 3.0 \times 10^{-16} = \gamma_{Zn^{2+}}[Zn^{2+}] \times \gamma_{OH^-}^2 [OH^-]^2$$

$$[Zn^{2+}][OH^-]^2 = \frac{3.00 \times 10^{-16}}{\gamma_{Zn^{2+}} \cdot \gamma_{OH^-}^2} = \frac{3.00 \times 10^{-16}}{0.675 (0.900)^2} = 5.49 \times 10^{-16} = K'_{sp}$$

Solubility $= S = [Zn^{2+}] = \frac{1}{2}[OH^-]$

$S(2S)^2 = 5.49 \times 10^{-16}$

$S = \underline{5.2 \times 10^{-6} M}$

(b) $\mu = \frac{1}{2}[2 \times 0.0167(1)^2 + 0.0167(2)^2] = 0.050$

$\gamma_{Zn^{2+}} = 0.48 \qquad \gamma_{OH^-} = 0.81$

$$K'_{sp} = \frac{3.0 \times 10^{-16}}{(0.48)(0.81)^2} = 9.53 \times 10^{-16}$$

$$S = [Zn^{2+}] = \left(\frac{9.53 \times 10^{-16}}{4}\right)^{1/3} = \underline{\underline{6.2 \times 10^{-6} \text{ M}}}$$

(c) Initial amounts

$20.0 \times 0.250 = 5.00$ mmol K^+

$ = 5.00$ mmol OH^-

$80.0 \times 0.0250 = 2.00$ mmol Zn^{2+}

$80.0 \times 0.0250 \times 2 = 4.00$ mmol Cl^-

$c_{K^+} = 5/100 = 0.0500$

$c_{OH^-} = (5.00 - 2 \times 2.00)/100 = 0.0100$

$c_{Cl^-} = 0.0400 \quad$ and $\quad c_{Zn^{2+}} = 0.00$

$\mu = \frac{1}{2}[0.0500(1)^2 + 0.0100(1)^2 + 0.0400(1)^2] = 0.05$

$\gamma_{Zn^{2+}} = 0.48 \quad$ and $\quad \gamma_{OH^-} = 0.81$

$K'_{sp} = 9.53 \times 10^{-16} \qquad$ (see part b)

$[Zn^{2+}][OH^-]^2 = [Zn^{2+}](0.0100)^2 = 9.5 \times 10^{-16}$

$[Zn^{2+}] = S = \underline{\underline{9.5 \times 10^{-12} \text{ M}}}$

(d) Initial amount

$20.0 \times 0.100 = 2.00$ mmol K^+

$ = 2.00$ mmol OH^-

Chapter 8

$$80.0 \times 0.0250 = 2.00 \text{ mmol Zn}^{2+}$$

$$80.0 \times 0.0250 \times 2 = 4.00 \text{ mmol Cl}^-$$

$$c_{K^+} = 2.00/100 = 0.0200 \text{ M}$$

$$c_{OH^-} = 0.00$$

$$c_{Zn^{2+}} = (2.00 - 1.00)/100 = 0.0100$$

$$c_{Cl^-} = 4.00/100 = 0.0400$$

$$\mu = \frac{1}{2}[0.0200(1)^2 + 0.0100(2)^2 + 0.0400(1)^2] = 0.0500$$

$$K'_{sp} = 9.53 \times 10^{-16} = 0.0100\,[OH^-]^2 \quad \text{(see part b)}$$

$$[OH^-] = \sqrt{9.53 \times 10^{-16}/0.0100} = 3.09 \times 10^{-7}$$

$$S = [OH^-]/2 = \underline{\underline{1.5 \times 10^{-7} \text{ M}}}$$

8-12 $\quad \mu = \frac{1}{2}\sum c_i Z_i^2 = \frac{1}{2}[0.0333(2)^2 + 0.0666(1)^2] = 0.100$ for all solutions

(a) (1) In Table 8-1 we find for $\mu = 0.100$,

$$\gamma_{Ag^+} = 0.75 \quad \text{and} \quad \gamma_{SCN^-} = 0.76$$

$$K_{sp} = a_{Ag^+} \times a_{SCN^-} = [Ag^+][SCN^-]\gamma_{Ag^+} \times \gamma_{SCN^-}$$

$$K'_{sp} = K_{sp}/(\gamma_{Ag^+} \times \gamma_{SCN^-}) = 1.1 \times 10^{-12}/(0.75 \times 0.76)$$

$$= 1.93 \times 10^{-12}$$

$$S = [Ag^+] = [SCN^-]$$

$$S = \sqrt{K'_{sp}} = \sqrt{1.93 \times 10^{-12}} = \underline{\underline{1.4 \times 10^{-6} \text{ M}}}$$

(2) $\quad S = \sqrt{K_{sp}} = \sqrt{1.1 \times 10^{-12}} = \underline{\underline{1.0 \times 10^{-6} \text{ M}}}$

Chapter 8

(b) (1) $\gamma_{Pb^{2+}} = 0.37$ and $\gamma_{I^-} = 0.76$

$$K_{sp} = [Pb^{2+}][I^-]\gamma_{Pb^{2+}} \cdot \gamma_{I^-}^2$$

$$K'_{sp} = K_{sp}/(\gamma_{Pb^{2+}} \cdot \gamma_{I^-}^2) = 7.9 \times 10^{-9}/[0.37 \times (0.76)^2] = 3.70 \times 10^{-8}$$

$[Pb^{2+}] = S$ and $[I^-] = 2S$

$S(2S)^2 = 3.70 \times 10^{-8}$

$S = (3.70 \times 10^{-8}/4)^{1/3} = \underline{\underline{2.1 \times 10^{-3} \text{ M}}}$

(2) $S = (7.9 \times 10^{-9}/4)^{1/3} = \underline{\underline{1.3 \times 10^{-3} \text{ M}}}$

(c) (1) Proceeding as in part (a),

$$K'_{sp} = 1.1 \times 10^{-10}/(0.38 \times 0.36) = 8.04 \times 10^{-10}$$

$S = [Ba^{2+}] = [SO_4^{2-}]$

$S = \sqrt{K'_{sp}} = \sqrt{8.04 \times 10^{-10}} = \underline{\underline{2.8 \times 10^{-5} \text{ M}}}$

(2) $S = \sqrt{1.1 \times 10^{-10}} = \underline{\underline{1.0 \times 10^{-5} \text{ M}}}$

(d) (1) Proceeding as in part (b),

$$K'_{sp} = 3.2 \times 10^{-17}/[(0.38)^2 \times 0.021] = 1.06 \times 10^{-14}$$

$[Cd^{2+}] = 2S$ and $[Fe(CN)_6^{4-}] = S$

$S(2S)^2 = 4S^3 = 1.06 \times 10^{-14}$

$S = (1.06 \times 10^{-14}/4)^{1/3} = \underline{\underline{1.4 \times 10^{-5} \text{ M}}}$

(2) $S = (3.2 \times 10^{-17}/4)^{1/3} = \underline{\underline{2.0 \times 10^{-6} \text{ M}}}$

Chapter 8

8-13 $\mu = \dfrac{1}{2}[0.0167(2)^2 + 2 \times 0.0167(1)^2] = 0.050$

(a) (1) $\gamma_{Ag^+} = 0.80 \qquad \gamma_{IO_3^-} = 0.82$

$$K'_{sp} = \dfrac{3.1 \times 10^{-8}}{0.80 \times 0.82} = 4.73 \times 10^{-8}$$

$$S = \sqrt{4.73 \times 10^{-8}} = \underline{\underline{2.2 \times 10^{-4} \text{ M}}}$$

(2) $S = \sqrt{3.1 \times 10^{-8}} = \underline{\underline{1.8 \times 10^{-4} \text{ M}}}$

(b) (1) $\gamma_{Mg^{2+}} = 0.52 \qquad \gamma_{OH^-} = 0.81$

$$K'_{sp} = \dfrac{7.1 \times 10^{-12}}{0.52\,(0.81)^2} = 2.08 \times 10^{-11}$$

$$S = [Mg^{2+}] = \dfrac{1}{2}[OH^-]$$

$$S(2S)^2 = 2.08 \times 10^{-11}$$

$$S = (2.08 \times 10^{-11}/4)^{1/3} = \underline{\underline{1.7 \times 10^{-4} \text{ M}}}$$

(2) $S = (7.1 \times 10^{-12}/4)^{1/3} = \underline{\underline{1.2 \times 10^{-4} \text{ M}}}$

(c) (1) $\gamma_{Ba^{2+}} = 0.46 \qquad \gamma_{SO_4^{2-}} = 0.44$

$$K'_{sp} = \dfrac{1.1 \times 10^{-10}}{0.46 \times 0.44} = 5.43 \times 10^{-10}$$

$$S = 5.43 \times 10^{-10}/0.0167 = \underline{\underline{3.3 \times 10^{-8} \text{ M}}}$$

(2) $S = 1.1 \times 10^{-10}/0.0167 = \underline{\underline{6.6 \times 10^{-9} \text{ M}}}$

(d) (1) $\gamma_{La^{3+}} = 0.24 \qquad \gamma_{IO_3^-} = 0.82$

$$K'_{sp} = \frac{1.0 \times 10^{-11}}{0.24\,(0.82)^3} = 7.56 \times 10^{-11}$$

$$S = [La^{3+}] = \frac{1}{3}[IO_3^-]$$

$$S(3S)^3 = 7.56 \times 10^{-11}$$

$$S = (7.56 \times 10^{-11}/27)^{1/4} = \underline{\underline{1.3 \times 10^{-3}\,M}}$$

(2) $S = (1.0 \times 10^{-11}/27)^{1/4} = \underline{\underline{8.0 \times 10^{-4}\,M}}$

CHAPTER 9

9-1 In the calculation of the molar solubility of Fe(OH)$_2$, the hydronium ion concentration may be assumed to be negligibly small. This assumption cannot be made in the case of Fe(OH)$_3$. In the latter case we assume [H$_3$O$^+$] = [OH$^-$] ≈ 1.00 × 10^{-7}.

9-2 The overall dissociation constant for H$_2$S is

$$\frac{[H_3O^+]^2[S^{2-}]}{[H_2S]} = K_1K_2$$

In a solution saturated with the gas, [H$_2$S] is constant; therefore

$$[S^{2-}] = \frac{[H_2S]}{[H_3O^+]^2}K_1K_2 \quad \text{or} \quad \frac{K}{[H_3O^+]^2}$$

9-3 The simplifications in equilibrium calculations involve assuming that the concentration of one or more species is 0.00 M. When 0.00 is inserted into equilibrium-constant expressions, the constant becomes equal to zero or infinity. Thus, the expression is meaningless.

9-4 A charge-balance equation is derived by relating the concentration of cations and anions in such a way that

  no. mol/L positive charge = no. mol/L negative charge

For a doubly charged ion such as Ba^{2+}, the concentration of electrons for each mole is twice the molar *concentration* of the Ba^{2+}. That is,

  mol/L positive charge = 2[Ba^{2+}]

Thus the molar concentration of all multiply charged species is always multiplied by the charge in a charge-balance equation.

9-5 (a) $0.10 = [H_3PO_4] + [H_2PO_4^-] + [HPO_4^{2-}] + [PO_4^{3-}]$

(b) $0.10 = [H_3PO_4] + [H_2PO_4^-] + [H_2PO_4^{2-}] + [PO_4^{3-}]$

$[Na^+] = 2c_{Na_2HPO_4} = 0.20$

(c) $0.100 + 0.0500 = [HNO_2] + [NO_2^-]$

$$[Na^+] = c_{NaNO_2} = 0.0500$$

(d) $[F^-] + [HF] = 0.025 + 2[Ca^{2+}]$

$[Na^+] = 0.025$

(e) $0.100 = [Na^+] = [OH^-] + 2[Zn(OH)_4^{2-}]$

(f) $[Mg^{2+}] = [H_2CO_3] + [HCO_3^-] + [CO_3^{2-}]$

(g) $[Ca^{2+}] = \dfrac{1}{2}([F^-] + [HF])$

9-6 (a) $[H_3O^+] = [OH^-] + [H_2PO_4^-] + 2[HPO_4^{2-}] + 3[PO_4^{3-}]$

(b) $[Na^+] + [H_3O^+] = [OH^-] + [H_2PO_4^-] + 2[HPO_4^{2-}] + 3[PO_4^{3-}]$

(c) $[Na^+] + [H_3O^+] = [OH^-] + [NO_2^-]$

(d) $[H_3O^+] + [Na^+] + 2[Ca^{2+}] = [OH^-] + [F^-]$

(e) $2[Zn^{2+}] + [H_3O^+] + [Na^+] = 2[Zn(OH)_4^{2-}] + [OH^-]$

(f) $2[Mg^{2+}] + [H_3O^+] = [OH^-] + 2[CO_3^{2-}] + [HCO_3^-]$

(g) $2[Ca^{2+}] + [H_3O^+] = [OH^-] + [F^-]$

9-7 (a) Following the systematic procedure shown in Figure 9-1 and Example 9-7, we write

Step 1 $Ag_2CO_3(s) \rightleftarrows 2Ag^+ + CO_3^{2-}$

$H_2CO_3 + H_2O \rightleftarrows H_3O^+ + HCO_3^-$

$HCO_3^- + H_2O \rightleftarrows H_3O^+ + CO_3^{2-}$

Step 2 $S = \text{solubility} = [Ag^+]/2$

Step 3 $[Ag^+]^2[CO_3^{2-}] = K_{sp} = 8.1 \times 10^{-12}$ \hfill (1)

Chapter 9

$$\frac{[H_3O^+][HCO_3^-]}{[H_2CO_3]} = K_1 = 4.45 \times 10^{-7} \qquad (2)$$

$$\frac{[H_3O^+][CO_3^{2-}]}{[HCO_3^-]} = K_2 = 4.69 \times 10^{-11} \qquad (3)$$

Step 4 $[Ag^+] = 2([CO_3^{2-}] + [HCO_3^-] + [H_2CO_3]) \qquad (4)$

$[H_3O^+] = 1.00 \times 10^{-6}$

Step 5 No charge-balance equation because a buffer of unknown composition is present.

Step 6 Unknowns: $[Ag^+]$, $[CO_3^{2-}]$, $[HCO_3^-]$, and $[H_2CO_3]$

Equations: (1), (2), (3), and (4)

Step 7 No approximations needed.

Step 8 Substituting $[H_3O^+] = 1.00 \times 10^{-6}$ into Equation (3) and rearranging gives

$$[HCO_3^-] = \frac{1.00 \times 10^{-6} [CO_3^{2-}]}{4.69 \times 10^{-11}} = 2.13 \times 10^4 [CO_3^{2-}]$$

Substituting this relationship and $[H_3O^+]$ into Equation (2) gives after rearranging

$$[H_2CO_3] = \frac{1.00 \times 10^{-6} \times 2.13 \times 10^4 [CO_3^{2-}]}{4.45 \times 10^{-7}} = 4.78 \times 10^4$$

Substituting these two relationships into Equation (4) gives

$[Ag^+] = 2[CO_3^{2-}] + 2 \times 2.13 \times 10^4 [CO_3^{2-}] + 2 \times 4.78 \times 10^4 [CO_3^{2-}]$

$ = 1.38 \times 10^5 [CO_3^{2-}]$

Substituting Equation (1) gives

$$[Ag^+] = \frac{1.38 \times 10^5 \times 8.1 \times 10^{-12}}{[Ag^+]^2}$$

or $[Ag^+] = (1.38 \times 10^5 \times 8.1 \times 10^{-12})^{1/3} = 1.04 \times 10^{-2}$

solubility $= 1.04 \times 10^{-2}/2 = \underline{\underline{5.2 \times 10^{-3} \text{ M}}}$

Substituting other values for [H$_3$O$^+$] yields the following solubility data:

	[H$_3$O$^+$]	Solubility, M
(a)	1.00×10^{-6}	5.2×10^{-3}
(b)	1.00×10^{-7}	1.7×10^{-3}
(c)	1.00×10^{-9}	3.6×10^{-4}
(d)	1.00×10^{-11}	1.3×10^{-4}

9-8 Proceeding as in Solution 9-7, we write

$$\text{BaSO}_4^{2-} \rightleftarrows \text{Ba}^{2+} + \text{SO}_4^{2-} \qquad K_{sp} = 1.1 \times 10^{-10}$$

$$\text{HSO}_4^- + \text{H}_2\text{O} \rightleftarrows \text{H}_3\text{O}^+ + \text{SO}_4^{2-} \qquad K_2 = 1.02 \times 10^{-2}$$

$$S = [\text{Ba}^{2+}]$$

$$[\text{Ba}^{2+}][\text{SO}_4^{2-}] = 1.1 \times 10^{-10} \qquad (1)$$

$$\frac{[\text{H}_3\text{O}^+][\text{SO}_4^{2-}]}{[\text{HSO}_4^-]} = 1.02 \times 10^{-2} \qquad (2)$$

Mass balance requires that

$$[\text{Ba}^{2+}] = [\text{SO}_4^{2-}] + [\text{HSO}_4^-] \qquad (3)$$

Since [H$_3$O$^+$] is known, we have 3 equations and 3 unknowns.

Substituting Equation (2) into Equation (3) yields

$$[\text{Ba}^{2+}] = [\text{SO}_4^{2-}] + \frac{[\text{H}_3\text{O}^+][\text{SO}_4^{2-}]}{1.02 \times 10^{-2}}$$

$$= [\text{SO}_4^{2-}](1 + [\text{H}_3\text{O}^+]/1.02 \times 10^{-2})$$

Substituting Equation (1) to eliminate [SO$_4^-$] gives

Chapter 9

$$[Ba^{2+}] = \frac{1.1 \times 10^{-10}}{[Ba^{2+}]} \times (1 + 98.0\,[H_3O^+])$$

$$[Ba^{2+}] = S = \sqrt{1.1 \times 10^{-10}\,(1 + 98.0\,[H_3O^+])}$$

$$= \sqrt{1.1 \times 10^{-10} + 1.078 \times 10^{-8}\,[H_3O^+]}$$

Substituting for $[H_3O^+]$ in this equation leads to

	$[H_3O^+]$	S, mol/L
(a)	2.00	1.47×10^{-4}
(b)	1.00	1.04×10^{-4}
(c)	0.500	7.42×10^{-5}
(d)	0.100	3.45×10^{-5}

9-9 The following derivation applies to this and the following two problems.

$$MS(s) \rightleftarrows M^{2+} + S^{2-}$$

$$H_2S + H_2O \rightleftarrows H_3O^+ + HS^- \qquad K_1 = 9.6 \times 10^{-8}$$

$$HS^- + H_2O \rightleftarrows H_3O^+ + S^{2-} \qquad K_2 = 1.3 \times 10^{-14}$$

$$H_2S + 2H_2O \rightleftarrows 2H_3O^+ + S^{2-} \qquad K_1K_2 = 9.6 \times 10^{-8} \times 1.3 \times 10^{-14} = 1.25 \times 10^{-21}$$

$$S = \text{solubility} = [M^{2+}] = [S^{2-}] + [HS^-] + [H_2S]$$

$$[M^{2+}][S^{2-}] = K_{sp} \qquad (1)$$

$$\frac{[H_3O^+][S^{2-}]}{[HS^-]} = K_2 = 1.3 \times 10^{-14} \qquad (2)$$

$$\frac{[H_3O^+]^2[S^{2-}]}{[H_2S]} = K_1K_2 = 1.25 \times 10^{-21} \qquad (3)$$

Chapter 9

From mass-balance consideration

$$[M^{2+}] = [S^{2-}] + [HS^-] + [H_2S] \qquad (4)$$

Substituting Equations (2) and (3) into (4) gives

$$[M^{2+}] = [S^{2-}] + \frac{[H_3O^+][S^{2-}]}{K_2} + \frac{[H_3O^+]^2[S^{2-}]}{K_1K_2}$$

$$= [S^{2-}]\left(1 + \frac{[H_3O^+]}{K_2} + \frac{[H_3O^+]^2}{K_1K_2}\right)$$

Substituting Equation (1) yields

$$[M^{2+}] = \frac{K_{sp}}{[M^{2+}]}\left(1 + \frac{[H_3O^+]}{K_2} + \frac{[H_3O^+]^2}{K_1K_2}\right)$$

$$[M^{2+}] = \sqrt{K_{sp}\left(1 + \frac{[H_3O^+]}{K_2} + \frac{[H_3O^+]^2}{K_1K_2}\right)} \qquad (5)$$

$$= \sqrt{K_{sp}\left(1 + \frac{[H_3O^+]}{1.3 \times 10^{-14}} + \frac{[H_3O^+]^2}{1.25 \times 10^{-21}}\right)}$$

(a) Substituting $K_{sp} = 8 \times 10^{-37}$ and $[H_3O^+] = 0.10$ into Equation (5) leads to

$$[Cu^{2+}] = \text{solubility} = \sqrt{8 \times 10^{-37}\left(1 + \frac{0.10}{1.3 \times 10^{-14}} + \frac{(0.10)^2}{1.25 \times 10^{-21}}\right)}$$

$$= \underline{\underline{2.5 \times 10^{-9} \text{ M}}}$$

(b) Substituting $K_{sp} = 8 \times 10^{-37}$ and $[H_3O^+] = 1.0 \times 10^{-4}$ into Equation (5) gives

$$\text{solubility} = \underline{\underline{2.5 \times 10^{-12} \text{ M}}}$$

9-10 $K_{sp} = [Cd^{2+}][S^{2-}] = 1 \times 10^{-27}$

Substituting this value into Equation (5), Solution 9-9 gives

(a) $S = [Cd^{2+}] = 8.9 \times 10^{-5} = \underline{\underline{9 \times 10^{-5} \text{ M}}}$

85

Chapter 9

(b) $S = [Cd^{2+}] = 8.9 \times 10^{-8} = \underline{\underline{9 \times 10^{-8}}}$

9-11 $K_{sp} = [Mn^{2+}][S^{2-}] = 3 \times 10^{-14}$

Substituting this value into Equation (5), Solution 9-9 gives

(a) $S = [Mn^{2+}] = 4.9 \times 10^{-2} = \underline{\underline{5 \times 10^{-2} \text{ M}}}$

(b) $S = [Mn^{2+}] = 6.9 \times 10^{-4} = \underline{\underline{7 \times 10^{-4} \text{ M}}}$

9-12 (a) Here we proceed as in Solution 9-8.

$$PbCO_3 \rightleftarrows Pb^{2+} + CO_3^{2-} \qquad K_{sp} = [Pb^{2+}][CO_3^{2-}] = 7.4 \times 10^{-14} \qquad (1)$$

$$H_2CO_3 + H_2O \rightleftarrows H_3O^+ + HCO_3^- \qquad K_1 = \frac{[H_3O^+][HCO_3^-]}{[H_2CO_3]} = 4.45 \times 10^{-7} \qquad (2)$$

$$HCO_3^- + H_2O \rightleftarrows H_3O^+ + CO_3^{2-} \qquad K_2 = \frac{[H_3O^+][CO_3^{2-}]}{[HCO_3^-]} = 4.69 \times 10^{-11} \qquad (3)$$

$$[Pb^{2+}] = [CO_3^{2-}] + [HCO_3^-] + [H_2CO_3] \qquad (4)$$

$$[H_3O^+] = 1.00 \times 10^{-7}$$

Proceeding as in Solution 9-9 we derive an equation analogous to Equation (5) in that solution. That is,

$$[Pb^{2+}] = \sqrt{K_{sp}\left(1 + \frac{[H_3O^+]}{K_2} + \frac{[H_3O^+]^2}{K_1 K_2}\right)}$$

$$= \sqrt{7.4 \times 10^{-14}\left(1 + \frac{1.0 \times 10^{-7}}{4.69 \times 10^{-11}} + \frac{1.0 \times 10^{-14}}{4.45 \times 10^{-7} \times 4.69 \times 10^{-11}}\right)}$$

$$\text{solubility} = [Pb^{2+}] = \underline{\underline{1.4 \times 10^{-5} \text{ M}}}$$

(b) In this case $[H_3O^+]$ is no longer fixed, and a more descriptive way of writing the equations is

$$PbCO_3(s) \rightleftarrows Pb^{2+} + CO_3^{2-} \qquad K_{sp} = [Pb^{2+}][CO_3^{2-}] = 7.4 \times 10^{-14} \qquad (1)$$

Chapter 9

$$CO_3^{2-} + H_2O \rightleftarrows HCO_3^- + OH^- \quad K_{b1} = \frac{K_w}{K_2} = \frac{1.00 \times 10^{-14}}{4.69 \times 10^{-11}} = 2.13 \times 10^{-4}$$

$$= \frac{[HCO_3^-][OH^-]}{[CO_3^{2-}]} \quad (2)$$

$$HCO_3^- + H_2O \rightleftarrows H_2CO_3 + OH^- \quad K_{b2} = \frac{K_w}{K_1} = \frac{1.00 \times 10^{-14}}{4.45 \times 10^{-7}} = 2.25 \times 10^{-8}$$

$$= \frac{[H_2CO_3][OH^-]}{[HCO_3^-]} \quad (3)$$

$$2H_2O \rightleftarrows H_3O^+ + OH^- \quad K_w = [H_3O^+][OH^-] = 1.00 \times 10^{-14} \quad (4)$$

Mass balance: $[Pb^{2+}] = [H_2CO_3] + [HCO_3^-] + [CO_3^{2-}]$ (5)

Charge balance: $2[Pb^{2+}] + [H_3O^+] = [OH^-] + [HCO_3^-] + 2[CO_3^{2-}]$ (6)

Sought: solubility = $[Pb^{2+}]$

We now have 6 equations and 6 unknowns. Therefore, in principle a solution is feasible, and we follow the procedure shown in Example 9-8.

To simplify we assume $[H_3O^+]$ is negligible with respect to $2[Pb^{2+}]$ in Equation (6) and $[H_2CO_3]$ is smaller than $[HCO_3^-]$ and $[H_2CO_3]$ in Equation (5), thus giving

$$[Pb^{2+}] \approx [HCO_3^-] + [CO_3^{2-}] \quad (7)$$

$$2[Pb^{2+}] = [OH^-] + [HCO_3^-] + 2[CO_3^{2-}] \quad (8)$$

Now we have 4 unknowns ($[Pb^{2+}]$, $[CO_3^-]$, $[HCO_3^-]$, and $[OH^-]$) and 4 independent equations (1), (2), (7), and (8).

If we multiply Equation (7) by (2) and subtract from Equation (8), we obtain

$$[OH^-] = [HCO_3^-]$$

which, as shown in Example 9-8, leads to an equation analogous to Equation 9-31. Thus,

Chapter 9

$$[Pb^{2+}]^2 - \sqrt{\frac{K_w}{K_2} K_{sp} [Pb^{2+}]} - K_{sp} = 0$$

Inserting numerical constants gives

$$[Pb^{2+}]^2 - 3.97 \times 10^{-9} \sqrt{[Pb^{2+}]} - 7.4 \times 10^{-14} = 0$$

Application of the method of successive approximation gives

$$[Pb^{2+}] = \text{solubility} = \underline{\underline{2.53 \times 10^{-6} \text{ M}}}$$

9-13 (a) In solution buffered to $[H_3O^+] = 1.00 \times 10^{-7}$, we proceed as in Example 9-7.

$$Ag_2SO_3(s) \rightleftharpoons 2Ag^+ + SO_3^{2-} \qquad [Ag^+]^2[SO_3^{2-}] = K_{sp} = 1.5 \times 10^{-14}$$

$$H_2SO_3 + H_2O \rightleftharpoons H_3O^+ + HSO_3^- \qquad \frac{[H_3O^+][HSO_3^-]}{[H_2SO_3]} = K_1 = 1.23 \times 10^{-2}$$

$$HSO_3^- + H_2O \rightleftharpoons H_3O^+ + SO_3^{2-} \qquad \frac{[H_3O^+][SO_3^{2-}]}{[HSO_3^-]} = K_2 = 6.6 \times 10^{-8}$$

Solubility = $[Ag^+]/2$

Mass balance:

$$\frac{[Ag^+]}{2} = [SO_3^{2-}] + [HSO_3^-] + [H_2SO_3] \qquad \text{and} \qquad [H_3O^+] = 1.00 \times 10^{-7}$$

Substituting the expression for K_1 and K_2 into the mass-balance equation gives

$$\frac{[Ag^+]}{2} = [SO_3^{2-}] + \frac{[H_3O^+][SO_3^{2-}]}{K_2} + \frac{[H_3O^+]^2[SO_3^{2-}]}{K_1 K_2}$$

$$\frac{[Ag^+]}{2} = [SO_3^{2-}]\left(1 + \frac{[H_3O^+]}{6.6 \times 10^{-8}} + \frac{[H_3O^+]^2}{1.23 \times 10^{-2} \times 6.6 \times 10^{-8}}\right)$$

$$= \frac{K_{sp}}{[Ag^+]^2}\left(1 + 1.52 \times 10^7 [H_3O^+] + 1.23 \times 10^9 [H_3O^+]^2\right)$$

$$[Ag^+] = \left[2 \times 1.5 \times 10^{-14}\left(1 + 1.52 \times 10^7 [H_3O^+] + 1.23 \times 10^9 [H_3O^+]^2\right)\right]^{1/3}$$

Substituting $[H_3O^+] = 1.00 \times 10^{-7}$ gives

$$[Ag^+] = 4.2 \times 10^{-5}$$

Solubility = $\underline{\underline{2.1 \times 10^{-5} \text{ M}}}$

(b) Here we proceed as in Example 9-7b and write

$$Ag_2SO_3(s) \rightleftarrows 2Ag^+ + SO_3^{2-} \qquad K_{sp} = [Ag^+]^2[SO_3^{2-}] = 1.5 \times 10^{-14} \qquad (1)$$

$$SO_3^{2-} + H_2O \rightleftarrows HSO_3^- + OH^- \qquad K_{b1} = \frac{K_w}{K_1} = \frac{1.00 \times 10^{-14}}{6.6 \times 10^{-8}}$$

$$= 1.52 \times 10^{-7} = \frac{[OH^-][HSO_3^-]}{[SO_3^{2-}]} \qquad (2)$$

$$HSO_3^- + H_2O \rightleftarrows H_2SO_3 + OH^- \qquad K_{b2} = \frac{K_w}{K_1} = \frac{1.00 \times 10^{-14}}{1.23 \times 10^{-2}}$$

$$= 8.13 \times 10^{-13} = \frac{[OH^-][H_2SO_3]}{[HSO_3^-]} \qquad (3)$$

$$2H_2O \rightleftarrows [H_3O^+] + [OH^-] \qquad K_w = [H_3O^+][OH^-] = 1.00 \times 10^{-14} \qquad (4)$$

We wish to find solubility = $[Ag^+]/2$.

Mass balance:

$$[Ag^+]/2 = \cancel{[H_2SO_3]} + [HSO_3^-] + [SO_3^{2-}] \qquad (5)$$

Charge balance:

$$[Ag^+] + \cancel{[H_3O^+]} = [OH^-] + [HSO_3^-] + 2[SO_3^{2-}] \qquad (6)$$

We assume

$$[H_2SO_3] \ll ([HSO_3^-] + [SO_3^{2-}]) \qquad \text{and} \qquad [H_3O^+] \ll [Ag^+]$$

Chapter 9

We then have 4 equations [(1), (2), (5), and (6)] and 4 unknowns. To solve the 4 equations, we first multiply Equation (5) by 2 and subtract from (6), which gives with rearrangement

$$[OH^-] = [HSO_3^-] \qquad (7)$$

Substituting Equation (7) into (2) and rearranging gives

$$[HSO_3^-] = \sqrt{1.52 \times 10^{-7} [SO_3^{2-}]} \qquad (8)$$

Substituting Equations (8) and (1) into (5) yields

$$\frac{[Ag^+]}{2} = \sqrt{\frac{1.52 \times 10^{-7} \times 1.5 \times 10^{-14}}{[Ag^+]^2}} + \frac{1.5 \times 10^{-14}}{[Ag^+]^2}$$

$$[Ag^+] = \frac{2 \times 4.77 \times 10^{-11}}{[Ag^+]} + \frac{2 \times 1.5 \times 10^{-14}}{[Ag^+]^2}$$

$$[Ag^+]^3 = 9.55 \times 10^{-11} [Ag^+] + 3.0 \times 10^{-14}$$

$$[Ag^+]^3 - 9.55 \times 10^{-11} [Ag^+] - 3.0 \times 10^{-14} = 0$$

The method of successive approximations gives

$$[Ag^+] = 3.21 \times 10^{-5} \quad \text{and} \quad \text{solubility} = \underline{\underline{1.61 \times 10^{-5} \text{ M}}}$$

9-14 $[Cu^{2+}][OH^-]^2 = 4.8 \times 10^{-20} \qquad [Mn^{2+}][OH^-]^2 = 2 \times 10^{-13}$

(a) $\underline{\underline{Cu(OH)_2 \text{ forms first}}}$

(b) Cu^{2+} begins to precipitate when

$$[OH^-] = \sqrt{4.8 \times 10^{-20} / 0.050} = \underline{\underline{9.8 \times 10^{-10}}}$$

(c) Mn^{2+} begins to precipitate when

$$[OH^-] = \sqrt{2 \times 10^{-13} / 0.0400} = 2.2 \times 10^{-6}$$

$$[Cu^{2+}] = 4.8 \times 10^{-20} / (2.2 \times 10^{-6})^2 = \underline{\underline{9.6 \times 10^{-9}}}$$

Chapter 9

9-15 $Ba(IO_3)_2(s) \rightleftharpoons Ba^{2+} + I$ $K_{sp} = 1.57 \times 10^{-9}$

$BaSO_4(s) \rightleftharpoons Pb^{2+} + SO_4^{2-}$ $K_{sp} = 1.3 \times 10^{-10}$

To initiate precipitation of $Ba(IO_3)_2$

$$[Ba^{2+}] = \frac{K_{sp}}{[IO_3^-]^2} = \frac{1.57 \times 10^{-9}}{(0.05)^2} = 6.28 \times 10^{-7} \approx 6.3 \times 10^{-7}$$

$$[Ba^{2+}] = \frac{K_{sp}}{[SO_4^{2-}]} = \frac{1.3 \times 10^{-10}}{0.040} = 3.25 \times 10^{-9} \approx 3.2 \times 10^{-9}$$

(a) $\underline{BaSO_4 \text{ precipitates first}}$.

(b) $[Ba^{2+}] = \underline{\underline{3.2 \times 10^{-9}}}$

(c) When $[Ba^{2+}] = 6.3 \times 10^{-7}$

$$[SO_4^{2-}] = \frac{K_{sp}}{[Ba^{2+}]} = \frac{1.1 \times 10^{-10}}{6.3 \times 10^{-7}} = 1.746 \times 10^{-4} \approx \underline{\underline{1.7 \times 10^{-4}}}$$

9-16 (a) $[Ag^+] = 8.3 \times 10^{-17} / 1.0 \times 10^{-6} = \underline{\underline{8.3 \times 10^{-11}}}$

(b) $[Ag^+] = 1.1 \times 10^{-12} / 0.070 = \underline{\underline{1.6 \times 10^{-11}}}$

(c) $[I^-] = 8.3 \times 10^{-17} / 1.6 \times 10^{-11} = \underline{\underline{5.2 \times 10^{-6}}}$

$[SCN^-]/[I^-] = 0.070/5.2 \times 10^{-6} = \underline{\underline{1.3 \times 10^4}}$

(d) $\dfrac{[SCN^-]}{[I^-]} = \dfrac{1.1 \times 10^{-12}/[Ag^+]}{8.3 \times 10^{-17}/[Ag^+]} = \underline{\underline{1.3 \times 10^4}}$

Note that this ratio is independent of $[Ag^+]$ as long as some $AgSCN(s)$ is present.

9-17 (a) $[Ba^{2+}][SO_4^{2-}] = 1.1 \times 10^{-10}$ $[Sr][SO_4^{2-}] = 3.2 \times 10^{-7}$

$BaSO_4$ precipitation complete when

$[SO_4^{2-}] = 1.1 \times 10^{-10}/(1.0 \times 10^{-6}) = 1.1 \times 10^{-4}$

Chapter 9

SrSO$_4$ precipitation begins when

$$[SO_4^{2-}] = 3.2 \times 10^{-7}/0.10 = 3.2 \times 10^{-6}$$

Thus, separation <u>not feasible</u>.

(b) $[Ag^+]^2[SO_4^{2-}] = 1.6 \times 10^{-5}$

BaSO$_4$ precipitation complete when $[SO_4^{2-}] = 1.1 \times 10^{-4}$. See part (a).

Ag$_2$SO$_4$ precipitates when $[Ag^+] = \sqrt{1.6 \times 10^{-5}/0.040} = 2.00 \times 10^{-2}$

Thus, separation theoretically <u>feasible</u>.

(c) Be precipitates when $[OH^-] = (7.0 \times 10^{-22}/0.020)^{1/2} = 1.9 \times 10^{-10}$

Hf precipitates when $[OH^-] = (4.0 \times 10^{-26}/0.010)^{1/4} = 1.4 \times 10^{-6}$

Be precipitation complete when $[OH^-] = \sqrt{7.0 \times 10^{-22}/(1.0 \times 10^{-6})} = 2.6 \times 10^{-8}$

Thus, separation theoretically <u>feasible</u>.

(d) Tl$^+$ begins to precipitate when $[IO_3^-] = 3.1 \times 10^{-6}/0.060 = 5.2 \times 10^{-5}$

In^{3+} begins to precipitate when $[IO_3^-] = (3.3 \times 10^{-11}/0.11)^{1/3} = 6.7 \times 10^{-4}$

Tl$^+$ precipitation complete when $[IO_3^-] = 3.1 \times 10^{-6}/(1.0 \times 10^{-6}) = 3.1$

Therefore, separation <u>not feasible</u>.

9-18 $AgBr(s) \rightleftarrows Ag^+ + Br^-$ $K_{sp} = 5.0 \times 10^{-13} = [Ag^+][Br^-]$ (1)

$Ag^+ + 2CN^- \rightleftarrows Ag(CN)_2^-$ $\beta_2 = \dfrac{[Ag(CN)_2^-]}{[Ag^+][CN^-]^2} = 1.3 \times 10^{21}$ (2)

It is readily shown that the reaction

$$CN^- + H_2O \rightleftarrows HCN + OH^-$$

proceeds to such a small extent that it can be neglected in formulating a solution to this problem. That is, HCN $<<$ [CN$^-$], and only the two equilibria shown need to be taken into account.

$$\text{Solubility} = [\text{Br}^-] = [\text{Ag}^+] + [\text{Ag(CN)}_2^-]$$

Mass balance requires that

$$[\text{Br}^-] = [\text{Ag}^+] + [\text{Ag(CN)}_2^-] \qquad (3)$$

$$0.100 = [\text{CN}^-] + 2[\text{Ag(CN)}_2^-] \qquad (4)$$

We now have 4 equations and 4 unknowns.

Based upon the large size of β_2, let us assume that

$$[\text{CN}^-] << [2\text{Ag(CN)}_2^-] \quad \text{and} \quad [\text{Ag}^+] << [\text{Ag(CN)}_2^-]$$

Equation (4) becomes

$$[\text{Ag(CN)}_2^-] = 0.100/2 = 0.0500$$

and Equation (3) becomes

$$[\text{Br}^-] = [\text{Ag(CN)}_2^-] = 0.0500$$

To check the assumption, we calculate [Ag$^+$] by substituting into Equation (1)

$$[\text{Ag}^+] = 5.0 \times 10^{-13}/0.0500 \approx 1 \times 10^{-11}$$

To obtain [CN$^-$] we substitute into Equation (2)

$$\frac{[\text{Ag(CN)}_2^-]}{[\text{Ag}^+][\text{CN}^-]^2} = \frac{0.0500}{1 \times 10^{-11} [\text{CN}^-]^2} = 1.3 \times 10^{21}$$

$$[\text{CN}^-] = 2.0 \times 10^{-6}$$

Thus, the two assumptions are valid and

$$\text{solubility} = [\text{Br}^-] = 0.0500 \text{ M}$$

Chapter 9

$$\text{mass AgBr}/200 \text{ mL} = 0.0500 \frac{\text{mmol}}{\text{mL}} \times 200 \text{ mL} \times \frac{0.1877 \text{ g AgBr}}{\text{mmol AgBr}}$$

$$= \underline{\underline{1.877 \text{ g}}}$$

9-19 Pertinent equilibria:

$$\text{CuCl}(s) \rightleftarrows \text{Cu}^+ + \text{Cl}^-$$

$$\text{Cu}^+ + 2\text{Cl}^- \rightleftarrows \text{CuCl}_2^-$$

$$[\text{Cu}^+][\text{Cl}^-] = K_{sp} = 1.9 \times 10^{-7} \qquad (1)$$

$$\frac{[\text{CuCl}_2^-]}{[\text{Cu}^+][\text{Cl}^-]^2} = 3.2 \times 10^5 \qquad (2)$$

It is convenient to multiply Equation (1) by Equation (2) to give

$$\frac{[\text{CuCl}_2^-]}{[\text{Cl}^-]} = 1.9 \times 10^{-7} \times 3.2 \times 10^5 = 6.08 \times 10^{-2} \qquad (3)$$

From charge-balance consideration, we can write

$$[\text{Cu}^+] + [\text{Na}^+] = [\text{Cl}^-] + [\text{CuCl}_2^-]$$

Here, we assume $[\text{H}_3\text{O}^+] = [\text{OH}^-]$. Furthermore, $[\text{Na}^+]$ is equal to the analytical concentration of NaCl. Thus, we can rearrange this equation to give

$$[\text{Cl}^-] = c_{\text{NaCl}} + [\text{Cu}^+] - [\text{CuCl}_2^-] \approx c_{\text{NaCl}} \qquad (4)$$

$$\text{Solubility} = S = [\text{Cu}^+] + [\text{CuCl}_2^-]$$

We now have 3 independent equations (1), (3), and (4) and 3 unknowns. To solve these we first substitute Equation (4) into Equations (1) and (3) to obtain

$$[\text{Cu}^+] = 1.9 \times 10^{-7}/c_{\text{NaCl}}$$

$$[\text{CuCl}_2^-] = 6.08 \times 10^{-2} \times c_{\text{NaCl}}$$

and $S = 1.9 \times 10^{-7}/c_{\text{NaCl}} + 6.08 \times 10^{-2} \, c_{\text{NaCl}} \qquad (5)$

The following solubilites are obtained with this equation

(a) 6.1×10^{-2} M   (d) 2.5×10^{-4} M

(b) 6.1×10^{-3} M   (e) 1.9×10^{-3} M

(c) 6.3×10^{-4} M

To check the assumption that $[Cu^+] - [CuCl_2^-] << c_{NaCl}$ in part (e), we substitute $[Cl^-] = 1.0 \times 10^{-4}$ into Equations (3) and (1) to give

$$[Cu^+] = 1.9 \times 10^{-9}/(1.0 \times 10^{-4}) = 1.9 \times 10^{-3}$$

$$[CuCl_2^-] = 1.0 \times 10^{-4} \times 6.08 \times 10^{-2} = 6.1 \times 10^{-6}$$

We see that the assumption that $[Cu^{2+}] - [CuCl_2^-] << c_{NaCl}$ in Equation (4) is not valid; thus we must derive a more rigorous relationship. Let us substitute Equations (1) and (3) into the unsimplified version of Equation (4) to give

$$[Cl^-] = c_{NaCl} + \frac{1.9 \times 10^{-7}}{[Cl^-]} - 6.08 \times 10^{-2} [Cl^-]$$

which rearranges to the quadratic

$$1.061 [Cl^-]^2 - c_{NaCl} [Cl^-] - 1.9 \times 10^{-7} = 0$$

$$[Cl^-]^2 - 0.942 \, c_{NaCl} [Cl^-] - 1.791 \times 10^{-7} = 0 \qquad (6)$$

Equation (6) permits the calculation of the *equilibrium* chloride concentration which can then be substituted for c_{NaCl} in Equation (5). That is,

$$S = 1.9 \times 10^{-7}/[Cl^-] + 6.08 \times 10^{-2} [Cl^-]$$

Thus when $c_{NaCl} = 10^{-4}$ Equation (6) becomes

$$[Cl^-]^2 - 0.942 \times 10^{-4} [Cl^-] - 1.791 \times 10^{-6} = 0$$

$$[Cl^-] = 4.73 \times 10^{-4}$$

Chapter 9

$$S = [Cu^+] + [CuCl_2^-]$$

$$= 1.9 \times 10^{-7}/(4.73 \times 10^{-4}) + 6.08 \times 10^{-2} \times 4.73 \times 10^{-4}$$

$$= \underline{\underline{4.3 \times 10^{-4}}}$$

Note that the equilibrium chloride concentration [Cl⁻] is *larger* than the analytical concentration c_{NaCl}. The reason for this apparent anomaly is that the dissolution of CuCl to give Cu⁺ and Cl⁻ contributes significantly to the equilibrium chloride concentration. The following data were obtained by the foregoing procedure:

	c_{NaCl}	[Cl⁻]	S, M
(a)	1.0	9.42×10^{-1}	$\underline{\underline{5.7 \times 10^{-2}}}$
(b)	1.0×10^{-1}	9.42×10^{-2}	$\underline{\underline{5.7 \times 10^{-3}}}$
(c)	1.0×10^{-2}	9.44×10^{-3}	$\underline{\underline{5.9 \times 10^{-4}}}$
(d)	1.0×10^{-4}	1.10×10^{-3}	$\underline{\underline{2.4 \times 10^{-4}}}$
(e)	1.0×10^{-4}	4.73×10^{-4}	$\underline{\underline{4.3 \times 10^{-4}}}$

9-20 (a) $\quad CaSO_4(s) \rightleftharpoons Ca^{2+} + SO_4^{2-} \qquad K_{sp} = 2.4 \times 10^{-5} = [Ca^{2+}][SO_4^{2-}] \qquad (1)$

$\quad CaSO_4(aq) \rightleftharpoons Ca^{2+} + SO_4^{2-} \qquad K_d = 5.2 \times 10^{-3} = \dfrac{[Ca^{2+}][SO_4^{2-}]}{[CaSO_4]_{aq}} \qquad (2)$

$CaSO_4(s) \rightleftharpoons CaSO_4(aq)$

$[Ca^{2+}] = [SO_4^{2-}] \qquad (3)$

Three equations and 3 unknowns ($[Ca^{2+}]$, $[SO_4^{2-}]$, and $[CaSO_4]_{aq}$).

To solve the equation we divide Equation (1) by (2) to give

$$[CaSO_4]_{aq} = K_{sp}/K_d = 2.4 \times 10^{-5}/(5.2 \times 10^{-3}) = 4.6 \times 10^{-3}$$

Chapter 9

Note that this is the equilibrium constant expression for the third equilibrium and indicates the concentration of un-ionized CaSO$_4$ *is always the same in a saturated solution of CaSO$_4$*. Substituting Equation (3) into (1) gives

$$[Ca^{2+}] = \sqrt{2.4 \times 10^{-5}} = 4.90 \times 10^{-3}$$

$$S = [CaSO_4]_{aq} + [Ca^{2+}]$$
$$= 4.6 \times 10^{-3} + 4.9 \times 10^{-3} = 9.5 \times 10^{-3} = \underline{\underline{0.0095 \text{ M}}}$$

$$\% \, CaSO_4(aq) = (4.6 \times 10^{-3}/9.5 \times 10^{-3}) \times 100\% = \underline{\underline{48\%}}$$

(b) Here, $[CaSO_4]_{aq}$ is again equal to 4.6×10^{-3}

$$[SO_4^{2-}] = 0.0100 + [Ca^{2+}] \approx 0.0100$$

$$[Ca^{2+}] = 2.4 \times 10^{-5}/0.0100 = 2.4 \times 10^{-3}$$

Thus, the approximation is not very good and we write

$$[Ca^{2+}] = K_{sp}/[SO_4^{2-}] = 2.4 \times 10^{-5}/(0.0100 + [Ca^{2+}])$$

We solve this equation by systematic approximations. Thus, if $[Ca^{2+}] = 0.00$

$$[Ca^{2+}]_1 = 2.4 \times 10^{-5}/(0.0100 + 0.00) = 2.4 \times 10^{-3}$$

$$[Ca^{2+}]_2 = 2.4 \times 10^{-5}/(0.0100 + 2.4 \times 10^{-3}) = 1.94 \times 10^{-3}$$

$$[Ca^{2+}]_3 = 2.4 \times 10^{-5}/(0.0100 + 1.94 \times 10^{-3}) = 2.01 \times 10^{-3}$$

$$[Ca^{2+}] = 2.4 \times 10^{-5}/(0.0100 + 2.01 \times 10^{-3}) = 2.00 \times 10^{-3}$$

$$S = 4.6 \times 10^{-3} + 2.00 \times 10^{-3} = \underline{\underline{6.6 \times 10^{-3} \text{ M}}}$$

$$\% \, CaSO_4(aq) = (4.6 \times 10^{-3}/6.6 \times 10^{-3}) \times 100\% = \underline{\underline{70\%}}$$

CHAPTER 10

10-1 (a) The initial pH of the NH_3 solution will be less than that for the solution containing NaOH. With the first addition of titrant, the pH of the NH_3 solution will decrease rapidly and then level off and become nearly constant throughout the middle part of the titration. In contrast, additions of standard acid to the NaOH solution will cause the pH of the NaOH solution to decrease gradually and nearly linearly until the equivalent point is approached. The equivalence point pH for the NH_3 solution will be well below 7, whereas for the NaOH solution it will be exactly 7.

(b) Beyond the equivalence point, the pH is determined by the excess titrant. Thus the curves become identical in this region.

10-2 Completeness of the reaction between analyte and reagent and concentration of analyte and reagent.

10-3 The limited sensitivity of the eye to small color differences requires that there be a roughly tenfold excess of one or the other form of the indicator to be present in order for the color change to be seen. This change corresponds to a pH range of ± 1 pH unit about the pK of the indicator.

10-4 Temperature, ionic strength, and the presence of organic solvents and colloidal particles.

10-5 The standard reagents in neutralization titrations are always strong acids or strong bases because the reactions with this type of reagent are more complete than with those of their weaker counterparts. Sharper end points are the consequence of this difference.

10-6 A *buffer* is a solution that resists changes in pH either when it is diluted or when small amounts of acids or bases are added to it. It generally consists of a mixture of a weak acid and its conjugate base or a weak base and its conjugate acid.

10-7 The *buffer capacity* of a solution is the number of moles of hydronium ion or hydroxide ion needed to cause 1.00 L of the buffer to undergo a unit change in pH.

10-8 Mixture (a) will have the greater buffer capacity because the concentration of buffering reagents is greater.

10-9 (a) $[H_3O^+] = K_a \times \dfrac{\text{mmol HOAc}}{\text{mmol OAc}^-} = K_a \times \dfrac{10.0}{8.00} = 2.50\, K_a$

(b) $$[H_3O^+] = K_a \times \frac{100 \times 0.175 - 100 \times 0.0500}{100 \times 0.0500} = K_a \times \frac{12.5}{5.00} = 2.50\, K_a$$

(c) $$[H_3O^+] = K_a \times \frac{40.0 \times 0.1200}{160.0 \times 0.0420 - 40.0 \times 0.1200} = K_a \times \frac{4.80}{1.92} = 2.50\, K_a$$

The three solutions will have the same pH since the ratios of the amounts of weak acid to conjugate base are identical. They will differ, however, in buffer capacity with (a) having the greatest and (c) the least.

10-10 (a) The ideal system would have a pK_a of 3.5 or a K_a of 3.2×10^{-4}.

Malic acid/sodium hydrogen malate with K_a of 3.48×10^{-4} could be used.

(b) The ideal system would have a $pK_a = 7.6$ or $K_a = 2.5 \times 10^{-8}$.

HOCl/NaOCl with $K_a = 3.0 \times 10^{-8}$.

(c) The ideal system would have a $pK_a = 9.3$ or $K_a = 5.0 \times 10^{-10}$.

NH_4Cl/NH_3 with $K_a = 5.7 \times 10^{-10}$

(d) The ideal system would have a $pK_a = 5.1$ or $K_a = 7.9 \times 10^{-6}$.

$C_5H_5NH^+/C_6H_5N$ with $K_a = 5.9 \times 10^{-6}$

10-11 The sharper end point will be observed with the solute having the larger K_b.

(a) For NaOCl, $$K_b = \frac{1.00 \times 10^{-14}}{3.0 \times 10^{-8}} = 3.3 \times 10^{-7}$$

For $HONH_2$, $$K_b = \frac{1.00 \times 10^{-14}}{1.1 \times 10^{-6}} = 9.1 \times 10^{-9} \qquad \text{Thus \underline{NaOCl}}$$

(b) For NH_3, $$K_b = \frac{1.00 \times 10^{-14}}{5.70 \times 10^{-10}} = 1.75 \times 10^{-5}$$

For sodium phenolate, $$K_b = \frac{1.00 \times 10^{-14}}{1.00 \times 10^{-10}} = 1.00 \times 10^{-4}$$

Thus <u>sodium phenolate</u>

Chapter 10

(c) For $HONH_2$, $K_b = 9.1 \times 10^{-9}$ (part a)

For CH_3NH_2, $K_b = \dfrac{1.00 \times 10^{-14}}{2.3 \times 10^{-11}} = 4.3 \times 10^{-4}$ Thus $\underline{\underline{CH_3NH_2}}$

(d) For NH_2NH_2, $K_b = \dfrac{1.00 \times 10^{-14}}{1.05 \times 10^{-8}} = 1.05 \times 10^{-7}$

For NaCN, $K_b = \dfrac{1.00 \times 10^{-14}}{6.2 \times 10^{-10}} = 1.6 \times 10^{-3}$ Thus $\underline{\underline{NaCN}}$

10-12 The sharper end point will be observed with the solute having the larger K_a.

(a) For HNO_2, $K_a = 7.1 \times 10^{-4}$

For HIO_3, $K_a = 1.7 \times 10^{-1}$ Thus $\underline{\underline{HIO_3}}$

(b) For $C_6H_5NH_3^+$, $K_a = 2.51 \times 10^{-5}$

For C_6H_5COOH, $K_a = 6.28 \times 10^{-5}$ Thus $\underline{\underline{C_6H_5COOH}}$

(c) For HOCl, $K_a = 3.0 \times 10^{-8}$

For $CH_3COCOOH$, $K_a = 3.2 \times 10^{-3}$ Thus $\underline{\underline{CH_3COCOOH}}$

(d) For $C_6H_4OHCOOH$, $K_a = 1.06 \times 10^{-3}$

For CH_3COOH, $K_a = 1.75 \times 10^{-5}$ Thus $\underline{\underline{C_6H_4OHCOOH}}$

10-13
$HIn + H_2O \rightleftharpoons H_3O^+ + In^-$ $\dfrac{[H_3O^+][In^-]}{[HIn]} = K_a$

$pK_a = -\log K_a = 7.10$ (Table 10–1)

$K_a = $ antilog $(-7.10) = 7.9 \times 10^{-8}$

$[HIn]/[In^-] = 1.43$

Chapter 10

$[H_3O^+] = 7.9 \times 10^{-8} [HIn]/[In^-] = 7.9 \times 10^{-8} \times 1.43 = 1.136 \times 10^{-7}$

$pH = -\log 1.136 \times 10^{-7} = \underline{6.94}$

10-14 For methyl orange, $pK_a = 3.46 = -\log K_a$

$K_a = \text{antilog}(-3.46) = 3.47 \times 10^{-4}$

$InH^+ + H_2O \rightleftharpoons In + H_3O^+$

$\dfrac{[In][H_3O^+]}{[InH^+]} = 3.47 \times 10^{-4}$

$[H_3O^+] = \dfrac{3.47 \times 10^{-4} [InH^+]}{[In]} = 3.47 \times 10^{-4} \times 1.64 = 5.69 \times 10^{-4}$

$pH = -\log 5.69 \times 10^{-4} = \underline{3.24}$

10-15 $[H_3O^+] = \sqrt{K_w}$ and $pH = -\log(K_w)^{1/2} = -\frac{1}{2}\log K_w$

At 0°C, $pH = -\frac{1}{2} \log (1.14 \times 10^{-15}) = \underline{7.47}$

At 50°C, $pH = -\frac{1}{2} \log (5.47 \times 10^{-14}) = \underline{6.63}$

At 100°C, $pH = -\frac{1}{2} \log (4.9 \times 10^{-13}) = \underline{6.16}$

10-16 At 0°C, $K_w = 1.14 \times 10^{-15}$ (see Problem 10-15)

$pK_w = -\log 1.14 \times 10^{-15} = \underline{14.94}$

At 50°C, $K_w = 5.47 \times 10^{-14}$

$pK_w = -\log 5.47 \times 10^{-14} = \underline{13.26}$

At 100°C, $K_w = 4.9 \times 10^{-13}$

$pK_w = -\log 4.9 \times 10^{-13} = \underline{12.31}$

Chapter 10

10-17 $[H_3O^+][OH^-] = K_w$

pH + pOH = pK_w

(a) pH = pK_w - pOH = 14.94 - 2.00 = <u>12.94</u>   (see Solution 10-16)

(b) pH = 13.26 - 2.00 = <u>11.26</u>

(c) pH = 12.31 - 2.00 = <u>10.31</u>

10-18 $\dfrac{14.0 \text{ g HCl}}{100 \text{ g soln}} \times \dfrac{1.054 \text{ g soln}}{\text{mL soln}} \times \dfrac{1 \text{ mmol HCl}}{0.03646/\text{g HCl}} = 4.047 \text{ M}$

$[H_3O^+] = 4.047 \text{ M}$ and   pH = $-\log 4.047$ = <u>−0.607</u>

10-19 $\dfrac{9.00 \text{ g NaOH}}{100 \text{ g soln}} \times \dfrac{1.098 \text{ g soln}}{\text{mL soln}} \times \dfrac{1 \text{ mmol NaOH}}{0.04000 \text{ g NaOH}} = 2.471 \text{ M} = [OH^-]$

pOH = $-\log 2.471 = -0.393$ and   pH = 14.00 - (-0.393) = <u>14.393</u>

10-20 The solution is so dilute that we must take into account the contribution of water to $[OH^-]$, which is equal to $[H_3O^+]$. Thus,

$[OH^-] = 2.0 \times 10^{-8} + [H_3O^+] = 2.0 \times 10^{-8} + \dfrac{1.00 \times 10^{-14}}{[OH^-]}$

$[OH^-]^2 - 2.0 \times 10^{-8}[OH^-] - 1.00 \times 10^{-14} = 0$

$[OH^-] = 1.105 \times 10^{-7}$

pOH = $-\log 1.105 \times 10^{-7} = 6.957$ and   pH = 14.00 - 6.957 = <u>7.04</u>

10-21 The solution is so dilute that we must take into account the contribution of water to $[H_3O^+]$, which is equal to $[OH^-]$. Thus,

$[H_3O^+] = 2 \times 10^{-8} + [OH^-] = 2 \times 10^{-8} + \dfrac{1.00 \times 10^{-14}}{[H_3O^+]}$

$[H_3O^+]^2 - 2 \times 10^{-8}[H_3O^+] - 1.00 \times 10^{-14} = 0$

$[H_3O^+] = 1.105 \times 10^{-7}$ and   pH = $-\log 1.105 \times 10^{-7}$ = <u>6.96</u>

Chapter 10

10-22 In each part, (20.0 mL HCl × 0.200 mmol HCl/mL) = 4.00 mmol HCl taken

(a) c_{HCl} = [H$_3$O$^+$] = $\dfrac{4.00 \text{ mmol HCl}}{(20.0 + 25.0) \text{ mL soln}}$ = 0.0889 M

pH = -log 0.0889 = <u>1.05</u>

(b) Same as in part (a); pH = <u>1.05</u>

(c) c_{HCl} = (4.00 − 25.0 × 0.132)/(20.0 + 25.0) = 1.556 × 10^{-2} M

[H$_3$O$^+$] = 1.556 × 10^{-2} M and pH = -log (1.556 × 10^{-2}) = <u>1.81</u>

(d) As in part (c), c_{HCl} = 1.556 × 10^{-2} and pH = <u>1.81</u>

(The presence of NH$_4^+$ will not alter the pH significantly.)

(e) c_{NaOH} = (25.0 × 0.232 − 4.00)/(45.0) = 4.00 × 10^{-2}

pOH = -log 4.00 × 10^{-2} = 1.398 and pH = 14.00 - 1.398 = <u>12.60</u>

10-23 In each part, $\dfrac{0.102 \text{ g Mg(OH)}_2}{0.05832 \text{ g Mg(OH)}_2/\text{mmol}}$ = 1.749 mmol Mg(OH)$_2$ taken

(a) c_{HCl} = (75.0 × 0.0600 − 1.749 × 2)/75.0 = 0.01336 M

[H$_3$O$^+$] = 0.01336 and pH = -log 0.01336 = <u>1.87</u>

(b) 15.0 × 0.0600 = 0.900 mmol HCl added. Solid Mg(OH)$_2$ remains and

[Mg^{2+}] = 0.900 mmol HCl × $\dfrac{1 \text{ mmol Mg}^+}{2 \text{ mmol HCl}}$ × $\dfrac{1}{15.0 \text{ mL soln}}$ = 0.0300 M

[M^{2+}][OH$^-$]^2 = 7.1 × 10^{-12} = K_{sp}

[OH$^-$] = (7.1 × 10^{-12}/0.0300)$^{1/2}$ = 1.54 × 10^{-5}

pH = 14.00 − (- log 1.54 × 10^{-5}) = <u>9.19</u>

(c) 30.00 × 0.0600 = 1.80 mmol HCl added, which forms 0.90 mmol Mg^{2+}.

[Mg^{2+}] = 0.90/30.0 = 3.00 × 10^{-2}

Chapter 10

$$[OH^-] = (7.1 \times 10^{-12}/0.0300)^{1/2} = 1.54 \times 10^{-5}$$

$$pH = 14.00 - (-\log 1.54 \times 10^{-5}) = \underline{9.19}$$

(d) $[Mg^{2+}] = 0.0600$

$$[OH^-] = (7.1 \times 10^{-12}/0.0600)^{1/2} = 1.09 \times 10^{-5}$$

$$pH = 14.00 - (-\log 1.09 \times 10^{-5}) = \underline{9.04}$$

10-24 (a) $[H_3O^+] = 0.050$ and $pH = -\log 0.0500 = \underline{1.30}$

(b) $\mu = (0.0500 \times 1^2 + 0.0500 \times 1^2)/2 = 0.0500$

$$\gamma_{H_3O^+} = 0.86 \quad \text{(Table 8-1)}$$

$$a_{H_3O^+} = 0.86 \times 0.0500 = 0.043$$

$$pH = -\log 0.043 = \underline{1.37}$$

10-25 $\mu = \dfrac{1}{2}[0.0167 \times (2)^2 + 2 \times 0.0167 \times (1)^2] = 0.0500$

(a) $[OH^-] = 2 \times 0.0167 = 0.0334$

$$pH = 14.00 - (-\log 0.0334) = \underline{12.52}$$

(b) At $\mu = 0.0500$, $\gamma_{OH^-} = 0.81$ (Table 8-1)

$$a_{OH^-} = 0.0334 \times 0.81 = 0.0271$$

$$a_{H^+} \times a_{OH^-} = 1.00 \times 10^{-14}$$

$$a_{H^+} = 1.00 \times 10^{-14}/0.0271 = 3.69 \times 10^{-13}$$

$$pH = -\log 3.69 \times 10^{-13} = \underline{12.43}$$

10-26 $\dfrac{[H_3O^+][OCl^-]}{[HOCl]} = K_a = 3.0 \times 10^{-8}$

$[H_3O^+] = [OCl^-]$ and $[HOCl] = c_{HOCl} - [H_3O^+] \cong c_{HOCl}$

$[H_3O^+]^2 / c_{HOCl} = 3.0 \times 10^{-8}$

$[H_3O^+] = \sqrt{c_{HOCl} \times 3.0 \times 10^{-8}}$

Thus when $c_{HOCl} =$

(a) 0.100 $[H_3O^+] = 5.477 \times 10^{-5}$ and pH = $\underline{4.26}$

(b) 0.0100 $[H_3O^+] = 1.732 \times 10^{-5}$ and pH = $\underline{4.76}$

(c) 1.00×10^{-4} $[H_3O^+] = 1.732 \times 10^{-6}$ and pH = $\underline{5.76}$

10-27 $OCl^- + H_2O \rightleftharpoons HOCl + OH^-$

$\dfrac{[HOCl][OH^-]}{[OCl^-]} = K_b = \dfrac{K_w}{K_a} = \dfrac{1.00 \times 10^{-14}}{3.00 \times 10^{-8}} = 3.333 \times 10^{-7}$

Substituting into the equation for K_b gives after rearranging

$[OH^-] = \sqrt{c_{NaOCl} \times K_b} = \sqrt{c_{NaOCl} \times 3.333 \times 10^{-7}}$

For $c_{NaOCl} =$

(a) 0.100, $[OH^-] = 1.826 \times 10^{-4}$ pOH = 3.74   pH = $\underline{10.26}$

(b) 0.010, $[OH^-] = 5.773 \times 10^{-5}$ pOH = 4.24   pH = $\underline{9.76}$

(c) 1.0×10^{-4}, $[OH^-] = 5.773 \times 10^{-6}$ pOH = 5.24   pH = $\underline{8.76}$

10-28 $NH_3 + H_2O \rightleftharpoons NH_4^+ + OH^-$ $K_b = \dfrac{1.00 \times 10^{-14}}{5.70 \times 10^{-10}} = 1.75 \times 10^{-5}$

(a) $\dfrac{[NH_4^+][OH^-]}{[NH_3]} = \dfrac{[OH^-]}{0.100 - [OH^-]} \approx \dfrac{[OH^-]^2}{0.100} = 1.75 \times 10^{-5}$

Chapter 10

$$[OH^-] = \sqrt{1.75 \times 10^{-5} \times 0.100} = 1.323 \times 10^{-3}$$

$$pH = 14.00 - (-\log 1.323 \times 10^{-3}) = \underline{11.12}$$

(b) Proceeding in the same way, we obtain pH = $\underline{10.62}$

(c) Proceeding in the same way, we obtain

$$[OH^-] = 4.19 \times 10^{-5} \quad \text{and} \quad pH = 9.62$$

To obtain a more exact solution, we write

$$\frac{[OH^-]^2}{1.00 \times 10^{-4} - [OH^-]} = 1.75 \times 10^{-5}$$

$$[OH^-]^2 + 1.75 \times 10^{-5}[OH^-] - 1.75 \times 10^{-9} = 0$$

$$[OH^-] = 3.40 \times 10^{-5}$$

$$pH = 14.00 - (-\log 3.40 \times 10^{-5}) = \underline{9.53}$$

10-29 $\dfrac{[H_3O^+][NH_3]}{[NH_4^+]} = K_a = 5.70 \times 10^{-10}$ $\quad [H_3O^+] = \sqrt{c_{NH_4Cl} \times 5.70 \times 10^{-10}}$

$$[H_3O^+] = \sqrt{c_{NH_4Cl} \times 5.70 \times 10^{-10}}$$

(a) $[H_3O^+] = \sqrt{0.100 \times 5.70 \times 10^{-10}} = 7.55 \times 10^{-6}$

$$pH = -\log 7.55 \times 10^{-6} = \underline{5.12}$$

Proceeding in the same way, we obtain at $c_{NH_4^+} =$

(b) 0.0100 $\quad [H_3O^+] = 2.387 \times 10^{-6}$ and pH = $\underline{5.62}$

(c) 1.00×10^{-4} $\quad [H_3O^+] = 2.387 \times 10^{-7}$ and pH = $\underline{6.62}$

10-30 $C_5H_{11}N + H_2O \rightleftarrows C_5H_{11}NH^+ + OH^- \quad K_b = \dfrac{1.00 \times 10^{-14}}{7.50 \times 10^{-12}} = 1.333 \times 10^{-3}$

(a) Because K_b is large we shall not make the usual approximations and will write

106

$$\frac{[OH^-]^2}{0.100 - [OH^-]} = 1.333 \times 10^{-3}$$

$$[OH^-]^2 + 1.333 \times 10^{-3}[OH^-] - 1.333 \times 10^{-4} = 0$$

$$[OH^-] = 1.090 \times 10^{-2}$$

$$pH = 14.00 - (-\log 1.090 \times 10^{-2}) = \underline{12.04}$$

Proceeding in the same way, we obtain

(b) $[OH^-] = 3.045 \times 10^{-3}$ and pH = $\underline{11.48}$

(c) $[OH^-] = 9.34 \times 10^{-5}$ and pH = $\underline{9.97}$

If the approximate method is used, the pH values are

(a) 12.06 (b) 11.56 (c) 10.56

10-31 $HIO_3 + H_2O \rightleftarrows H_3O^+ + IO_3^-$ $K_a = 1.7 \times 10^{-1}$

$[H_3O^+] = [IO_3^-]$ and $[HIO_3] = c_{HIO_3} - [H_3O^+]$

Here it cannot be assumed that $[H_3O^+] \ll 0.100$ and we must solve the exact quadratic equation. Thus,

$$\frac{[H_3O^+]^2}{c_{HIO_3} - [H_3O^+]} = K_a = 0.17$$

(a) $[H_3O^+]^2 + K_a[H_3O^+] - K_a c_{HIO_3} = 0$

$[H_3O^+] + 1.7 \times 10^{-1}[H_3O^+] - 1.7 \times 10^{-1} \times 0.100 = 0$

Solving this equation leads to

$[H_3O^+] = 7.064 \times 10^{-2}$ and pH = $-\log 7.064 \times 10^{-2} = \underline{1.15}$

Proceeding as in part (a), we obtain

(b) $[H_3O^+] = 9.472 \times 10^{-3}$ and pH = $\underline{2.02}$

(c) $[H_3O^+] = 9.994 \times 10^{-5}$ and pH = $\underline{4.00}$

Chapter 10

10-32 (a) $\mathcal{M}_{HA} = 90.079$

$$c_{HA} = 43.0 \text{ g HA} \times \frac{1 \text{ mmol HA}}{0.090079 \text{ g HA}} \times \frac{1}{500 \text{ mL soln}} = 0.9547 \text{ M HA}$$

$$\frac{[H_3O^+][A^-]}{[HA]} = 1.38 \times 10^{-4}$$

$[H_3O^+] = [A^-]$ and $[HA] = 0.9547 - [H_3O^+] \approx 0.9547$

$[H_3O^+]^2/0.9547 = 1.38 \times 10^{-4}$

$[H_3O^+] = \sqrt{1.38 \times 10^{-4} \times 0.9547} = 1.148 \times 10^{-2}$

pH = -log 1.148×10^{-2} = <u>1.94</u>

(b) $c_{HA} = 0.9547 \times 25/250 = 0.09547 \text{ M}$

Proceeding as in part (a) we obtain

$[H_3O^+] = 3.63 \times 10^{-3}$ and pH = <u>2.44</u>

To obtain a more exact answer,

$[H_3O^+] = [A^-]$ and $[HA] = 0.09547 - [H_3O^+]$

$$\frac{[H_3O^+]}{0.09547 - [H_3O^+]} = 1.38 \times 10^{-4}$$

$[H_3O^+]^2 + 1.38 \times 10^{-4}[H_3O^+] - 1.317 \times 10^{-5} = 0$

$[H_3O^+] = 3.561 \times 10^{-3}$ and pH = <u>2.45</u>

(c) $c_{HA} = 0.09547 \times 10/1000 = 9.547 \times 10^{-4}$

Here we cannot assume that $[H_3O^+] \ll 9.547 \times 10^{-4}$. Thus,

$[H_3O^+] = [A^-]$ and $[HA] = 9.547 \times 10^{-4} - [H_3O^+]$

$$\frac{[H_3O^+]^2}{9.547 \times 10^{-4} - [H_3O^+]} = 1.38 \times 10^{-4}$$

$$[H_3O^+]^2 + 1.38 \times 10^{-4}[H_3O^+] - 1.317 \times 10^{-7} = 0$$

$$[H_3O^+] = 3.004 \times 10^{-4} \text{ and } pH = -\log 3.004 \times 10^{-4} = \underline{3.52}$$

10-33 (a) $c_{HA} = 1.05 \text{ g HA} \times \dfrac{1 \text{ mmol HA}}{0.22911 \text{ g HA}} \times \dfrac{1}{100 \text{ mL}} = 0.04583 \text{ M}$

$$\frac{[H_3O^+][A^-]}{[HA]} = 0.43 = \frac{[H_3O^+]^2}{0.04583 - [H_3O^+]}$$

$$[H_3O^+]^2 + 0.43[H_3O^+] - 0.01971 = 0$$

$$[H_3O^+] = 4.18 \times 10^{-2} \text{ and } pH = \underline{1.38}$$

Proceeding in the same way

(b) $c_{HA} = 4.58 \times 10^{-3} \text{ M}$, $[H_3O^+] = 4.53 \times 10^{-3}$, and $pH = \underline{2.34}$

(c) $c_{HA} = 4.58 \times 10^{-4} \text{ M}$, $[H_3O^+] = 4.58 \times 10^{-4}$, and $pH = \underline{3.34}$

10-34 For formic acid, HA $K_{HA} = [H_3O^+][A^-]/[HA] = 1.80 \times 10^{-4}$

Throughout Solution 10-34

$$\text{amount HA taken} = 20.00 \text{ mL} \times \frac{0.200 \text{ mmol}}{\text{mL}} = 4.00 \text{ mmol}$$

(a) $c_{HA} = 4.00/45.0 = 8.89 \times 10^{-2}$

$[H_3O^+] = [A^-]$ and $[HA] = 8.89 \times 10^{-2} - [H_3O^+]$

$[H_3O^+]^2/(8.89 \times 10^{-2} - [H_3O^+]) = 1.80 \times 10^{-4}$

$[H_3O^+]^2 + 1.80 \times 10^{-4}[H_3O^+] - 8.89 \times 10^{-2} \times 1.80 \times 10^{-4} = 0$

$[H_3O^+] = 3.91 \times 10^{-3}$ and $pH = \underline{2.41}$

Chapter 10

(b) Amount NaOH added = $25.0 \times 0.160 = 4.00$ mmol.

Therefore, we have a solution of NaA.

c_{NaA} = $4.00/(20.0 + 25.0)$ = 8.89×10^{-2}

$A^- + H_2O \rightleftharpoons HA + OH^-$

K_b = $1.00 \times 10^{-14}/1.80 \times 10^{-4}$ = 5.56×10^{-11}

$[OH^-]$ = $[HA]$ $[A^-]$ = $8.89 \times 10^{-2} - [OH^-]$ ≈ 8.89×10^{-2}

$[OH^-]^2/(8.89 \times 10^{-2})$ = 5.56×10^{-11}

$[OH^-]$ = $\sqrt{8.89 \times 10^{-2} \times 5.56 \times 10^{-11}}$ = 2.22×10^{-6}

pH = $14.00 - (-\log 2.22 \times 10^{-6})$ = __8.35__

(c) Amount NaOH added = $25.0 \times 0.200 = 5.00$ mmol.

Therefore, we have an excess of NaOH and the pH is determined by its concentration.

$[OH^-]$ = c_{NaOH} = $(5.00 - 4.00)/45.0$ = 2.222×10^{-2}

pH = $14.00 - (-\log 2.222 \times 10^{-2})$ = __12.35__

(d) Amount NaA added = $25.0 \times 0.200 = 5.00$ mmol

c_{NaA} = $5.00/45.00$ = 0.1111 M ≈ $[A^-]$

c_{HA} = $4.00/45.00$ = 0.0889 M ≈ $[HA]$

$[H_3O^+] \, 0.111/0.0889 = 1.80 \times 10^{-4}$

$[H_3O^+]$ = 1.440×10^{-4} and pH = __3.84__

10-35
$NH_3 + H_2O \rightleftharpoons NH_4^+ + OH^-$ $K_b = \dfrac{1.00 \times 10^{-14}}{5.70 \times 10^{-10}} = 1.75 \times 10^{-5}$

Throughout this problem the amount of NH_3 taken is 4.00 mmol.

(a) $$c_{NH_3} = \frac{4.00 \text{ mmol NH}_3}{60.0 \text{ mL}} = 0.0667 \text{ M}$$

$$[OH^-] = \sqrt{1.75 \times 10^{-5} \times 0.0667} = 1.08 \times 10^{-3}$$

$$pH = 14.00 - (-\log 1.08 \times 10^{-3}) = \underline{11.03}$$

(b) Amount HCl added = $20.0 \text{ mL} \times 0.200 \text{ mmol/mL} = 4.00$

Thus we have 4.00 mmol NH$_4$Cl per 60.0 mL solution.

$$c_{NH_4Cl} = 4.00/60.0 = 0.0667 \text{ M}$$

$$NH_4^+ + H_2O \rightleftharpoons H_3O^+ + NH_3 \qquad K_a = 5.70 \times 10^{-10}$$

$$[H_3O^+] = \sqrt{K_a c_{NH_4Cl}} = \sqrt{5.70 \times 10^{-10} \times 0.0667} = 6.17 \times 10^{-6}$$

$$pH = -\log 6.17 \times 10^{-5} = \underline{5.21}$$

(c) Amount HCl added = $20.0 \times 0.250 = 5.00$ mmol

$$c_{HCl} = \frac{5.00 \text{ mmol} - 4.00 \text{ mmol}}{60.0 \text{ mL}} = 1.667 \times 10^{-2} = [H_3O^+]$$

$$pH = -\log 1.667 \times 10^{-2} = \underline{1.78}$$

(d) Amount NH$_4$Cl added = $20.0 \times 0.200 = 4.00$ mmol

$$\frac{[H_3O^+][NH_3]}{[NH_4^+]} = 5.70 \times 10^{-10} = \frac{[H_3O^+] \, 4.00/60.0}{4.00/60.0}$$

$[H_3O^+] = 5.70 \times 10^{-10}$ and pH = $-\log 5.70 \times 10^{-10} = \underline{9.24}$

(e) Amount HCl added = 2.00 mmol

Amount NH$_3$ remaining = 4.00 - 2.00 = 2.00 mmol

$$[H_3O^+] \frac{2.00/60.0}{2.00/60.0} = 5.70 \times 10^{-10}$$

$[H_3O^+] = 5.70 \times 10^{-10}$ and pH = $-\log 5.70 \times 10^{-10} = \underline{9.24}$

Chapter 10

10-36 $\dfrac{[H_3O^+][NH_3]}{[NH_4^+]} = 5.70 \times 10^{-10}$

(a) $[NH_3] \approx 0.0300$ and $[NH_4^+] \approx 0.0500$

$[H_3O^+] = 5.70 \times 10^{-10} \times 0.0500/0.0300 = 9.50 \times 10^{-10}$

$[OH^-] = 1.00 \times 10^{-14}/9.50 \times 10^{-10} = \underline{\underline{1.05 \times 10^{-5}}}$

$pH = -\log 9.50 \times 10^{-10} = \underline{\underline{9.022}}$

(b) $\mu = (0.0500 \times 1^2 + 0.0500 \times 1^2)/2 = 0.0500$

From Table 8-1, $\gamma_{NH_4^+} = 0.80$ and $\gamma_{NH_3} = 1.00$

$\dfrac{a_{H_3O^+} \times a_{NH_3}}{a_{NH_4^+}} = \dfrac{a_{H_3O^+} \times 1.00\,[NH_3]}{0.80\,[NH_4^+]} = 5.70 \times 10^{-10}$

$a_{H_3O^+} = \dfrac{5.70 \times 10^{-10} \times 0.80 \times 0.0500}{1.00 \times 0.0300} = 7.60 \times 10^{-10}$

$pH = -\log a_{H_3O^+} = -\log 7.60 \times 10^{-10} = \underline{\underline{9.12}}$

10-37 In each part of this problem, a buffer mixture of a weak acid HA and its conjugate base NaA is formed and Equations 10-6 and 10-7 (page 201) apply.

That is,

$[HA] = c_{HA} - [H_3O^+] + [OH^-]$

$[A^-] = c_{NaA} + [H_3O^+] - [OH^-]$

In each case, we will assume initially that $[H_3O^+]$ and $[OH^-]$ are much smaller than the molar concentration of the acid and conjugate so that $[A^-] \approx c_{NaA}$ and $[HA] \approx c_{HA}$. These assumptions then lead to Equation 10-10.

$[H_3O^+] = K_a c_{HA}/c_{NaA}$

112

(a) c_{HA} = 9.20 g HA $\times \dfrac{1 \text{ mol HA}}{90.08 \text{ g HA}} \times \dfrac{1}{1.00 \text{ L}}$ = 0.1021 M

c_{NaA} = 11.15 g NaA $\times \dfrac{1 \text{ mol NaA}}{112.06 \text{ g NaA}} \times \dfrac{1}{1.00 \text{ L}}$ = 0.0995 M

$[H_3O^+]$ = $1.38 \times 10^{-4} \times 0.1021/0.0995$ = 1.416×10^{-4}

Note that $[H_3O^+]$ (and $[OH^-]$) << $c_{HA} c_{NaA}$ as assumed. Therefore,

pH = $-\log 1.416 \times 10^{-4}$ = <u>3.85</u>

(b) c_{HA} = 0.0550 and c_A = 0.0110

$[H_3O^+]$ = $1.75 \times 10^{-5} \times 0.0550/0.0110$ = 8.75×10^{-5} and pH = <u>4.06</u>

(c) Original amount HA = 3.00 g $\times \dfrac{\text{mmol HA}}{0.13812 \text{ g}}$ = 21.72 mmol

Original amount NaOH = 50.0 mL $\times \dfrac{0.1130 \text{ mmol}}{\text{mL}}$ = 5.65 mmol

c_{HA} = $(21.72 - 5.65)/500$ = 3.214×10^{-2}

c_{NaA} = $5.65/500$ = 1.130×10^{-2}

$[H_3O^+]$ = $1.06 \times 10^{-3} \times 3.214 \times 10^{-2} / 1.130 \times 10^{-2}$ = 3.015×10^{-3}

pH = 2.521

Note, however, that $[H_3O^+]$ is *not* << c_{HA} and c_{NaA}.

Thus, Equations 10-6 and 10-7 must be used.

$[A^-]$ = $1.130 \times 10^{-2} + [H_3O^+] - \cancel{[OH^-]}$

$[HA]$ = $3.214 \times 10^{-2} - [H_3O^+] + \cancel{[OH^-]}$

Certainly, $[OH^-]$ will be negligible since the solution is acidic.

Substituting into the dissociation-constant expression gives

Chapter 10

$$\frac{[H_3O^+](1.130 \times 10^{-2} + [H_3O^+])}{3.214 \times 10^{-2} - [H_3O^+]} = 1.06 \times 10^{-3}$$

Rearranging gives

$$[H_3O^+]^2 + 1.236 \times 10^{-2} [H_3O^+] - 3.407 \times 10^{-5} = 0$$

$[H_3O^+] = 2.321 \times 10^{-3}$ and pH = <u>2.63</u>

(d) Here, Equation 10-10 gives $[H_3O^+] = 4.3 \times 10^{-2}$, which is clearly not $\ll c_{HA}$ or c_{NaA} and we must proceed as in part (c).

This leads to

$$\frac{[H_3O^+](0.100 + [H_3O^+])}{0.0100 - [H_3O^+]} = 4.3 \times 10^{-1}$$

$$[H_3O^+]^2 + 0.53 [H_3O^+] - 4.3 \times 10^{-3} = 0$$

$[H_3O^+] = 7.99 \times 10^{-3}$ and pH = <u>2.10</u>

10-38 In each of the parts of this problem, we are dealing with a weak base B and its conjugate acid BHCl or $(BH)_2SO_4$. The pH-determining equilibrium can then be written as

$$BH^+ + H_2O \rightleftarrows H_3O^+ + B$$

The equilibrium concentration of BH^+ and B are given by Equations 10-6 and 10-7. That is,

$$[BH^+] = c_{BHCl} + \cancel{[OH^-]} - [H_3O^+] \qquad (1)$$

$$[B] = c_B - \cancel{[OH^-]} + [H_3O^+] \qquad (2)$$

In many cases $[OH^-]$ and $[H_3O^+]$ will be much smaller than c_B and c_{BHCl} and $[BH^+] \approx c_{BHCl}$ and $[B] \approx c_B$ so that

$$[H_3O^+] = K_a \times \frac{c_{BHCl}}{c_B} \qquad (3)$$

(a)
$$\text{Amount NH}_4^+ = 3.30 \text{ g (NH}_4)_2\text{SO}_4 \times \frac{1 \text{ mmol (NH}_4)_2\text{SO}_4}{0.13214 \text{ g (NH}_4)_2\text{SO}_4} \times \frac{2 \text{ mmol NH}_4^+}{\text{mmol (NH}_4)_2\text{SO}_4} =$$

$$49.95 \text{ mmol}$$

Amount NaOH = 125.0 mL × 0.1011 mmol/mL = 12.64 mmol

$$c_{\text{NH}_3} = 12.64 \text{ mmol NaOH} \times \frac{1 \text{ mmol NH}_3}{\text{mmol NaOH}} \times \frac{1}{500.0 \text{ mL}} = 2.528 \times 10^{-2} \text{ M}$$

$$c_{\text{NH}_4^+} = (49.95 - 12.64) \text{ mmol NH}_4^+ \times \frac{1}{500.0 \text{ mL}} = 7.462 \times 10^{-2} \text{ M}$$

Substituting these relationships in equation (3) gives

$$[\text{H}_3\text{O}^+] = K_a \times \frac{c_{\text{NH}_4\text{Cl}}}{c_{\text{NH}_3}} = 5.70 \times 10^{-10} \times 7.462 \times 10^{-2} / (2.528 \times 10^{-2})$$

$$= 1.682 \times 10^{-9}$$

pH = -log 1.682 × 10^{-9} = <u>8.77</u>

(b) Substituting into equation (3) gives

$$[\text{H}_3\text{O}^+] = 7.5 \times 10^{-12} \times \frac{c_{\text{BHCl}}}{c_{\text{B}}} = 7.50 \times 10^{-12} \times 0.080/0.120 = 5.00 \times 10^{-12}$$

pH = -log 5.00 × 10^{-12} = <u>11.30</u>

If the assumption that [OH$^-$] ≪ c_B and c_{BCl} is not made the pH is found to be 11.28.

(c) c_B = 0.050 and c_{BHCl} = 0.167

[H$_3$O$^+$] = 2.31 × 10^{-11} × 0.167/0.050 = 7.715 × 10^{-11}

pH = -log 7.715 × 10^{-11} = <u>10.11</u>

(d)
$$\text{Original amount B} = 2.32 \text{ g B} \times \frac{1 \text{ mmol B}}{0.09313 \text{ g B}} = 24.91 \text{ mmol}$$

Amount HCl added = 100 mL × 0.0200 mmol/mL = 2.00 mmol

Chapter 10

$$c_B = (24.91 - 2.00)/250.0 = 9.164 \times 10^{-2}$$

$$c_{BH^+} = 2.00/250.0 = 8.00 \times 10^{-3}$$

$$[H_3O^+] = 2.51 \times 10^{-5} \times 8.00 \times 10^{-3} / 9.164 \times 10^{-2} = 2.191 \times 10^{-6}$$

$$pH = -\log 2.191 \times 10^{-6} = \underline{5.66}$$

10-39 (a) $\Delta pH = \underline{0.00}$

(b) $[H_3O^+]$ changes to 0.00500 M from 0.05000 M

$$\Delta pH = -\log 0.00500 - (-\log 0.0500) = 2.301 - 1.301 = \underline{1.000}$$

(c) pH diluted solution $= 14.000 - (-\log 0.00500) = 11.699$

pH undiluted solution $= 14.000 - (-\log 0.0500) = \underline{12.699}$

$$\Delta pH = \underline{-1.000}$$

(d) In order to get a better picture of the pH change with dilution, we will dispense with the usual approximations and write

$$\frac{[H_3O^+][OAc^-]}{[HOAc]} = \frac{[H_3O^+]^2}{0.0500 - [H_3O^+]} = 1.75 \times 10^{-5}$$

$$[H_3O^+]^2 + 1.75 \times 10^{-5}[H_3O^+] - 0.0500 \times 1.75 \times 10^{-5} = 0$$

Solving by the quadratic formula or by successive approximations gives

$$[H_3O^+] = 9.267 \times 10^{-4} \text{ and } pH = -\log 9.267 \times 10^{-4} = 3.033$$

For the diluted solution, the quadratic becomes

$$[H_3O^+]^2 + 1.75 \times 10^{-5} - 0.00500 \times 1.75 \times 10^{-5}$$

$$[H_3O^+] = 2.872 \times 10^{-4} \text{ and } pH = 3.542$$

$$\Delta pH = 3.033 - 3.542 = \underline{-0.509}$$

(e) $OAc^- + H_2O \rightleftarrows HOAc + OH^-$

$$\frac{[HOAc][OH^-]}{[OAc^-]} = \frac{1.00 \times 10^{-14}}{1.75 \times 10^{-5}} = 5.71 \times 10^{-10} = K_b$$

Here we can use an approximation solution because K_b is so very small. For the undiluted sample

$$\frac{[OH^-]^2}{0.0500} = 5.71 \times 10^{-10}$$

$$[OH^-] = \sqrt{5.71 \times 10^{-10} \times 0.0500} = 5.343 \times 10^{-6}$$

$$pH = 14.00 - (-\log 5.345 \times 10^{-6}) = 8.728$$

For the diluted sample

$$[OH^-] = \sqrt{5.71 \times 10^{-10} \times 0.00500} = 1.690 \times 10^{-6}$$

$$pH = 14.00 - (-\log 1.690 \times 10^{-6}) = 8.228$$

$$\Delta pH = 8.228 - 8.728 = \underline{\underline{-0.500}}$$

(f) Here we must avoid the approximate Equation 10-10 because it will not reveal the small pH change resulting from dilution. Instead write Equations 10-6 and 10-7 as

$$[HOAc] = c_{HOAc} + \cancel{[OH^-]} - [H_3O^+] \approx c_{HOAc} - [H_3O^+]$$

$$[OAc^-] = c_{NaOAc} - \cancel{[OH^-]} + [H_3O^+] \approx c_{NaOAc} + [H_3O^+]$$

$$K_a = 1.75 \times 10^{-5} = \frac{[H_3O^+](0.0500 + [H_3O^+])}{0.0500 - [H_3O^+]}$$

Rearranging gives

$$[H_3O^+]^2 + 5.0018 \times 10^{-2}[H_3O^+] - 8.75 \times 10^{-7} = 0$$

$$[H_3O^+] = 1.749 \times 10^{-5} \text{ and } pH = 4.757$$

Chapter 10

Proceeding in the same way we obtain for the diluted sample

$$1.75 \times 10^{-5} = \frac{[H_3O^+](0.00500 + [H_3O^+])}{0.00500 - [H_3O^+]}$$

$$[H_3O^+]^2 + 5.0175 \times 10^{-3} [H_3O^+] - 8.75 \times 10^{-8} = 0$$

$$[H_3O^+] = 1.738 \times 10^{-5} \text{ and } pH = 4.760$$

$$\Delta pH = 4.760 - 4.757 = \underline{0.003}$$

(g) Proceeding as in part (f), a 10-fold dilution of this solution results in a pH change that is less than 1 in the third decimal place. Thus for all practical puposes,

$$\Delta pH = \underline{0.000}$$

10-40 (a) After addition of acid, $[H_3O^+]$ = 1 mmol/100 mL = 0.0100 M and pH = 2.00. Since original pH = 7.00

$$\Delta pH = 2.00 - 7.00 = \underline{-5.00}$$

(b) After addition of acid,

$$c_{HCl} = (100 \times 0.0500 + 1.00)/100 = 0.0600 \text{ M}$$

$$\Delta pH = -\log 0.0600 - (-\log 0.0500) = 1.222 - 1.301 = \underline{-0.079}$$

(c) After addition of acid,

$$c_{NaOH} = (100 \times 0.0500 - 1.00)/100 = 0.0400$$

$$[OH^-] = 0.0400 \quad \text{and} \quad pH = 14.00 - (-\log 0.0400) = 12.602$$

From Problem 10-39(c), original pH $= \underline{12.699}$

$$\Delta pH = \underline{-0.097}$$

Chapter 10

(d) From Solution 10-39(d), original pH = 3.033

Upon adding 1 mmol HCl to the 0.0500 M HCl, we produce a mixture that is 0.0500 M in HOAc and 1.00/100 = 0.0100 M in HCl. The pH of this solution is approximately that of a 0.0100 M HCl solution, or 2.000. Thus,

$\Delta pH = 2.000 - 3.033 = \underline{\underline{-1.033}}$

(If the contribution of dissociation of HOAc to the pH is taken into account, a pH of 1.996 is obtained and ΔpH = -1.037 is obtained.)

(e) From Solution 10-39(e), original pH = 8.728

Upon adding 1.00 mmol HCl we form a buffer having the composition

$c_{HOAc} = 1.00/100 = 0.0100$

$c_{NaOAc} = (0.0500 \times 100 - 1.00)/100 = 0.0400$

Applying Equation 10-10 gives

$[H_3O^+] = 1.75 \times 10^{-5} \times 0.0100/0.0400 = 4.575 \times 10^{-6}$

$pH = -\log 4.575 \times 10^{-6} = 5.359$

$\Delta pH = 5.359 - 8.728 = \underline{\underline{-3.369}}$

(f) From Solution 10-39(f), original pH = 4.757

With the addition of 1.00 mmol of HCl we have a buffer whose concentrations are

$c_{HOAc} = 0.0500 + 1.00/100 = 0.0600$

$c_{NaOAc} = 0.0500 - 1.00/100 = 0.0400$

Proceeding as in part (e) we obtain

$[H_3O^+] = 2.625 \times 10^{-5}$ and pH = 4.581

$\Delta pH = 4.581 - 4.757 = \underline{\underline{-0.176}}$

Chapter 10

(g) For the original solution

$$[H_3O^+] = 1.75 \times 10^{-5} \times 0.500/0.500 = 1.75 \times 10^{-5}$$

$$pH = -\log 1.75 \times 10^{-5} = 4.757$$

After addition of 1.00 mmol HCl

$$c_{HOAc} = 0.500 + 1.00/100 = 0.510$$

$$c_{NaOAc} = 0.500 - 1.00/100 = 0.490$$

Proceeding as in part (e), we obtain

$$[H_3O^+] = 1.75 \times 10^{-5} \times 0.510/0.490 = 1.821 \times 10^{-5}$$

$$pH = -\log 1.821 \times 10^{-5} = 4.740$$

$$\Delta pH = 4.740 - 4.757 = \underline{-0.017}$$

10-41 (a) $c_{NaOH} = 1.00/100 = 0.0100 = [OH^-]$

$$pH = 14.00 - (-\log 0.0100) = 12.00$$

Original pH = 7.00 and $\Delta pH = 12.00 - 7.00 = \underline{5.00}$

(b) Original pH = 1.301 [see Problem 10-39(b)]

After addition of base, $c_{HCl} = (100 \times 0.0500 - 1.00)/100 = 0.0400$

$$\Delta pH = -\log 0.0400 - 1.301 = 1.398 - 1.301 = \underline{0.097}$$

(c) Original pH = 12.699 [see Problem 10-39(c)]

After addition of base, $c_{NaOH} = (100 \times 0.0500 + 1.00)/100 = 0.0600$

$$pH = 14.00 - (-\log 0.0600) = 12.778$$

$$\Delta pH = 12.778 - 12.699 = \underline{0.079}$$

(d) Original pH = 3.033   [see Problem 10-39(d)]

Addition of strong base gives a buffer of HOAc and NaOAc.

c_{NaOAc} = 1.00 mmol/100 = 0.0100

c_{HOAc} = 0.0500 – 1.00/100 = 0.0400

Proceeding as in Solution 10-40(e) we obtain

$[H_3O^+]$ = 1.75 × 10^{-5} × 0.0400/0.0100 = 7.00 × 10^{-5}

pH = – log 7.00 × 10^{-5} = 4.155

ΔpH = 4.155 – 3.033 = 1.122

(e) Original pH = 8.728   [see Problem 10-39(e)]

Here, we have a mixture of NaOAc and NaOH and the pH is determined by the excess NaOH.

c_{NaOH} = 1.00 mmol/100 = 0.0100

pH = 14.00 – (–log 0.0100) = 12.00

ΔpH = 12.00 – 8.728 = 3.272

(f) Original pH = 4.757   [see Problem 10-39(f)]

c_{NaOAc} = 0.0500 + 1.00/100 = 0.0600

c_{HOAc} = 0.0500 – 1.00/100 = 0.0400

Proceeding as in Solution 10-40(e) we obtain

$[H_3O^+]$ = 1.167 × 10^{-5} and pH = 4.933

ΔpH = 4.933 – 4.757 = 0.176

Chapter 10

(g) Original pH = 4.757   [see Problem 10-40(f)]

$$c_{HOAc} = 0.500 - 1.00/100 = 0.490$$

$$c_{NaOAc} = 0.500 + 1.00/100 = 0.510$$

Substituting into Equation 10-10 gives

$$[H_3O^+] = 1.75 \times 10^{-5} \times 0.400/0.510 = 1.681 \times 10^{-5}$$

$$pH = -\log 1.681 \times 10^{-5} = 4.774$$

$$\Delta pH = 4.774 - 4.757 = \underline{0.017}$$

10-42 For lactic acid, $K_a = 1.38 \times 10^{-4} = [H_3O^+][A^-]/[HA]$

Throughout this problem we will base calculations on Equations 10-6 and 10-7. That is,

$$[A^-] = c_{NaA} + [H_3O^+] - \cancel{[OH^-]}$$

$$[HA] = c_{HA} - [H_3O^+] + \cancel{[OH^-]}$$

$$\frac{[H_3O^+](c_{NaA} + [H_3O^+])}{c_{HA} - [H_3O^+]} = 1.38 \times 10^{-4}$$

This equation rearranges to

$$[H_3O^+]^2 + (1.38 \times 10^{-4} + c_{NaA})[H_3O^+] - 1.38 \times 10^{-4} c_{HA} = 0$$

(a) Before addition of acid

$$[H_3O^+]^2 + (1.38 \times 10^{-4} + 0.0800)[H_3O^+] - 1.38 \times 10^{-4} \times 0.0200 = 0$$

$$[H_3O^+] = 3.443 \times 10^{-5} \text{ and } pH = 4.463$$

Upon adding 0.500 mmol of strong acid

$$c_{HA} = (100 \times 0.0200 + 0.500)/100 = 0.0250 \text{ M}$$

$$c_{NaA} = (100 \times 0.0800 - 0.500)/100 = 0.0750 \text{ M}$$

$[H_3O^+]^2 + (1.38 \times 10^{-4} + 0.0750)[H_3O^+] - 1.38 \times 10^{-4} \times 0.0250 = 0$

$[H_3O^+] = 4.589 \times 10^{-5}$ and pH = 4.338

$\Delta pH = 4.338 - 4.463 = \underline{\underline{-0.125}}$

(b) Before addition of acid

$[H_3O^+]^2 + (1.38 \times 10^{-4} + 0.0200)[H_3O^+] - 1.38 \times 10^{-4} \times 0.0800 = 0$

$[H_3O^+] = 5.341 \times 10^{-4}$ and pH = 3.272

After adding acid

$c_{HA} = (100 \times 0.0800 + 0.500)/100 = 0.085$ M

$c_{NaA} = (100 \times 0.0200 - 0.500)/100 = 0.015$ M

$[H_3O^+]^2 + (1.38 \times 10^{-4} + 0.015) - 1.38 \times 10^{-4} \times 0.085 = 0$

$[H_3O^+] = 7.388 \times 10^{-4}$ and pH = 3.131

$\Delta pH = 3.131 - 3.272 = \underline{\underline{-0.141}}$

(c) Before addition of acid

$[H_3O^+]^2 + (1.38 \times 10^{-4} + 0.0500)[H_3O^+] - 1.38 \times 10^{-4} \times 0.0500 = 0$

$[H_3O^+] = 1.372 \times 10^{-4}$ and pH = 3.863

After adding acid

$c_{HA} = (100 \times 0.0500 + 0.500)/100 = 0.0550$

$c_{NaA} = (100 \times 0.0500 - 0.500)/100 = 0.0450$

$[H_3O^+]^2 + (1.38 \times 10^{-4} + 0.0450)[H_3O^+] - 1.38 \times 10^{-4} \times 0.0550 = 0$

$[H_3O^+] = 1.675 \times 10^{-4}$ and pH = 3.776

$\Delta pH = 3.776 - 3.863 = \underline{\underline{-0.087}}$

Chapter 10

10-43 pH = 3.50 and $[H_3O^+]$ = antilog (-3.50) = 3.162×10^{-4}

$[H_3O^+][A^-]/[HA]$ = 1.80×10^{-4}

$[A^-]/[HA]$ = $1.80 \times 10^{-4}/3.162 \times 10^{-4}$ = 0.5693

[HA] = 1.00 and $[A^-]$ = 0.5693×1.00 = 0.5693 M

mass HCOONa = $0.5693 \dfrac{\text{mmol NaA}}{\text{mL}} \times 400 \text{ mL} \times 0.06801 \dfrac{1}{\text{mmol NaA}}$

= 15.5 g sodium formate

10-44 Proceeding as in Problem 10-43, we find that 43.2 g of sodium glycolate are needed.

10-45 pH = 3.37 and $[H_3O^+]$ = antilog (-3.37) = 4.266×10^{-4}

4.0×10^{-4} = $[H_3O^+][A^-]/[HA]$ = $4.266 \times 10^{-4} [A^-]/[HA]$

$[A^-]/[HA]$ = 0.9377

Let V_{HCl} = mL 0.200 M HCl added.

amount HA formed = amount HCl added = $V_{HCl} \times 0.200$ mmol

amount NaA remaining = original amount NaA − amount V_{HCl} added

= $(250 \times 0.300 - V_{HCl} \times 0.200)$ mmol

Total volume of solution = $(250 + V_{HCl})$ mL

c_{HA} = $0.200 \, V_{HCl}/(250 + V_{HCl})$ ≅ [HA]

c_{NaA} = $(75.0 - 0.200 \, V_{HCl})/(250 + V_{HCl})$ ≅ $[A^-]$

Substituting

$\dfrac{[A^-]}{[HA]} = \dfrac{(75.0 - 0.200 \, V_{HCl})/\cancel{(250 + V_{HCl})}}{0.200 \, V_{HCl}/\cancel{(250 + V_{HCl})}} = 0.9377$

$$75.0 - 0.200\, V_{HCl} = 0.200\, V_{HCl} \times 0.9377 = 0.1875\, V_{HCl}$$

$$V_{HCl} = 75.0/(0.200 + 0.1875) = \underline{\underline{194 \text{ mL HCl}}}$$

10-46 Proceeding as in Problem 10-44, we find that $\underline{89.3}$ mL of 2.00 M NaOH are required.

10-47 0.00 to 49.00 mL reagent. Excess base present and we may write

$$c_{NaOH} = [OH^-] = \frac{\text{original amount base} - \text{amount HCl added}}{50.0 + V_{HCl}}$$

$$= \frac{50.00 \times 0.1000 - V_{HCl} \times 0.1000}{50.0 + V_{HCl}} \quad (1)$$

Substituting $V_{HCl} = 0.00$ mL into equation (1) gives

$$[OH^-] = 0.1000 \quad \text{and} \quad pH = 14.00 - (-\log 0.1000) = \underline{13.00}$$

Substituting $V_{HCl} = 10.00$ mL into the equation yields

$$[OH^-] = \frac{5.000 - 10.00 \times 0.1000}{50.00 + 10.00} = 6.67 \times 10^{-2}$$

$$pH = 14.00 - (-\log 6.67 \times 10^{-2}) = \underline{12.82}$$

The data from additional calculations of this type are given in the table that follows.

50.00 mL reagent. This is equivalence point where

$$[H_3O^+] = [OH^-] = 1.00 \times 10^{-7} \quad \text{and} \quad pH = \underline{7.00}$$

51.00, 55.00, and 60.00 mL reagent. Here we have an excess of HCl and

$$c_{HCl} = [H_3O^+] = \frac{\text{amount HCl added} - \text{original amount NaOH}}{50.00 + V_{HCl}}$$

$$= \frac{V_{HCl} \times 0.100 - 50.00 \times 0.1000}{50.00 + V_{HCl}}$$

Chapter 10

Thus at 51.00 mL

$$[H_3O^+] = \frac{51.00 \times 0.1000 - 50.00 \times 0.1000}{101.0} = 9.90 \times 10^{-4}$$

pH = -log 9.90×10^{-4} = <u>3.00</u>

V_{HCl} (mL)	pH	V_{HCl} (mL)	pH
0.00	13.00	49.00	11.00
10.00	12.82	50.00	7.00
25.00	12.52	51.00	3.00
40.00	12.05	55.00	2.32
45.00	11.72	60.00	2.04

10-48 Let us calculate pH when 24.95 and 25.05 mL of reagent have been added.

24.95 mL reagent

$$c_{A^-} \cong \frac{\text{amount KOH added}}{\text{total volume soln}} = \frac{24.95 \times 0.1000 \text{ mmol KOH}}{74.95 \text{ mL soln}} = \frac{2.495}{74.95}$$

$$c_{HA} \cong [HA] = \frac{\text{original amount HA} - \text{amount KOH added}}{\text{total volume soln}}$$

$$= \frac{(50.00 \times 0.0500 - 24.95 \times 0.1000) \text{ mmol HA}}{74.95 \text{ mL soln}}$$

$$= \frac{2.500 - 2.495}{74.95} = \frac{0.005}{74.95}$$

Substituting into Equation 10-10

$$[H_3O^+] = K_a \frac{c_{HA}}{c_{A^-}} = \frac{1.80 \times 10^{-4} \times 0.005/74.95}{2.495/74.95} = 3.607 \times 10^{-7}$$

pH = -log 3.607×10^{-7} = <u>6.44</u>

Chapter 10

25.05 mL KOH

$$c_{KOH} = \frac{\text{amount KOH added} - \text{initial amount HA}}{(25.05 + 50.00) \text{ mL}}$$

$$= \frac{25.05 \times 0.1000 - 50.00 \times 0.05000}{75.05} = 6.66 \times 10^{-5} = [OH^-]$$

$$pH = 14.00 - (-\log 6.66 \times 10^{-5}) = \underline{9.82}$$

Thus, the indicator should change color in the range of pH <u>6.5 to 9.8</u>. Cresol purple (range: 7.6 to 9.2) (Table 10-1) would be quite suitable.

10-49 (See Solution 10-48) Let us calculate the pH when 49.95 and 50.05 mL of $HClO_4$ have been added.

49.95 mL $HClO_4$

$$B = C_2H_5NH_2 \qquad BH^+ = C_2H_5NH_3^+$$

$$c_{BH^+} = \frac{\text{no. mmol } HClO_4}{\text{total volume soln}} = \frac{49.95 \times 0.1000}{99.95} = \frac{4.995}{99.95} \approx [BH^+]$$

$$c_B = \frac{(50.00 \times 0.1000 - 49.95 \times 0.1000)}{99.95} = \frac{0.00500}{99.95} \approx [B]$$

$$\frac{[H_3O^+][B]}{[BH^+]} = 2.31 \times 10^{-11} = \frac{[H_3O^+](0.005/\cancel{99.95})}{4.995/\cancel{99.95}}$$

$$[H_3O^+] = 2.31 \times 10^{-11} \times 4.995/0.005 = 2.308 \times 10^{-8}$$

$$pH = -\log 2.308 \times 10^{-8} = 7.64$$

50.05 mL $HClO_4$

$$c_{HClO_4} = \frac{50.05 \times 0.1000 - 50.00 \times 0.1000}{100.05} = 4.998 \times 10^{-5} = [H_3O^+]$$

$$pH = -\log 4.998 \times 10^{-5} = 4.30$$

Indicator should change color in the pH range of 7.64 to 4.30. Bromocresol purple would be suitable.

Chapter 10

10-50 (a) **0.00 mL NaOH**

$$c_{HA} = 0.1000 \text{ M} \quad [H_3O^+] = [A^-] \quad [HA] = 0.1000 - [H_3O^+]$$

$$K_a = 7.1 \times 10^{-4} = \frac{[H_3O^+][A^-]}{[HA]} = \frac{[H_3O^+]^2}{0.1000 - [H_3O^+]}$$

$$[H_3O^+]^2 + 7.1 \times 10^{-4} [H_3O^+] - 0.1000 \times 7.1 \times 10^{-4} = 0$$

$[H_3O^+] = 8.08 \times 10^{-3}$ and pH = 2.09

Assuming $[H_3O^+] \ll 0.1000$ gives

$[H_3O^+] = 8.43 \times 10^{-3}$ and pH = 2.07

5.00 mL NaOH

$$c_{NaA} = (5.00 \times 0.1000)/(50.00 + 5.00) = 9.09 \times 10^{-3} \text{ M}$$

$$c_{HA} = (50.00 \times 0.1000 - 5.00 \times 0.1000)/(55.00) = 8.18 \times 10^{-2} \text{ M}$$

We now have a buffer mixture and employ Equation 10-10.

$$[H_3O^+] = 7.1 \times 10^{-4} \times 8.18 \times 10^{-2} / (9.09 \times 10^{-3}) = 6.39 \times 10^{-3}$$

pH = 2.19

Note, however, that $[H_3O^+]$ is not $\ll c_{NaA}$ and c_{HA}.

Therefore, we must use Equations 10-6 and 10-7.

$$K_a = \frac{[H_3O^+](c_{NaA} + [H_3O^+] - [OH^-])}{(c_{HA} - [H_3O^+] + [OH^-])}$$

which arranges to

$$[H_3O^+]^2 + (K_a + c_{NaA})[H_3O^+] - K_a c_{HA} = 0 \quad (1)$$

$$[H_3O^+]^2 + (7.1 \times 10^{-4} + 9.09 \times 10^{-3})[H_3O^+] - 5.808 \times 10^{-5} = 0$$

$$[H_3O^+]^2 + 9.80 \times 10^{-3} [H_3O^+] - 5.808 \times 10^{-5} = 0$$

$[H_3O^+] = 4.160 \times 10^{-3}$ and pH = 2.38

Chapter 10

The other preequivalence point calculations are performed by substitution into equation (1). The results are shown in the table that follows.

50.00 mL NaOH

Here we are at the equivalence point and have a solution of NaA.

$$c_{NaA} = \frac{50.00 \times 0.1000}{50.00 + 50.00} = 0.05000 \text{ M}$$

$$A^- + H_2O \rightleftharpoons HA + OH^-$$

$$K_b = \frac{[OH^-][HA]}{[A^-]} = \frac{1.00 \times 10^{-14}}{7.1 \times 10^{-4}} = 1.408 \times 10^{-11}$$

$[HA] = [OH^-]$ and $[A^-] = 0.0500 - [OH^-] \approx 0.05000$

$[OH^-]^2/0.05000 = 1.408 \times 10^{-11}$

$[OH^-] = \sqrt{0.05000 \times 1.408 \times 10^{-11}} = 8.390 \times 10^{-7}$

$\text{pH} = 14.00 - (-\log 8.390 \times 10^{-7}) = \underline{7.92}$

51.00 mL NaOH. We now have an excess of NaOH.

$[OH^-] \approx c_{NaOH} = (51.00 \times 0.1000 - 50.00 \times 0.1000)/101.0$

$= 9.901 \times 10^{-4}$

$\text{pH} = 14.00 - (-\log 9.901 \times 10^{-4}) = \underline{11.00}$

The other post-equivalence point data are treated in the same way.

(b) The calculations here are identical to those in part (a) except that $K_a = 1.38 \times 10^{-4}$. The answers are found in the table that follows.

(c) Here the acid is $C_5H_{11}^+$.

$$C_5H_5NH^+ + H_2O \rightleftharpoons C_5HN_5 + 4H_3O^+$$

$$K_a = 5.90 \times 10^{-6}$$

The calculations for this titration can be performed exactly as those in part (a). K_a is small enough, however, so that the approximate method can be used instead. Thus, the initial hydronium ion concentration is given by

$$[H_3O^+] = \sqrt{c_{HA} K_a}$$

For the preequivalence point data,

$$[H_3O^+] = K_a c_{C_5H_5NH^+} / c_{C_5H_5N} \qquad \text{(Equation 10-10)}$$

At the equivalence point, the hydroxide ion concentration is given by

$$[OH^-] = \sqrt{c_{C_5H_5N} \times K_b} = \sqrt{0.0500 \times 1.00 \times 10^{-14} / (5.90 \times 10^{-6})} = 9.21 \times 10^{-6}$$

$$pH = 14.00 - (-\log 9.21 \times 10^{-6}) = \underline{8.96}$$

The data for this titration curve are given in the table that follows.

Chapter 10

Vol, mL	(a) pH	(b) pH	(c) pH
0.00	2.09	2.44	3.12
5.00	2.38	2.96	4.28
15.00	2.82	3.50	4.86
25.00	3.17	3.86	5.23
40.00	3.76	4.46	5.83
45.00	4.11	4.82	6.18
49.00	4.85	5.55	6.92
50.00	7.92	8.28	8.96
51.00	11.00	11.00	11.00
55.00	11.68	11.68	11.68
60.00	11.96	11.96	11.96

10-51 (a) $NH_4^+ + H_2O \rightleftharpoons H_3O^+ + NH_3 \quad K_a = 5.70 \times 10^{-10}$

$$NH_3 + H_2O \rightleftharpoons NH_4^+ + OH^- \quad K_b = \frac{K_w}{K_a} = 1.75 \times 10^{-5} = [OH^-][NH_4^+]/[NH_3]$$

0.00 mL 0.1000 M HCl

$[NH_4^+] = [OH^-] \quad [NH_3] = 0.1000 - [OH^-] = 0.1000$

$[OH^-] = \sqrt{c_{NH_3} K_b} = \sqrt{0.1000 \times 1.75 \times 10^{-5}} = 1.323 \times 10^{-3}$

$pH = 14.00 - (-\log 1.323 \times 10^{-3}) = \underline{11.12}$

5.00 mL 0.1000 M HCl

$c_{NH_4^+} = 5.00 \times 0.1000/(50.00 + 5.00) = 9.091 \times 10^{-3}$

Chapter 10

$$c_{NH_3} = (50.00 \times 0.1000 - 5.00 \times 0.1000)/55.00 = 8.182 \times 10^{-2}$$

$$[NH_3] \approx c_{NH_3} \qquad [NH_4^+] \approx c_{NH_4^+}$$

Substituting these equalities into the expression for K_a gives upon rearranging

$$[H_3O^+] = K_a c_{NH_4^+}/c_{NH_3}$$

$$= 5.70 \times 10^{-10} \times 9.091 \times 10^{-3}/(8.182 \times 10^{-2})$$

$$= 6.33 \times 10^{-11}$$

$$pH = -\log 6.33 \times 10^{-11} = \underline{10.20}$$

The other preequivalence point data are treated in the same way.

The answers appear in the table that follows.

50.00 mL 0.1000 M HCl. This is the equivalence point at which

$$c_{NH_4^+} = 50.00 \times 0.1000/100.0 = 0.05000$$

$$[H_3O^+] = [NH_3] \qquad \text{and} \qquad [NH_4^+] = 0.05000 - [H_3O^+] \approx 0.05000$$

$$[H_3O^+]^2/0.05000 = 5.70 \times 10^{-10}$$

$$[H_3O^+] = \sqrt{0.05000 \times 5.70 \times 10^{-10}} = 5.34 \times 10^{-6}$$

$$pH = -\log 5.34 \times 10^{-6} = \underline{5.27}$$

51.00 mL HCl

$$c_{HCl} = [H_3O^+] = (51.00 \times 0.1000 - 50.00 \times 0.1000)/101.0$$

$$= 9.901 \times 10^{-4} \quad \text{and} \quad pH = -\log 9.901 \times 10^{-4} = \underline{3.00}$$

Other post-equivalence point data are obtained in the same way.

Chapter 10

(b) These calculations are identical to those in part (a) except that $K_a = 1.05 \times 10^{-8}$ and $K_b = 9.52 \times 10^{-7}$. The data are found in the table that follows.

(c) $CN^- + H_2O \rightleftharpoons HCN + OH^-$

$$K_b = \frac{[OH^-][HCN]}{[CN^-]} = \frac{K_w}{K_a} = \frac{1.00 \times 10^{-14}}{6.2 \times 10^{-10}} = 1.61 \times 10^{-5}$$

The calculations are performed in exactly the same way as those in part (a).

0.00 mL

$$[OH^-] = \sqrt{c_{CN^-} \times K_b} = \sqrt{0.1000 \times 1.61 \times 10^{-5}} = 1.27 \times 10^{-3}$$

$$pH = 14.00 - (-\log 1.27 \times 10^{-3}) = \underline{11.10}$$

5.00 mL

$$c_B = c_{NaCN} = 8.182 \times 10^{-2} \quad \text{[see part (a), 5.00 mL]}$$

$$c_{BH^+} = c_{HCN} = 9.091 \times 10^{-3}$$

$$[H_3O^+] = K_a c_{HCN} / c_{CN^-}$$

$$= 6.2 \times 10^{-10} \times 9.091 \times 10^{-3} / 8.182 \times 10^{-2} = 6.89 \times 10^{-11}$$

$$pH = -\log 6.89 \times 10^{-11} = \underline{10.16}$$

50.00 mL. At equivalence we have a 0.05000 M solution of HCN.

$$[H_3O^+] = \sqrt{0.0500 \times K_a} = \sqrt{0.0500 \times 6.2 \times 10^{-18}}$$

$$= 5.568 \times 10^{-6} \quad \text{and} \quad pH = \underline{5.25}$$

The post-equivalence point calculations are identical to those in part (a).

Chapter 10

	(a)	(b)	(c)
Vol HCl, mL	pH	pH	pH
0.00	11.12	10.49	11.10
5.00	10.20	8.93	10.16
15.00	9.61	8.35	9.58
25.00	9.24	7.98	9.21
40.00	8.64	7.38	8.61
45.00	8.29	7.02	8.25
49.00	7.55	6.29	7.52
50.00	5.27	4.64	5.25
51.00	3.00	3.00	3.00
55.00	2.32	2.32	2.32
60.00	2.04	2.04	2.04

10-52 (a) $C_6H_5NH_3^+ + H_2O \rightleftarrows H_3O^+ + C_6H_5NH_2$

$$K_a = 2.51 \times 10^{-5} = \frac{[H_3O^+][C_6H_5NH_2]}{[C_6H_5NH_3^+]}$$

0.00 mL

$[H_3O^+] = [C_6H_5NH_2]$ and $[C_6H_5NH_3^+] = 0.1000 - [H_3O^+] \approx 0.1000$

$[H_3O^+]^2 = 2.51 \times 10^{-5} \times 0.1000$

$[H_3O^+] = \sqrt{2.51 \times 10^{-6}} = 1.584 \times 10^{-3}$

pH = $-\log 1.584 \times 10^{-3}$ = <u>2.80</u>

Chapter 10

5.00 mL

$$c_{C_6H_5NH_2} = 5.00 \times 0.1000/55.00 = 9.091 \times 10^{-3}$$

$$c_{C_6H_5NH_3^+} = (50.00 \times 0.1000 - 5.00 \times 0.1000)/55.00 = 8.182 \times 10^{-2}$$

Substituting into Equation 10-10

$$[H_3O^+] = 2.51 \times 10^{-5} \times 8.182 \times 10^{-2}/9.091 \times 10^{-3}$$

$$= 2.259 \times 10^{-4} \quad \text{and} \quad pH = \underline{3.65}$$

The other preequivalence data are treated in the same way. The answers are in the table that follows.

50.00 mL

Here we have 0.05000 M solution of the weak base aniline.

$$C_6H_5NH_2 \rightleftharpoons C_6H_5NH_3^+ + OH^- \quad K_b = \frac{1.00 \times 10^{-14}}{2.51 \times 10^{-5}} = 3.98 \times 10^{-10}$$

$$[OH^-] = [C_6H_5NH_3^+] \quad \text{and} \quad [C_6H_5NH_2] = 0.05000 - [OH^-] \approx 0.05000$$

$$[OH^-]^2/0.05000 = 3.98 \times 10^{-10}$$

$$[OH^-] = \sqrt{0.05000 \times 3.98 \times 10^{-10}} = 4.461 \times 10^{-6}$$

$$pH = 14.00 - (-\log 4.461 \times 10^{-6}) = \underline{8.65}$$

51.00 mL

Here, the pH is determined by the concentration of NaOH. That is,

$$c_{NaOH} = (51.00 \times 0.1000 - 50.00 \times 0.1000)/101.00$$

$$= 9.901 \times 10^{-4} \text{ M}$$

$$pH = 14.00 - (-\log 9.901 \times 10^{-4}) = 11.00$$

55.00 and 60.00 mL. Proceeding in the same way, we obtain pH = 11.68 and 11.96.

(b) $\dfrac{[H_3O^+][A^-]}{[HA]} = 1.36 \times 10^{-3}$

0.00 mL. 0.0100 M NaOH

$$\dfrac{[H_3O^+]^2}{0.0100 - [H_3O^+]} = K_a = 1.36 \times 10^{-3}$$

$[H_3O^+]^2 + 1.36 \times 10^{-3}[H_3O^+] - 1.36 \times 10^{-5} = 0$

$[H_3O^+] = 3.07 \times 10^{-3}$ and $[H_3O^+] = -\log 9.78 \times 10^{-3} = \underline{2.51}$

5.00 mL

$c_{A^-} = 5.00 \times 0.0100/55.00 = 9.091 \times 10^{-4}$

$c_{HA} = (50.00 \times 0.0100 - 5.00 \times 0.0100)/55.00 = 8.182 \times 10^{-3}$

Because K_a is relatively large we must use Equations 10-6 and 10-7. Thus,

$[A^-] = c_A + [H_3O^+] - [OH^-] \approx c_A + [H_3O^+]$

$[HA] = c_{HA} - [H_3O^+] + [OH^-] = c_A - [H_3O^+]$

$$\dfrac{[H_3O^+](9.091 \times 10^{-4} + [H_3O^+])}{8.182 \times 10^{-3} - [H_3O^+]} = 1.36 \times 10^{-3}$$

$[H_3O^+]^2 + 2.269 \times 10^{-3}[H_3O^+] - 1.113 \times 10^{-5} = 0$

Solving the quadratic gives

$[H_3O^+] = 2.389 \times 10^{-3}$ and pH = $-\log 2.389 \times 10^{-3} = \underline{2.62}$

Chapter 10

The remaining data up to the equivalence point are derived in the same way and are found in the accompanying table.

50.00 mL. $c_{NaA} = 0.005$

$$A^- + H_2O \rightleftharpoons HA + OH^- \qquad K_b = \frac{1.00 \times 10^{-14}}{1.36 \times 10^{-3}} = 7.353 \times 10^{-12}$$

$$[OH^-] = \sqrt{0.00500 \times 7.353 \times 10^{-12}} = 1.917 \times 10^{-7}$$

$$pH = 14.00 - (-\log 1.917 \times 10^{-7}) = \underline{7.28}$$

51.00 mL. 0.0100 M NaOH

$$c_{NaOH} = \frac{51.00 \times 0.0100 - 50.00 \times 0.0100}{101.00} = 9.901 \times 10^{-5} = [OH^-]$$

$$pH = 14.00 - (-\log 9.901 \times 10^{-5}) = \underline{10.00}$$

The pH values at 55.00 and 60.00 mL are obtained in the same way and are given in the accompanying table.

(c) Here, $K_a = 3.0 \times 10^{-8}$. The calculations are identical to those in part (a) of this problem. The answers are found in the table that follows.

(d)
$$HONH_2 + H_2O \rightleftharpoons HONH_3^+ + OH^- \qquad K_b = \frac{1.00 \times 10^{-14}}{1.10 \times 10^{-6}} = 9.09 \times 10^{-9}$$

Proceeding as in the solution to Problem 10-51(a), we obtain the data shown in the last column of the table that follows.

Chapter 10

Vol, mL	(a) pH	(b) pH	(c) pH	(d) pH
0.00	2.80	2.51	4.26	9.48
5.00	3.65	2.62	6.57	6.91
15.00	4.23	2.84	7.15	6.33
25.00	4.60	3.09	7.52	5.96
40.00	5.20	3.60	8.12	5.36
49.00	6.29	4.66	9.21	4.27
50.00	8.65	7.28	10.11	3.63
51.00	11.00	10.00	11.00	3.00
55.00	11.68	10.68	11.68	2.32
60.00	11.96	10.96	11.96	2.04

10-53 Substituting into Equations 10-16 and 10-17

$$\alpha_0 = \frac{[H_3O^+]}{[H_3O^+] + K_a} \qquad \alpha_1 = \frac{K_a}{[H_3O^+] + K_a}$$

(a) $[H_3O^+] = 4.786 \times 10^{-6}$ and $K_a = 1.75 \times 10^{-5}$

$$\alpha_0 = \frac{4.786 \times 10^{-6}}{4.786 \times 10^{-6} + 1.75 \times 10^{-5}} = \underline{\underline{0.215}}$$

$\alpha_1 = 1.000 - 0.215 = \underline{\underline{0.785}}$

Proceeding in the same way, we obtain

(b) $[H_3O^+] = 5.598 \times 10^{-2}$; $\alpha_0 = \underline{\underline{0.116}}$; $\alpha_1 = \underline{\underline{0.884}}$

(c) $[H_3O^+] = 1.000 \times 10^{-7}$; $\alpha_0 = \underline{\underline{0.769}}$; $\alpha_1 = \underline{\underline{0.231}}$

(d) $[H_3O^+] = 7.586 \times 10^{-6}$; $\alpha_0 = \underline{\underline{0.873}}$; $\alpha_1 = \underline{\underline{0.127}}$

(e) $[H_3O^+] = 8.318 \times 10^{-11}$; $\alpha_0 = \underline{0.917}$; $\alpha_1 = \underline{0.083}$

10-54 $[H_3O^+] = 6.310 \times 10^{-4}$.   Substituting into Equation 10-16 gives

$$\alpha_0 = \frac{6.310 \times 10^{-4}}{6.310 \times 10^{-4} + 1.80 \times 10^{-4}} = 0.778$$

$$\frac{HCOOH}{c_T} = \frac{[HCOOH]}{0.0850} = \alpha_0$$

$$[HCOOH] = 0.778 \times 0.0850 = \underline{6.61 \times 10^{-2} \text{ M}}$$

10-55 $[H_3O^+] = 3.38 \times 10^{-12}$

For $CH_3NH_3^+$, Equation 10-17 takes the form

$$\alpha_1 = \frac{[CH_3NH_2]}{c_T} = \frac{K_a}{[H_3O^+] + K_a} = \frac{2.3 \times 10^{-11}}{3.38 \times 10^{-12} + 2.3 \times 10^{-11}}$$

$$= 0.872 = \frac{[CH_3NH_2]}{0.120}$$

$$[CH_3NH_2] = 0.872 \times 0.120 = 0.105 \text{ M}$$

10-56 For lactic acid, $K_a = 1.38 \times 10^{-4}$

$$\alpha_0 = \frac{[H_3O^+]}{K_a + [H_3O^+]} = \frac{[H_3O^+]}{1.38 \times 10^{-4} + [H_3O^+]}$$

$$= 0.640 = \frac{[HA]}{c_T} = \frac{[HA]}{0.120}$$

$$[HA] = 0.640 \times 0.120 = \underline{0.0768}$$

$$\alpha_1 = 1.000 - 0.640 = \underline{0.360}$$

$$[A^-] = \alpha_1 \times 0.120 = (1.000 - 0.640) \times 0.120) = \underline{0.0432}$$

Chapter 10

$$[H_3O^+] = \frac{0.640 \times 1.38 \times 10^{-4}}{1-0.640} = 2.453 \times 10^{-4}$$

$$pH = -\log 2.453 \times 10^{-4} = \underline{\underline{3.61}}$$

The remaining data are obtained in the same way.

Acid	c_T	pH	[HA]	[A$^-$]	α_0	α_1
Lactic	0.120	3.61	0.0768	0.0432	0.640	0.360
Iodic	0.200	1.28	0.0470	0.153	0.235	0.765
Butanoic	0.162	5.00	0.0644	0.0979	0.397	0.604
Nitrous	0.179	3.30	0.0739	0.105	0.413	0.587
HCN	0.366	9.39	0.145	0.221	0.396	0.604
Sulfamic	0.250	1.20	0.095	0.155	0.380	0.620

CHAPTER 11

11-1 Not only is NaHA a proton donor, it is also the conjugate base of the parent acid H_2A

$$HA^- + H_2O \rightleftarrows \begin{array}{c} H_3O^+ + A^{2-} \\ H_2A + OH^- \end{array}$$

Solutions of acid salts are acidic or alkaline, depending upon which of these equilibrium predominates. In order to compute the pH of solutions of this type, it is necessary to take both equilibria into account.

11-2 A solution of NaHA will be acidic or basic depending upon which of the following equilibrium constants is larger.

$$HA^- + H_2O \rightleftarrows H_3O^+ + A^{2-} \quad K_2$$

$$HA^- + H_2O \rightleftarrows H_2A + OH^- \quad K_b = \frac{K_w}{K_1}$$

(a) For the acid salt of fumaric acid

$$K_2 = 3.21 \times 10^{-5}$$

$$K_b = \frac{1.00 \times 10^{-14}}{8.85 \times 10^{-4}} = 1.13 \times 10^{-11}$$

Because K_2 is much larger than K_b, the solution will be acidic.

(b) For $NaHCO_3$

$$K_2 = 4.69 \times 10^{-11}$$

$$K_b = \frac{1.00 \times 10^{-14}}{4.45 \times 10^{-7}} = 2.25 \times 10^{-8}$$

Because $K_b \gg K_2$, the solution is basic.

11-3 The HPO_4^{2-} ion is such a weak acid ($K_a = 4.5 \times 10^{-13}$) that the change in pH in the vicinity of the third equivalence point is too small to be observable.

Chapter 11

11-4 (a) $NH_4^+ + H_2O \rightleftharpoons NH_3 + H_3O^+$ $\quad K_{NH_4^+} = 5.70 \times 10^{-10}$

$OAc^- + H_2O \rightleftharpoons HOAc + OH^-$ $\quad K_{OAc^-} = \dfrac{K_w}{K_{HOAc}} = \dfrac{1.00 \times 10^{-14}}{5.70 \times 10^{-10}} = 5.71 \times 10^{-10}$

Since the K's are essentially identical, the solution should be approximately <u>neutral</u>.

(b) $NO_2^- + H_2O \rightleftharpoons HNO_2 + OH^-$ $\quad$ Solution will be <u>basic</u>.

(c) Neither Na^+ nor NO_3^- reacts with H_2O. Solution will be <u>neutral</u>.

(d) $HC_2O_4^- + H_2O \rightleftharpoons H_3O^+ + C_2O_4^{2-}$ $\quad K_2 = 5.42 \times 10^{-5}$

$HC_2O_4^- + H_2O \rightleftharpoons OH^- + H_2C_2O_4$ $\quad K_b = \dfrac{1.00 \times 10^{-14}}{5.60 \times 10^{-2}} = 1.79 \times 10^{-13}$

$\qquad\qquad\qquad\qquad\qquad\qquad\qquad\qquad\qquad\qquad\qquad$ Solution will be <u>acidic</u>.

(e) $C_2O_4^{2-} + H_2O \rightleftharpoons HC_2O_4^- + OH^-$ $\quad K_b = \dfrac{1.00 \times 10^{-14}}{5.42 \times 19^{-5}} = 1.84 \times 10^{-10}$

$\qquad\qquad\qquad\qquad\qquad\qquad\qquad\qquad\qquad\qquad\qquad$ Solution will be <u>basic</u>.

(f) $HPO_4^{2-} + H_2O \rightleftharpoons H_3O^+ + PO_4^{3-}$ $\quad K_3 = 4.5 \times 10^{-13}$

$HPO_4^{2-} + H_2O \rightleftharpoons OH^- + H_2PO_4^-$ $\quad K_b = \dfrac{1.00 \times 10^{-14}}{6.32 \times 10^{-8}} = 1.58 \times 10^{-7}$

$\qquad\qquad\qquad\qquad\qquad\qquad\qquad\qquad\qquad\qquad\qquad$ Solution will be <u>basic</u>.

(g) $H_2PO_4^- + H_2O \rightleftharpoons H_3O^+ + HPO_4^-$ $\quad K_2 = 6.32 \times 10^{-8}$

$H_2PO_4^- + H_2O \rightleftharpoons OH^- + H_3PO_4$ $\quad K_b = \dfrac{1.00 \times 10^{-14}}{7.11 \times 10^{-3}} = 1.41 \times 10^{-12}$

$\qquad\qquad\qquad\qquad\qquad\qquad\qquad\qquad\qquad\qquad\qquad$ Solution will be <u>acidic</u>.

Chapter 11

(h) $PO_4^{3-} + H_2O \rightleftarrows HPO_4^{2-} + OH^-$ $K_b = \dfrac{1.00 \times 10^{-14}}{4.5 \times 10^{-13}} = 2.2 \times 10^{-2}$

Solution will be <u>basic</u>.

11-5 $H_3AsO_4 + H_2O \rightleftarrows H_2AsO_4^- + H_3O^+$ $K_1 = 5.8 \times 10^{-3}$

$H_2AsO_4^- + H_2O \rightleftarrows HAsO_4^{2-} + H_3O^+$ $K_2 = 1.1 \times 10^{-7}$

Substituting into Equation 11-12

$[H_3O^+] = \sqrt{5.8 \times 10^{-3} \times 1.1 \times 10^{-7}} = 2.53 \times 10^{-5}$

$pH = -\log(2.53 \times 10^{-5}) = 4.60$

<u>Bromocresol green</u> would be satisfactory.

11-6 Let us assume that $c_{Na_2HAsO_4} = 0.033$ M at the equivalence point and that $K_2 = 1.1 \times 10^{-7}$ and $K_3 = 3.2 \times 10^{-12}$. Substituting into Equation 11-10 gives

$[H_3O^+] = \sqrt{\dfrac{3.2 \times 10^{-12} \times 0.033 + 1 \times 10^{-14}}{1 + 0.033/1.1 \times 10^{-7}}} = 6.2 \times 10^{-10}$

$pH = -\log(6.2 \times 10^{-10}) = 9.2$ <u>Phenolphthalein</u> would be suitable.

11-7 Curve A in Figure 11-4 (page 238) is the titration curve for H_3PO_4. Note that one end point occurs at about pH 4.5 and a second at about pH 9.5. Thus H_3PO_4 would be determined by titration with bromocresol green as an indicator (pH 3.8 to 5.4). A titration to the second end point with phenolphthalein would give the number of millimoles of NaH_2PO_4 plus twice the number of millimoles of H_3PO_4. Thus the concentration of NaH_2PO_4 is obtained from the difference in volume for the two titrations.

11-8 (a) To obtain the approximate equivalence point pH, we will employ Equation 11-11

$[H_3O^+] = \sqrt{K_1K_2} = \sqrt{4.45 \times 10^{-7} \times 4.69 \times 10^{-11}}$

$= 4.57 \times 10^{-9}$ and $pH = -\log(4.57 \times 10^{-9}) = 8.34$

<u>Cresol purple</u>. (7.6 to 9.2)   (See Table 10-1)

Chapter 11

(b) $P^{2-} + H_2O \rightleftarrows HP^- + OH^-$

$$K_b = \frac{[OH^-][HP^-]}{[P^{2-}]} = \frac{K_w}{K_2} = \frac{1.00 \times 10^{-14}}{3.91 \times 10^{-6}} = 2.56 \times 10^{-9}$$

$[OH^-] = [HP^-]$ and we assume $[P^{2-}] = 0.05 - [OH^-] \approx 0.050$

$[OH^-] = \sqrt{0.050 \times 2.56 \times 10^{-9}} = 1.13 \times 10^{-5}$

$pH = 14.00 - (-\log(1.13 \times 10^{-5})$ Phenolphthalein. (8.3 to 10)

(c) Proceeding as in part (b), we obtain

$[OH^-] = \sqrt{0.0500 \times 1.00 \times 10^{-14}/4.31 \times 10^{-5}} = 3.41 \times 10^{-6}$

$pH = 14.00 - (-\log 3.41 \times 10^{-6}) = 8.5$ Cresol purple. (7.6 to 9.2)

(d) Here we will use Equation 11-10 to give

$$[H_3O^+] = \sqrt{\frac{0.0500 \times 1.18 \times 10^{-11} + 1.00 \times 10^{-14}}{1 + 0.0500/1.42 \times 10^{-7}}} = 1.31 \times 10^{-9}$$

$pH = -\log(1.52 \times 10^{-9}) = 8.88$ Cresol purple. (7.6 to 9.2)

(e) $^+H_3NCH_2CH_2NH_3^+ + H_2O \rightarrow H_2NCH_2CH_2NH_3^+ + H_3O^+$ $K_1 = 1.42 \times 10^{-7}$

$\frac{[H_3O^+]^2}{0.0500} = 1.42 \times 10^{-7}$

$[H_3O^+] = \sqrt{1.42 \times 10^{-7} \times 0.0500} = 8.43 \times 10^{-5}$

$pH = -\log(8.43 \times 10^{-5}) = 4.07$ Bromocresol green. (3.8 to 5.4)

(f) Substituting into Equation 11-10 gives

$$[H_3O^+] = \sqrt{\frac{0.0500 \times 6.6 \times 10^{-8} + 1.00 \times 10^{-14}}{1 + 0.0500/1.23 \times 10^{-2}}} = 2.55 \times 10^{-3}$$

$pH = 4.59$ Bromocresol green. (3.8 to 5.4)

(g) Proceeding as in part (b), we obtain pH = 9.94.

Phenolphthalein. (8.5 to 10.0)

11-9 (a) $H_3PO_4 + H_2O \rightleftarrows H_3O^+ + H_2PO_4^-$ $K_1 = 7.11 \times 10^{-3}$

$$\frac{[H_3O^+][H_2PO_4^-]}{[H_3PO_4]} = \frac{[H_3O^+]^2}{0.04 - [H_3O^+]} = 7.11 \times 10^{-3}$$

$$[H_3O^+]^2 + 7.11 \times 10^{-3}[H_3O^+] - 0.0400 \times 7.11 \times 10^{-3} = 0$$

Solving by systematic approximation or by the quadratic formula gives

$$[H_3O^+] = 1.37 \times 10^{-2} \text{ and } pH = -\log(1.37 \times 10^{-2}) = \underline{1.86}$$

Proceeding in the same way, we obtain

(b) pH = $\underline{1.57}$

(c) pH = $\underline{1.64}$

(d) pH = $\underline{1.77}$

(e) pH = $\underline{4.21}$

(f) $H_2NCH_2CH_2NH_2 + H_2O \rightleftarrows H_2NCH_2CH_2NH_3^+ + OH^-$ $K_b = \dfrac{1.00 \times 10^{-14}}{1.18 \times 10^{-10}}$

$$= 8.47 \times 10^{-5}$$

$$\frac{[OH^-]^2}{0.0400} = 8.47 \times 10^{-5}$$

$$[OH^-] = \sqrt{0.0400 \times 8.47 \times 10^{-5}} = 1.84 \times 10^{-3}$$

$$pH = 14.00 - (-\log 1.84 \times 10^{-3}) = \underline{11.26}$$

11-10 In this problem, we will use Equation 11-10 or one of its simplifications.

(a) $[H_3O^+] = \sqrt{\dfrac{0.0400 \times 6.32 \times 10^{-8}}{1 + 0.0400/7.11 \times 10^{-3}}}$ $K_2 c \gg K_w$

$$= 1.95 \times 10^{-5} \quad \text{and} \quad pH = -\log(1.95 \times 10^{-5}) = \underline{4.71}$$

145

Chapter 11

Proceeding as in part (a), we obtain

(b) pH = 2.95

(c) pH = 4.28

(d) pH = 4.60

(e) pH = 9.80   Here, $c_{HA}/K_1 \gg 1$.

(f) pH = 8.39   Here, $c_{HA}/K_1 \gg 1$.

11-11 (a)
$$PO_4^{3-} + H_2O \rightleftarrows HPO_4^{2-} + OH^- \quad K_b = \frac{K_w}{K_3} = \frac{1.00 \times 10^{-14}}{4.5 \times 10^{-13}} = 2.22 \times 10^{-2}$$

$$\frac{[OH^-]^2}{0.0400 - [OH^-]} = 2.22 \times 10^{-2}$$

$$[OH^-]^2 + 2.22 \times 10^{-2}[OH^-] - 8.88 \times 10^{-4} = 0$$

Solving gives $[OH^-] = 2.07 \times 10^{-2}$

$$pH = 14.00 - (-\log 2.07 \times 10^{-2}) = \underline{12.32}$$

(b)
$$C_2O_4^{2-} + H_2O \rightleftarrows HC_2O_4^- + OH^- \quad K_b = \frac{1.00 \times 10^{-14}}{5.42 \times 10^{-5}} = 1.84 \times 10^{-10}$$

$$[OH^-] = \sqrt{0.0400 \times 1.84 \times 10^{-10}} = 2.72 \times 10^{-6} \quad \text{and} \quad pH = \underline{8.43}$$

Proceeding as in part (b), we obtain

(c) pH = 9.70

(d) pH = 9.89

(e) Proceeding as in part (a) gives pH = 12.58

(f) $BH_2^+ + H_2O \rightleftarrows BH^+ + H_3O^+$ where $BH_2^+ = {}^+H_3NCH_2CH_2NH_3^+$

$$[H_3O^+] = \sqrt{0.0400 \times 1.42 \times 10^{-7}} = 7.54 \times 10^{-5} \quad \text{and} \quad pH = \underline{4.12}$$

Chapter 11

11-12 (a) $[H_3O^+][H_2AsO_4^-]/[H_3AsO_4] = 5.8 \times 10^{-3}$

Here we must use Equations 10-6 and 10-7 (page 201) because K_a is relatively large. Thus,

$[H_2AsO_4^-] = 0.0200 + [H_3O^+] - \cancel{[OH^-]}$

$[H_3AsO_4] = 0.0500 - [H_3O^+] + \cancel{[OH^-]}$

Since $[OH^-]$ is negligible, we can write

$[H_3O^+] (0.0200) + [H_3O^+]) / (0.0500 - [H_3O^+]) = 5.8 \times 10^{-3}$

$[H_3O^+]^2 + 2.58 \times 10^{-2} [H_3O^+] - 2.90 \times 10^{-4} = 0$

$[H_3O^+] = 8.46 \times 10^{-3}$ and pH = <u>2.07</u>

(b) $H_2AsO_4^- + H_2O \rightleftarrows H_3O^+ + HAsO_4^{2-}$

$[H_3O^+][HAsO_4^{2-}]/[H_2AsO_4^-] = 1.11 \times 10^{-7}$

$[HAsO_4^{2-}] \approx 0.0500$ and $[H_2AsO_4^-] \approx 0.0300$

$[H_3O^+] = 1.1 \times 10^{-7} \times 0.0300/0.0500 = 6.6 \times 10^{-8}$

pH = $-\log(6.6 \times 10^{-8})$ = <u>7.18</u>

(c) Proceeding as in part (b), we obtain pH = <u>10.63</u>.

(d) $H_3PO_4 + HPO_4^{2-} \rightarrow 2H_2PO_4^-$

For each milliliter of solution, 0.0200 mmol Na_2HPO_4 reacts with 0.0200 mmol H_3PO_4 to give 0.0400 mmol NaH_2PO_4 and to leave 0.0200 mmol H_3PO_4. Thus we have a buffer that is 0.0200 M in H_3PO_4 and 0.0400 M in NaH_2PO_4. Since K_1 for H_3PO_4 is relatively large, we proceed as in part (a), and obtain pH = <u>2.55</u>.

(e) Proceeding as in part (a), we obtain pH = <u>2.06</u>.

11-13 (a) Proceeding as in Problem 11-12(a) and using K_1 for H_3PO_4, we obtain pH = <u>2.46</u>.

Chapter 11

(b) Proceeding as in Problem 11-12(b), we obtain pH = 7.51

(c) $HOC_2H_4NH_3^+ + H_2O \rightleftarrows HOC_2H_4NH_2 + H_3O^+ \quad K_a = 3.18 \times 10^{-10}$

$[H_3O^+]0.640/0.750 = 3.18 \times 10^{-10} = 3.73 \times 10^{-10}$

$pH = -\log(3.73 \times 10^{-10}) = 9.43$

(d) For each milliliter of solution, 0.0240 mmol of $H_2C_2O_4$ will have reacted with 0.0240 mmol of $Na_2C_2O_4$ to give 0.480 mmol of $NaHC_2O_4$. That is,

$H_2C_2O_4 + Na_2C_2O_4 \rightarrow 2NaHC_2O_4$

The reaction leaves $0.0360 - 0.0240 = 0.0120$ mmol of $Na_2C_2O_4$. Thus a buffer is formed that is 0.0120 M in $Na_2C_2O_4$ and 0.0480 M in $NaHC_2O_4$. Proceeding as in Problem 11-12(b), we obtain pH = 3.66

(e) Proceeding as in Problem 11-12(b), we obtain pH = 3.66.

11-14 (a) $HA + H_2O \rightleftarrows H_3O^+ + A^- \quad K_a = 4.3 \times 10^{-1}$

Picric acid is a strong enough weak acid so that it contributes significantly to $[H_3O^+]$. Its contribution is given by $[A^-]$ and the total $[H_3O^+]$ is then given by

$[H_3O^+] = c_{HCl} + [A^-] = 0.0100 + [A^-]$

$[A^-] = [H_3O^+] - 0.0100$

Mass balance requires that $[HA] + [A^-] = 0.0200$, or

$[HA] = 0.0200 - [A^-] = 0.0200 - [H_3O^+] + 0.0100$

$= 0.0300 - [H_3O^+]$

$\dfrac{[H_3O^+][A^-]}{[HA]} = 4.3 \times 10^{-1} = \dfrac{[H_3O^+]([H_3O^+] - 0.0100)}{0.0300 - [H_3O^+]}$

$[H_3O^+]^2 + 0.42[H_3O^+] - 1.29 \times 10^{-2} = 0$

$[H_3O^+] = 2.87 \times 10^{-2}$ and pH = 1.54

Chapter 11

(b) Proceeding in the same way, we obtain for benzoic acid, HBz

$$\frac{[H_3O^+]([H_3O^+] - 0.0100)}{0.0300 - [H_3O^+]} = 6.28 \times 10^{-5}$$

$$[H_3O^+]^2 - 9.94 \times 10^{-3}[H_3O^+] - 6.28 \times 10^{-5} \times 0.0300 = 0$$

$$[H_3O^+] = 1.013 \times 10^{-2} \text{ and } pH = \underline{1.99}$$

Note that if we assumed that the HCl completely repressed the dissociation of HBz, the calculated pH would be 2.00.

(c) $CO_3^{2-} + H_2O \rightleftharpoons HCO_3^- + OH^-$ $K_b = \dfrac{K_w}{K_2} = \dfrac{1.00 \times 10^{-14}}{4.69 \times 10^{-11}} = 2.13 \times 10^{-4}$

Hydroxide ions are introduced from two sources; NaOH and Na_2CO_3. The concentration from the latter source is equal to $[HCO_3^-]$. Thus,

$$[OH^-] = c_{NaOH} + [HCO_3^-]$$

$$= 0.0100 + [HCO_3^-]$$

$$[HCO_3^-] = [OH^-] - 0.0100$$

Since the solution is 0.100 M in Na_2CO_3, we may write

$$[CO_3^{2-}] = 0.100 - [HCO_3^-] - [H_2CO_3] \approx 0.100 - [HCO_3^-]$$

Substituting the previous equation gives

$$[CO_3^{2-}] = 0.100 - ([OH^-] - 0.0100) = 0.110 - [OH^-]$$

$$\frac{[OH^-][HCO_3^-]}{[CO_3^{2-}]} = 2.13 \times 10^{-4} = \frac{[OH^-]([OH^-] - 0.0100)}{0.110 - [OH^-]}$$

$$2.343 \times 10^{-5} - 2.13 \times 10^{-4}[OH^-] = [OH^-]^2 - 0.0100[OH^-]$$

$$[OH^-]^2 - 9.79 \times 10^{-3}[OH^-] - 2.343 \times 10^{-5} = 0$$

$$[OH^-] = 1.178 \times 10^{-2}$$

Chapter 11

$$\text{pH} = 14.00 - (-\log 1.178 \times 10^{-2}) = \underline{12.07}$$

(d)
$$NH_3 + H_2O \rightleftarrows NH_4^+ + OH^- \quad K_b = \frac{[NH_4^+][OH^-]}{[NH_3]} = \frac{1.00 \times 10^{-14}}{5.70 \times 10^{-10}} = 1.75 \times 10^{-5}$$

Proceeding in the same way as part (c), we write

$$[OH^-] = 0.0100 + [NH_4^+] \quad \text{or} \quad [NH_4^+] = [OH^-] - 0.0100$$

$$[NH_4^+] + [NH_3] = 0.100 \quad \text{or} \quad [NH_3] = 0.100 - [NH_4^+]$$

$$\frac{[OH^-]([OH^-] - 0.0100)}{0.110 - [OH^-]} = 1.75 \times 10^{-5}$$

$$[OH^-]^2 - 9.983 \times 10^{-3}[OH^-] - 1.925 \times 10^{-6} = 0$$

$$[OH^-] = 1.0172 \times 10^{-2} \quad \text{and} \quad \text{pH} = 14.00 - \log(1.0172 \times 10^{-2}) = \underline{12.01}$$

11-15 (a) H_3O^+ ions are produced from both HCl and chloroacetic acid (HA).

The concentration from the latter is equal to $[A^-]$. Thus,

$$[H_3O^+] = 0.0100 + [A^-] \quad \text{or} \quad [A^-] = [H_3O^+] - 0.0100$$

From mass-balance consideration

$$[A^-] + [HA] = 0.0300$$

Substituting the previous equation and rearranging yield

$$[HA] = 0.0300 - [H_3O^+] + 0.0100 = 0.0400 - [H_3O^+]$$

Substituting these relationships for [HA] and $[A^-]$ into the expression for K_a gives

$$K_a = 1.36 \times 10^{-3} = \frac{[H_3O^+]([H_3O^+] - 0.0100)}{0.0400 - [H_3O^+]}$$

$$[H_3O^+]^2 - 8.64 \times 10^{-3}[H_3O^+] - 5.44 \times 10^{-5} = 0$$

$$[H_3O^+] = 1.287 \times 10^{-2} \quad \text{and} \quad \text{pH} = \underline{1.89}$$

(b) Here, $[H_3O^+]$ from HCl is 0.0100 and from H_2SO_4 is $0.0150 + [SO_4^{2-}]$.

Therefore,

$$[H_3O^+] = 0.0250 + [SO_4^{2-}] \quad \text{or} \quad [SO_4^{2-}] = [H_3O^+] - 0.0250$$

$$[SO_4^{2-}] + [HSO_4^-] = 0.0150$$

$$[HSO_4^-] = 0.0150 - [SO_4^{2-}] = 0.0150 - ([H_3O^+] - 0.0250)$$

$$= 0.0400 - [H_3O^+]$$

Substituting into the dissociation constant expression for HSO_4^- gives

$$K = 1.02 \times 10^{-2} = \frac{[H_3O^+]([H_3O^+] - 0.0250)}{0.0400 - [H_3O^+]}$$

$$[H_3O^+]^2 - 1.48 \times 10^{-2}[H_3O^+] - 4.08 \times 10^{-4} = 0$$

$$[H_3O^+] = 2.89 \times 10^{-2} \quad \text{and} \quad pH = \underline{1.54}$$

(c) Proceeding as in the solution to Problem 11-14(c), we write

$$S^{2-} + H_2O \rightleftarrows HS^- + OH^- \qquad K_b = \frac{1.00 \times 10^{-14}}{1.3 \times 10^{-14}} = 0.769$$

$$[OH^-] = 0.0100 + [HS^-] \quad \text{or} \quad [HS^-] = [OH^-] - 0.0100$$

$$0.0300 = [S^{2-}] + [HS^-] + [H_2S] \approx [S^{2-}] + [HS^-]$$

Replacing $[HS^-]$ with $([OH^-] - 0.0100)$ and rearranging gives

$$[S^{2-}] = 0.0300 - [OH^-] + 0.0100 = 0.0400 - [OH^-]$$

Substituting into the expression for K_b gives

$$0.769 = \frac{[OH^-]([OH^-] - 0.0100)}{0.0400 - [OH^-]}$$

$$[OH^-]^2 + 0.759[OH^-] - 0.0308 = 0 \qquad [OH^-] = 3.86 \times 10^{-2}$$

Chapter 11

$$pH = 14.00 - (-\log 3.86 \times 10^{-2}) = \underline{12.59}$$

(d)
$$OAc^- + H_2O \rightleftharpoons HOAc + OH^- \quad K_b = \frac{1.00 \times 10^{-14}}{1.75 \times 10^{-5}} = 5.7 \times 10^{-10}$$

OAc⁻ is such a weak base that it makes no significant contribution to [OH⁻] and we may write

$$[OH^-] = 0.0100 - [HOAc] \approx 0.0100$$

$$pH = 14.00 - (-\log 0.0100) = \underline{12.00}$$

11-16 (a) Let us compare the ratio $[H_2SO_3]/[HSO_3^-]$ with that of $[SO_3^{2-}]/[HSO_3^-]$.

The larger will contain the predominant acid/base pair. The first is obtained by inverting the numerical value for K_1 and substituting $[H_3O^+] = 1.00 \times 10^{-6}$.

$$\frac{[H_2SO_3]}{[H_3O^+][HSO_3^-]} = \frac{1}{K_1} \quad \text{and} \quad \frac{[H_3O^+][SO_3^{2-}]}{[HSO_3^-]} = K_2$$

$$\frac{[H_2SO_3]}{[HSO_3^-]} = \frac{1.00 \times 10^{-6}}{1.23 \times 10^{-2}} = 8.1 \times 10^{-5} \qquad \frac{[SO_3^{2-}]}{[HSO_3^-]} = \frac{6.6 \times 10^{-8}}{1.00 \times 10^{-6}} = 0.066$$

Clearly the predominant pair is SO_3^{2-}/HSO_3^- and its acid/base ratio is

$$1/0.066 \approx \underline{15.2}.$$

(b) Substituting $[H_3O^+] = 1.00 \times 10^{-6}$ into the expressions for K_1, K_2, and K_3 yields

$$\frac{[H_2Cit^-]}{[H_3Cit]} = 745 \qquad \frac{[HCit^{2-}]}{[H_2Cit^-]} = 17.3 \qquad \frac{[Cit^{3-}]}{[HCit^{2-}]} = 0.40$$

The large size of the first two ratios and the small size of the third indicate that $HCit^{2-}$ is a predominant species in this solution. To compare $[Cit^{3-}]$ and $[H_2Cit^-]$ we invert the second ratio. Then

$$[H_2Cit^-]/[HCit^{2-}] = 1/17.3 = 0.058$$

Thus, the predominant acid/base system involves [Cit^{3-}] and [HCit^{2-}] and their acid/base ratio is

$$[\text{HCit}^{2-}]/[\text{Cit}^{3-}] \ = \ 1/0.40 \ = \ \underline{\underline{2.5}}$$

(c) Proceeding as in part (a), we obtain [HM$^-$]/[M^{2-}] = $\underline{\underline{0.498}}$

(d) Proceeding as in part (a), we obtain [HT$^-$]/[T^{2-}] = $\underline{\underline{0.0232}}$

11-17 (a) Proceeding as in Problem 11-16(a) with [H$_3$O$^+$] = 1.00 × 10^{-9}, we obtain

[H$_2$S]/[HS$^-$] = $\underline{\underline{0.010}}$

(b) Formulating the three species as BH$_2^{2+}$, BH$^+$ and B, where B is the symbol for NH$_2$C$_2$H$_5$NH$_2$.

$$\frac{[\text{BH}^+][\text{H}_3\text{O}^+]}{[\text{BH}_2^{2+}]} \ = \ K_1 \ = \ 1.42 \times 10^{-7} \quad \text{and} \quad \frac{[\text{B}][\text{H}_3\text{O}^+]}{[\text{BH}^+]} \ = \ K_2 \ = \ 1.18 \times 10^{-10}$$

$$[\text{BH}_2^{2+}]/[\text{BH}^+] \ = \ \frac{1.00 \times 10^{-9}}{1.42 \times 10^{-7}} \ = \ 0.0070$$

$$[\text{B}]/[\text{BH}^+] \ = \ \frac{1.18 \times 10^{-10}}{1.00 \times 10^{-9}} \ = \ 0.118$$

[BH$_2^{2+}$] is clearly < [B] and [BH$^+$]/[B] = 1.00/0.118 = $\underline{\underline{8.5}}$

(c) Proceeding as in Problem 11-16(b), we obtain

[H$_2$AsO$_4^-$]/[HAsO$_4^{2-}$] = $\underline{\underline{9.1 \times 10^{-3}}}$

(d) Proceeding as in Problem 11-16(a), we find

[HCO$_3^-$]/[CO$_3^{2-}$] = $\underline{\underline{21}}$

11-18 pH = 7.30   [H$_3$O$^+$] = antilog (−7.30) = 5.012 × 10^{-8}

[H$_3$O$^+$][HPO$_4^{2-}$]/[H$_2$PO$_4^-$] = 6.32 × 10^{-8}

[HPO$_4^{2-}$]/[H$_2$PO$_4^-$] = 6.32 × 10^{-8}/(5.012 × 10^{-8}) = 1.261

Chapter 11

$$HPO_4^{2-} + H_3PO_4 \rightarrow 2H_2PO_4^-$$

no. mmol H_3PO_4 present = 400×0.200 = 80.0

no. mmol $H_2PO_4^{2-}$ in the buffer = 2×80.0 = 160.0

no. mmol HPO_4^{2-} needed for the buffer = 1.261×160.0 = 201.8

Thus we need 80.0 mL of Na_2HPO_4 to react with the H_3PO_4 and an additional 201.8 to provide the needed concentration of HPO_4^{2-} or 281.8 mmol.

$$\text{mass } Na_2HPO_4 \cdot 2H_2O = 281.8 \text{ mmol} \times 0.17799 \text{ g/mmol} = \underline{\underline{50.2 \text{ g}}}$$

11-19 pH = 5.75 $[H_3O^+]$ = antilog(-5.75) = 1.778×10^{-6}

$[H_3O^+][P^{2-}]/[HP^-]$ = 3.91×10^{-6}

$[P^{2-}]/[HP^-]$ = $3.91 \times 10^{-6}/(1.778 \times 10^{-6})$ = 2.199

$H_2P + P^{2-} \rightarrow 2HP^-$

no. mmol H_2P present = 750×0.0500 = 37.5

no. mmol HP^- in the buffer = 2×37.5 = 75.0

no. mmol P^{2-} needed in the buffer = 2.199×75.0 = 164.9

Thus, we need $37.5 + 164.9 = 202.4$ mmol of K_2P.

$$\text{mass } K_2P = 202.4 \text{ mmol} \times 0.24232 \text{ g } K_2P/\text{mmol} = \underline{\underline{49.0 \text{ g } K_2P}}$$

11-20 no. mmol NaH_2PO_4 = 50.0×0.200 = 10.0

(a) no. mmol H_3PO_4 formed = no. mmol HCl added = 50.0×0.120 = 6.00

$c_{H_3PO_4}$ = 6.00/100 = 0.0600 M

$c_{NaH_2PO_4}$ = (10.0 – 6.00)/100 = 0.0400

Proceeding as in Problem 11-12(a), we obtain pH = $\underline{\underline{2.11}}$.

154

Chapter 11

(b) $c_{Na_2HPO_4}$ = 6.00/100 = 0.0600 M

$c_{NaH_2PO_4}$ = (10.00 – 6.00)/100 = 0.0400

Proceeding as in Problem 11-12(b), using K_2, gives pH = 7.38.

11-21 no. mmol KHP = 100 × 0.150 = 15.0

(a) no. mmol P^{2-} = 100 × 0.0800 = 8.00

no. mmol KHP = 15.0 - 8.00 = 7.00

c_{HP^-} = 7.00/200 = 0.0350 $c_{P^{2-}}$ = 8.00/200 = 0.0400

Proceeding as in Problem 11-12(b), we obtain pH = 5.47.

(b) c_{H_2P} = 8.00/200 = 0.0400 c_{HP^-} = (15.00 – 8.00)/200 = 0.0350

Proceeding as in Problem 11-12(a), we obtain pH = 2.92.

11-22 pH = 9.60   [H$_3$O$^+$] = antilog (-9.60) = 2.512 × 10^{-10}

[H$_3$O$^+$][CO$_3^{2-}$]/[HCO$_3^-$] = 4.69 × 10^{-11}

[CO$_3^{2-}$]/[HCO$_3^-$] = 4.69 × 10^{-11}/2.512 × 10^{-10} = 0.1867

Let V_{HCl} = mL 0.200 M HCl and $V_{Na_2CO_3}$ = mL 0.300 M Na$_2$CO$_3$

Since the solutions are dilute, the volumes will be additive

$V_{HCl} + V_{Na_2CO_3}$ = 1000 mL

Assume

[CO$_3^{2-}$] ≈ $c_{Na_2CO_3}$ = ($V_{Na_2CO_3}$ × 0.300 – V_{HCl} × 0.200)/1000

[HCO$_3^-$] ≈ $c_{HCO_3^-}$ = V_{HCl} × 0.200/1000

Substituting these relationships into the ratio [CO$_3^{2-}$]/[HCO$_3^-$] gives

155

Chapter 11

$$\frac{0.300\, V_{Na_2CO_3} - 0.200\, V_{HCl}}{0.200\, V_{HCl}} = 0.1867$$

$$0.300\, V_{Na_2CO_3} - 0.200\, V_{HCl} = 0.03734\, V_{HCl}$$

$$0.300\,(1000 - V_{HCl}) = 0.23734\, V_{HCl}$$

$$V_{HCl} = 300/0.5373 = \underline{\underline{558\ \text{mL}}}$$

$$V_{Na_2CO_3} = 1000 - 558 = \underline{\underline{442\ \text{mL}}}$$

Thus mix 442 mL of 0.300 M Na_2CO_3 with (1000 - 442) = 558 mL of 0.200 M HCl.

11-23 $[H_3O^+][HPO_4^{2-}]/[H_2PO_4^-] = 6.32 \times 10^{-8}$

$$\frac{[HPO_4^{2-}]}{[H_2PO_4^-]} = \frac{6.32 \times 10^{-8}}{1.00 \times 10^{-7}} = 0.632 \qquad (1)$$

Let $V_{H_3PO_4}$ and V_{NaOH} be the volume in milliliters of the two reagents. Then

$$V_{H_3PO_4} + V_{NaOH} = 1000\ \text{mL} \qquad (2)$$

From mass-balance consideration, we may write that in the 1000 mL,

$$\text{no. mmol } NaH_2PO_4 + \text{no. mmol } Na_2HPO_4 = 0.200 \times V_{H_3PO_4} \qquad (3)$$

$$\text{no. mmol } NaH_2PO_4 + 2 \times \text{no. mmol } Na_2HPO_4 = 0.160 \times V_{NaOH} \qquad (4)$$

Equation (1) can be written as

$$\frac{\text{no. mmol } Na_2HPO_4 / 1000}{\text{no. mmol } NaH_2PO_4 / 1000} = \frac{\text{no. mmol } Na_2HPO_4}{\text{no. mmol } NaH_2PO_4} = 0.632 \qquad (5)$$

Thus we have four equations: (2), (3), (4), and (5) and four unknowns:
$V_{H_3PO_4}$, V_{NaOH}, no. mmol NaH_2PO_4. and no. mmol Na_2HPO_4.

Subtracting Equation (3) from (4) yields

$$\text{no. mmol } Na_2HPO_4 = 0.160\, V_{NaOH} - 0.200\, V_{H_3PO_4} \qquad (6)$$

Substituting Equation (6) into (3) gives

no. mmol NaH_2PO_4 + 0.160 V_{NaOH} − 0.200 $V_{H_3PO_4}$ = 0.200 $V_{H_3PO_4}$

no. mmol NaH_2PO_4 = −0.160 V_{NaOH} + 0.400 $V_{H_3PO_4}$ (7)

Substituting Equations (6) and (7) into (5) gives

$$\frac{0.160\ V_{NaOH} - 0.200\ V_{H_3PO_4}}{0.400\ V_{H_3PO_4} - 0.160\ V_{NaOH}} = 0.632$$

This equation rearranges to

0.2611 V_{NaOH} = 0.4528 $V_{H_3PO_4}$

Substituting Equation (2) gives

0.2611 (1000 − $V_{H_3PO_4}$) = 0.4528 $V_{H_3PO_4}$

$$V_{H_3PO_4} = \frac{261.1}{0.7139} = \underline{\underline{366\ mL}} \qquad V_{NaOH} = 1000 - 366 = \underline{\underline{634\ mL}}$$

Thus mix 366 mL H_3PO_4 with (1000 − 366) = 634 mL NaOH.

11-24 $[H_3O^+][HAsO_4^{2-}]/[H_2AsO_4^-] = 1.1 \times 10^{-7}$

$[HAsO_4^{2-}]/[H_2AsO_4^-] = 1.1 \times 10^{-7}/(1.00 \times 10^{-6}) = 0.11$ (1)

As in Problem 11-23 we now develop four independent equations that allow calculation of V_{HCl}, $V_{Na_3AsO_4}$, no. mmol of $HAsO_4^{2-}$, and no. mmol of $H_2AsO_4^-$.

$V_{Na_3AsO_4} + V_{HCl} = 1000\ mL$ (2)

no. mmol NaH_2AsO_4 + no. mmol Na_2HAsO_4 = 0.500 $V_{Na_3AsO_4}$ (3)

2 × no. mmol NaH_2AsO_4 + no. mmol Na_2HAsO_4 = 0.400 V_{HCl} (4)

no. mmol Na_2HAsO_4 / no. mmol NaH_2AsO_4 = 0.11   (5)

Proceeding as in Problem 11-23, we solve equations (2), (3), (4), and (5) and conclude that $\underline{704\ mL}$ of the HCl should be mixed with $\underline{296\ mL}$ of the Na_3AsO_4.

Chapter 11

11-25 (a) For Na_2M, $K_1 = 1.3 \times 10^{-2}$ and $K_2 = 5.9 \times 10^{-7}$. This initial pH is high and two equally spaced end points will be encountered. Thus curve C.

(b) Curve A.

(c) Curve C for the reasons given in part (a).

11-26 (a) Titration with NaOH of a solution containing a mixture of two weak acids HA_1 and HA_2. HA_1 is present in a greater concentration and has a dissociation constant that is larger by a factor of about 10^4.

(b) Titration of a typical monoprotic weak acid.

(c) Titration of a mixture of a weak base, such as Na_2CO_3 and an acid salt, such as $NaHCO_3$.

11-27 For the titration of a mixture of H_3PO_4 and $H_2PO_4^-$, the volume to the first end point would have to be smaller than one half the total volume to the second end point because in the titration from the first to second end points both analytes are titrated, whereas to the first end point only the H_3PO_4 is titrated.

11-28 (a)
$$K_1 = \frac{[H_3O^+][HCO_3^-]}{[H_2CO_3]} = 4.45 \times 10^{-7} \qquad K_2 = \frac{[H_3O^+][CO_3^{2-}]}{[HCO_3^-]} = 4.69 \times 10^{-11}$$

0.00 mL

$$CO_3^{2-} + H_2O \rightleftharpoons HCO_3^- + OH^-$$

$$K_b = 1.00 \times 10^{-14} / 4.69 \times 10^{-11} = 2.132 \times 10^{-4}$$

$$[OH^-][HCO_3^-]/[CO_3^{2-}] = 2.132 \times 10^{-4}$$

$$[OH^-] = [HCO_3^-] \quad \text{and} \quad [CO_3^{2-}] = 0.1000 - [OH^-] \approx 0.100$$

$$[OH^-]^2 / 0.1000 = 2.132 \times 10^{-4}$$

$$[OH^-] = \sqrt{0.1000 \times 2.132 \times 10^{-4}} = 4.618 \times 10^{-3}$$

$$pH = 14.00 - (-\log 4.618 \times 10^{-3}) = \underline{11.66}$$

Chapter 11

12.50 mL

We now form a Na_2CO_3 / $NaHCO_3$ buffer in which

$[HCO_3^-] \approx c_{NaHCO_3} = 12.50 \times 0.2000/62.50 = 4.000 \times 10^{-2}$

$[CO_3^{2-}] \approx c_{Na_2CO_3} = (50.0 \times 0.1000 - 12.50 \times 0.2000)/62.5$

$= 4.000 \times 10^{-2}$

Substituting into the expressions for K_2 and rearranging give

$[H_3O^+] = 4.000 \times 10^{-2} \times 4.69 \times 10^{-11} / (4.000 \times 10^{-2}) = 4.69 \times 10^{-11}$

pH = <u>10.33</u>

20.00 and 24.00 mL

These data are treated in the same way. The results are found in the table that follows.

25.00 mL

Here we have a 0.05000 M solution of HCO_3^- and Equation 11-11 (page 233) applies. That is,

$[H_3O^+] = \sqrt{4.45 \times 10^{-7} \times 4.69 \times 10^{-11}} = 4.568 \times 10^{-9}$

pH = <u>8.34</u>

26.00 mL

We are now dealing with a new buffer mixture of H_2CO_3 / $NaHCO_3$.

no. mmol HCl added = 26.00×0.2000 = 5.200

no. mmol HCl consumed to form HCO_3^- = 50.00×0.1000 = <u>5.000</u>

no. mmol H_2CO_3 formed = 0.200

no. mmol HCO_3^- remaining = $50.00 \times 0.1000 - 0.2000$ = 4.800

$c_{H_2CO_3} = 0.2000/76.00 = 2.632 \times 10^{-3} \approx [H_2CO_3]$

Chapter 11

$$c_{NaHCO_3} = 4.800/76.00 = 6.316 \times 10^{-2} \approx [HCO_3^-]$$

$$[H_3O^+] = 4.45 \times 10^{-7} \times 2.632 \times 10^{-3}/6.316 \times 10^{-2} = 1.854 \times 10^{-8}$$

$$pH = \underline{7.73}$$

37.50, 45.00, and 49.00 mL

These data are treated in the same way. Answers in the table that follows.

50.00 mL

Here we have a solution that is 0.0500 M in H_2CO_3 and the pH is calculated on the basis of K_1.

$$[H_3O^+][HCO_3^-]/[H_2CO_3] = 4.45 \times 10^{-7}$$

$$[H_3O^+] = [HCO_3^-] \text{ and } [H_2CO_3] = 0.0500 - [H_3O^+] \approx 0.05000$$

$$[H_3O^+] = \sqrt{K_1 c_{H_2CO_3}} = \sqrt{4.45 \times 10^{-7} \times 0.0500} = 1.49 \times 10^{-4}$$

$$pH = \underline{3.83}$$

51.00 mL

1.00 mL 0.2000 M HCl in excess and

$$c_{HCl} \approx [H_3O^+] = 1.00 \times 0.2000/101.0 = 1.980 \times 10^{-3}$$

$$pH = -\log(1.00 \times 10^{-3}) = \underline{2.70}$$

60.00 mL

Using calculations similar to the previous one, pH = $\underline{1.74}$

(b) Let BH_2^{2+} symbolize $^+H_3NCH_2CH_2NH_3^+$

$$BH_2^{2+} + H_2O \rightleftarrows BH^+ + H_3O^+ \quad K_{a1} = 1.42 \times 10^{-7}$$

$$BH^+ + H_2O \rightleftarrows B + H_3O^+ \quad K_{a2} = 1.18 \times 10^{-10}$$

160

$$B + H_2O \rightleftarrows BH^+ + OH^- \qquad K_{b1} = \frac{K_w}{K_{a2}} = 8.47 \times 10^{-5}$$

0.00 mL

As in part (a)

$$[OH^-] = \sqrt{c_B \times 8.47 \times 10^{-5}} = \sqrt{0.1000 \times 8.47 \times 10^{-5}} = 2.910 \times 10^{-3}$$

$$pH = 14.00 - (-\log 2.910 \times 10^{-3}) = \underline{11.46}$$

12.50 mL

$$c_{BH^+} \approx [BH^+] = 12.50 \times 0.2000/62.5 = 0.04000$$

$$c_B \approx [B] = (50.00 \times 0.1000 - 12.50 \times 0.2000) = 0.04000$$

$$K_a = 1.18 \times 10^{-10} = [H_3O^+][B]/[BH^+] = [H_3O^+] \times 0.04000/0.04000$$

$$[H_3O^+] = 1.18 \times 10^{-10} \text{ and } pH = \underline{9.93}$$

20.00 and 24.00 mL

Calculations performed in the same way. Results are shown in the table that follows.

25.00 mL

$$c_{BHCl} = 50.00 \times 0.1000/75.00 = 0.06667$$

Here, we employ Equation 11-10

$$[H_3O^+] = \sqrt{\frac{1.18 \times 10^{-10} \times 0.06667 + 1.000 \times 10^{-14}}{1 + 0.06667/1.42 \times 10^{-7}}} = 4.096 \times 10^{-9}$$

$$pH = \underline{8.39}$$

26.00 mL

Proceeding as in part (a), for 26.00 mL

$$c_{BH_2^{2+}} \approx [BH_2^{2+}] = 0.2000/76.00 = 2.632 \times 10^{-3}$$

Chapter 11

$$c_{BH^+} \approx [BH^+] = 4.800/76.00 = 6.316 \times 10^{-2}$$

$$\frac{[H_3O^+] \times 6.316 \times 10^{-2}}{2.632 \times 10^{-3}} = 1.42 \times 10^{-7}$$

$$[H_3O^+] = 5.917 \times 10^{-9} \text{ and } pH = \underline{8.23}$$

37.50, 45.00, and 49.00 mL

These data are treated in the same way giving pH values shown in the table at the end of this solution.

50.00 mL

Here we have a solution that is 0.0500 M in BH_2^+ and the pH is calculated in the same way as at 50.00 mL in part (a). pH = $\underline{4.07}$

51.00 and 60.00 mL

These calculations are performed in the same way as for 51.00 mL in part (a). The results are shown in the table that follows.

(c) $HSO_4^- + H_2O \rightleftarrows H_3O^+ + SO_4^{2-}$ $[H_3O^+][SO_4^{2-}]/[HSO_4^-] = 1.02 \times 10^{-2}$

0.00 mL

Proceeding as in Feature 11-1 (page 239), we obtain

$$[H_3O^+] = 1.086 \times 10^{-1} \text{ and } pH = \underline{0.96}$$

12.50 mL

$$c_{NaHSO_4} = 12.50 \times 0.1000/62.50 = 0.04000$$

$$c_{H_2SO_4} = (50.00 \times 0.1000 - 12.50 \times 0.1000)/62.50 = 0.04000$$

$$[H_3O^+] = c_{H_2SO_4} + [SO_4^{2-}]$$

where $c_{H_2SO_4}$ represents the concentration of H_3O^+ from the complete dissociation of the H_2SO_4 to give HSO_4^-, and $[SO_4^{2-}]$ is equal to $[H_3O^+]$ from the partial dissociation of HSO_4^-.

Rearranging gives

$$[SO_4^{2-}] = [H_3O^+] - c_{H_2SO_4} \qquad (1)$$

Material balance requires that

$$[SO_4^{2-}] + [HSO_4^-] = c_{H_2SO_4} + c_{NaHSO_4} \qquad (2)$$

Substituting Equation (1) into (2) gives upon rearranging

$$[HSO_4^-] = c_{H_2SO_4} + c_{NaHSO_4} + c_{H_2SO_4} - [H_3O^+]$$

$$= 2c_{H_2SO_4} + c_{NaHSO_4} - [H_3O^+] \qquad (3)$$

Substituting Equations (1) and (3) into the acid dissociation-constant expression gives

$$\frac{[H_3O^+]\left([H_3O^+] - c_{H_2SO_4}\right)}{\left(2c_{H_2SO_4} + c_{NaHSO_4} - [H_3O^+]\right)} = 1.02 \times 10^{-2} = 0.0102$$

$$[H_3O^+]^2 + (0.0102 - c_{H_2SO_4})[H_3O^+] - 0.0102(2c_{H_2SO_4} + c_{NaHSO_4}) = 0$$

Substituting the numerical data leads to

$$[H_3O^+]^2 - 0.0298[H_3O^+] - 1.224 \times 10^{-3} = 0$$

$$[H_3O^+] = 5.293 \times 10^{-2} \text{ and } pH = \underline{1.28}$$

20.00 and 24.00 mL

These calculations are performed in the same way and lead to pH = 1.50 and pH = 1.63.

25.00 mL

$$c_{NaHSO_4} = 50.00 \times 0.1000/75.00 = 6.667 \times 10^{-2}$$

$$[H_3O^+] = [SO_4^{2-}] \text{ and } [HSO_4^-] = 6.667 \times 10^{-2} - [H_3O^+]$$

$$[H_3O^+]^2 / (6.67 \times 10^{-2} - [H_3O^+]) = 1.02 \times 10^{-2}$$

Chapter 11

$$[H_3O^+]^2 + 1.02 \times 10^{-2} [H_3O^+] - 6.803 \times 10^{-4} = 0$$

$$[H_3O^+] = 2.148 \times 10^{-2} \text{ and } pH = \underline{1.67}$$

26.00 mL

Proceeding as in part (a) for the concentration calculation of 26.00 mL, we find

$$c_{Na_2SO_4} = 2.632 \times 10^{-3} \text{ and } c_{NaHSO_4} = 6.316 \times 10^{-2}$$

HSO_4^- is a sufficiently strong acid that we must employ Equations 10-6 and 10-7, page 201.

$$[SO_4^{2-}] = 2.632 \times 10^{-3} + [H_3O^+] - [OH^-]$$

$$[HSO_4^-] = 6.316 \times 10^{-2} - [H_3O^+] + [OH^-]$$

Since $[OH^-]$ is negligible, we can write

$$[H_3O^+](2.632 \times 10^{-3} + [H_3O^+])/(6.316 \times 10^{-2} - [H_3O^+]) = 1.02 \times 10^{-2}$$

$$[H_3O^+]^2 + 1.283 \times 10^{-2} [H_3O^+] - 6.442 \times 10^{-4} = 0$$

$$[H_3O^+] = 1.976 \times 10^{-2} \text{ and } pH = \underline{1.70}$$

37.50, 45.00, and 49.00 mL

These calculations are performed in the same way as that for 26.00 mL. The results are located in the table that follows.

50.00 mL

Here, $c_{Na_2SO_4} = 0.05000$

Proceeding as in Problem 11-11(b), we obtain $pH = \underline{7.35}$

51.00 mL

Here,

$$c_{NaOH} = [OH^-] = (51.00 \times 0.2000 - 2 \times 50.00 \times 0.1000)/101.00$$

$$= 1.980 \times 10^{-3}$$

$$pH = 14.00 - (-\log 1.98 \times 10^{-3}) = \underline{11.30}$$

60.00 mL

Proceeding in the same way, we obtain pH = $\underline{12.26}$

(d) $[H_3O^+][HSO_3^-]/[H_2SO_3] = 1.23 \times 10^{-2}$

$[H_3O^+][SO_3^{2-}]/[HSO_3^-] = 6.6 \times 10^{-8}$

0.00 mL

Proceeding as in Problem 11-9(a), we obtain pH = $\underline{1.53}$

12.50 mL

$$c_{HSO_3^-} = 12.50 \times 0.2000/62.50 = 4.000 \times 10^{-2}$$

$$c_{H_2SO_3} = (50.00 \times 0.1000 - 12.50 \times 0.2000)/62.50 = 4.000 \times 10^{-2}$$

Because H_2SO_3 is relative strong, we must use Equations 10-6 and 10-7 (page 201)

$$[HSO_3^-] = 4.00 \times 10^{-2} + [H_3O^+] - [OH^-] \approx 4.00 \times 10^{-2} + [H_3O^+]$$

$$[H_2SO_3] = 4.00 \times 10^{-2} - [H_3O^+] + [OH^-] \approx 4.00 \times 10^{-2} - [H_3O^+]$$

$$[H_3O^+](4.000 \times 10^{-2} + [H_3O^+])/(4.000 \times 10^{-2} - [H_3O^+]) = 1.23 \times 10^{-2}$$

$$[H_3O^+]^2 + 5.23 \times 10^{-2}[H_3O^+] - 4.920 \times 10^{-4} = 0$$

$$[H_3O^+] = 8.14 \times 10^{-2} \text{ and } pH = \underline{2.09}$$

20.00 and 24.00 mL

Proceeding in the same way, we obtain pH = $\underline{2.61}$ and pH = $\underline{3.36}$

Chapter 11

25.00 mL

c_{NaHSO_3} = 50.00 × 0.1000/75.0 = 0.06667

Proceeding as in Problem 11-10, we obtain pH = <u>4.58</u>

26.00 mL

Proceeding as in part (a) for the concentration calculations at 26.00 mL.

$c_{Na_2SO_3}$ = 2.632 × 10^{-3} and c_{NaHSO_3} = 6.316 × 10^{-2}

[H$_3$O$^+$] (2.632 × 10^{-3})/(6.316 × 10^{-2}) = 6.6 × 10^{-8}

[H$_3$O$^+$] = 1.585 × 10^{-6} and pH = 5.80

37.50, 45.00, and 49.00 mL

Proceeding in the same way, we obtain pH = <u>7.18</u>, <u>7.78</u>, and <u>8.56</u> respectively.

50.00 mL

Here we have a 0.05000 M solution of Na$_2$SO$_3$. Proceeding as in Problem 11-11(b), we write

[OH$^-$] = $\sqrt{0.0500 \times K_b}$ = $\sqrt{0.0500 \times 1.00 \times 10^{-14}/6.6 \times 10^{-8}}$ =

8.70 × 10^{-4}

pH = 14.00 − (−log 8.70 × 10^{-4}) = <u>9.94</u>

51.00 mL

c_{NaOH} = (51.00 × 0.2000 − 50.00 × 0.1000 × 2)/101.0

= 1.98 × 10^{-3} = [OH$^-$]

pH = 14.00 − (−log 1.98 × 10^{-3}) = <u>11.30</u>

60.00 mL

Proceeding in the same way, we obtain pH = <u>12.26</u>

Chapter 11

Vol reagent, mL	(a) pH	(b) pH	(c) pH	(d) pH
0.00	11.66	11.46	0.96	1.53
12.50	10.33	9.93	1.28	2.09
20.00	9.73	9.33	1.50	2.61
24.00	8.95	8.55	1.63	3.37
25.00	8.34	8.39	1.67	4.58
26.00	7.73	8.23	1.70	5.80
37.50	6.35	6.85	2.19	7.18
45.00	5.75	6.25	2.70	7.78
49.00	4.97	5.47	3.46	8.56
50.00	3.83	4.07	7.35	9.94
51.00	2.70	2.70	11.30	11.30
60.00	1.74	1.74	12.26	12.26

11-29 $H_2NNH_3^+ + H_2O \rightleftharpoons H_3O^+ + H_2NNH_2 \quad K_a = 1.05 \times 10^{-8}$

$H_2NNH_2 + H_2O \rightleftharpoons H_2NNH_3^+ + OH^- \quad K_b = \dfrac{1.00 \times 10^{-14}}{1.05 \times 10^{-8}} = 9.52 \times 10^{-7}$

0.00 mL

$c_{NaOH} = 0.1000 \approx [OH^-]$

$pH = 14.00 - (-\log 0.1000) = \underline{13.00}$

This calculation assumes that H_2NNH_2 contributes essentially no OH^- to the solution.

10.00 mL

$c_{NaOH} = (50.00 \times 0.1000 - 10.00 \times 0.200)/60.0 = 5.00 \times 10^{-2}$

Chapter 11

Assume $[OH^-] \approx c_{NaOH} = 5.00 \times 10^{-2}$

pH = $14.00 - (-\log 5.00 \times 10^{-2})$ = <u>12.70</u>

20.00 and 24.00 mL

Proceeding in the same way, we obtain pH = <u>12.15</u> and <u>11.43</u>.

25.00 mL

$c_{H_2NNH_2} = 50.00 \times 0.08000/75.00 = 5.333 \times 10^{-2} - [OH^-]$

$c_{NaOH} = 0.000$

$[OH^-][H_2NNH_3^+]/[H_2NNH_2] = 9.52 \times 10^{-7}$

$[OH^-] = [H_2NNH_3^+]$

$[H_2NNH_2] = 5.333 \times 10^{-2} - [OH^-] \approx 5.333 \times 10^{-2}$

$[OH^-] = \sqrt{(5.333 \times 10^{-2}) \times (9.52 \times 10^{-7})} = 2.253 \times 10^{-4}$

pH = <u>10.35</u>

26.00 mL

no. mmol $HClO_4$ added = $26.00 \times 0.2000 = 5.2000$

initial no. mmol NaOH = 50.00×0.100 = <u>5.0000</u>

no. mmol $H_2NNH_3^+$ formed = 0.2000

initial no. mmol $H_2NNH_2 = 50.00 \times 0.0800 = 4.000$

no. mmol H_2NNH_2 present = 4.000 - 0.2000 = 3.800

$c_{H_2NNH_3^+} = 0.2000/76.00 = 2.632 \times 10^{-3} \approx [H_2NNH_3^+]$

$c_{H_2NNH_2} = 3.800/76.00 = 5.000 \times 10^{-2} \approx [H_2NNH_2]$

$[H_3O^+] = 1.05 \times 10^{-8} \times 2.632 \times 10^{-3}/(5.00 \times 10^{-2}) = 5.53 \times 10^{-10}$

Chapter 11

pH = <u>9.26</u>

35.00 and 44.00 mL

Proceeding in this same way, we obtain pH = <u>7.98</u> and <u>6.70</u> respectively.

45.00 mL

$$c_{H_2NNH_3^+} = 50.00 \times 0.0800/95.00 = 0.04211$$

$$[H_3O^+] = \sqrt{(0.04211) \times (1.05 \times 10^{-8})} = 2.103 \times 10^{-5}$$

$$pH = \underline{4.68}$$

46.00 mL

$$c_{HClO_4} \approx [H_3O^+] = 1.000 \times 0.2000/96.00 = 2.083 \times 10^{-3}$$

$$pH = -\log(2.083 \times 10^{-3}) = \underline{2.68}$$

50.00 mL

Proceeding in the same way gives pH = <u>2.00</u>

11-30 $HA + H_2O \rightleftharpoons H_3O^+ + A^-$ $[H_3O^+][A^-]/[HA] = 1.80 \times 10^{-4}$

For 0.00 and 10.00 mL, it is possible to assume that the formic acid does not contribute significantly to the $[H_3O^+]$.

0.00 mL

$$[H_3O^+] = c_{HClO_4} = 0.1000 \quad \text{and} \quad pH = \underline{1.00}$$

10.00 mL

$$[H_3O^+] = (50.00 \times 0.1000 - 10.00 \times 0.2000)/60.00 = 0.05000$$

$$pH = \underline{1.30}$$

20.00 mL

$$c_{HClO_4} = (50.00 \times 0.1000 - 20.00 \times 0.2000)/70.00 = 0.01429$$

Chapter 11

Here, we must take into account the H_3O^+ produced from the dissociation of HA.

$$[H_3O^+] = 0.01429 + [A^-] \quad \text{or} \quad [A^-] = [H_3O^+] - 0.01429$$

where $[A^-]$ accounts for the H_3O^+ produced by dissociation of HA.

$$c_{HA} = 50.00 \times 0.08000/70.0 = 0.05714$$

$$[HA] = 0.05714 - [A^-] = 0.05714 - [H_3O^+] + 0.01429$$

$$= 0.07143 - [H_3O^+]$$

$$[H_3O^+]([H_3O^+] - 0.01429)/(0.07143 - [H_3O^+]) = 1.77 \times 10^{-4}$$

$$[H_3O^+]^2 - 0.01411\,[H_3O^+] - 1.264 \times 10^{-5} = 0$$

$$[H_3O^+] = 1.496 \times 10^{-2} \quad \text{and} \quad pH = \underline{1.83}$$

24.00 mL

Proceeding as for 20.00 mL, we obtain $pH = \underline{2.33}$.

(If dissociation of HA is neglected, $pH = 2.57$.)

25.00 mL

$$c_{HClO_4} = 0.00$$

$$c_{HA} = 50.00 \times 0.08000/75.00 = 5.333 \times 10^{-2}$$

$$[H_3O^+] = [A^-] \quad \text{and} \quad [HA] = 5.333 \times 10^{-2} - [H_3O^+]$$

$$[H_3O^+]^2/(5.333 \times 10^{-2} - [H_3O^+]) = 1.80 \times 10^{-4}$$

Rearranging and solving the quadratic equation give

$$[H_3O^+] = 3.010 \times 10^{-3} \quad \text{and} \quad pH = \underline{2.52}$$

26.00 mL

total no. mmol KOH taken = 26.00×0.2000 = 5.200

original no. mmol HClO$_4$ = 50.00×0.1000 = 5.000

no. mmol KA formed = 0.200

no. mmol HA remaining = $50.00 \times 0.08000 - 0.2000$ = 3.800

$[H_3O^+][A^-]/[HA]$ = 1.80×10^{-4}

$[A^-]$ = $0.2000/76.00 + [H_3O^+] - \cancel{[OH^-]}$ Equation 10 – 7 (page 201)

$[HA]$ = $3.8000/76.00 - [H_3O^+] + \cancel{[OH^-]}$ Equation 10 – 6 (page 201)

$[H_3O^+](2.632 \times 10^{-3} + [H_3O^+])/(0.0500 - [H_3O^+])$ = 1.80×10^{-4}

$[H_3O^+]^2 + 2.812 \times 10^{-3}[H_3O^+] - 9.00 \times 10^{-6} = 0$

$[H_3O^+] = 1.907 \times 10^{-3}$ and pH = 2.72

35.00 and 44.00 mL

Proceeding in the same way, we obtain pH = 3.75 and 5.03.

45.00 mL

c_A = $50.00 \times 0.08000/95.00$ = 4.211×10^{-2} ≈ $[A^-]$

K_b = $1.00 \times 10^{-14}/1.80 \times 10^{-4}$ = 5.556×10^{-11} and $[OH^-] = [HA]$

$[OH^-]$ = $\sqrt{(4.211 \times 10^{-2}) \times (5.556 \times 10^{-11})}$ = 1.530×10^{-6}

pH = $14.00 - (-\log 1.530 \times 10^{-6})$ = 8.19

46.00 mL

1.00 mL 0.2000 M KOH in excess

c_{KOH} = $0.2000/96.00$ = 2.083×10^{-3}

pH = $14.00 - (-\log 2.083 \times 10^{-3})$ = 11.32

Chapter 11

50.00 mL

Proceeding in the same way, we obtain pH = <u>12.00</u>.

11-31 (a) $2H_2AsO_4^- \rightleftharpoons H_3AsO_4 + HAsO_4^{2-}$

$$K_1 = \frac{[H_3O^+][H_2AsO_4^-]}{[H_3AsO_4]} = 5.8 \times 10^{-3} \quad (1)$$

$$K_2 = \frac{[H_3O^+][HAsO_4^{2-}]}{[H_2AsO_4^-]} = 1.1 \times 10^{-7} \quad (2)$$

$$K_3 = \frac{[H_3O^+][AsO_4^{3-}]}{[HAsO_4^{2-}]} = 3.2 \times 10^{-12} \quad (3)$$

Dividing Equation (2) by Equation (1) leads to

$$\frac{K_2}{K_1} = \frac{[H_3AsO_4][HAsO_4^{2-}]}{[H_2AsO_4^-]^2} = \underline{\underline{1.9 \times 10^{-5}}}$$

which is the desired equilibrium-constant expression.

(b) $2HAsO_4^{2-} \rightleftharpoons AsO_4^{3-} + H_2AsO_4^-$

Here we divide Equation (3) by Equation (2)

$$\frac{K_3}{K_2} = \frac{[AsO_4^{3-}][H_2AsO_4^-]}{[HAsO_4^{2-}]^2} = \underline{\underline{2.9 \times 10^{-5}}}$$

11-32 $HOAc + H_2O \rightleftharpoons H_3O^+ + OAc^- \qquad K_{HOAc} = 1.75 \times 10^{-5}$

$NH_4^+ + H_2O \rightleftharpoons H_3O^+ + NH_3 \qquad K_{NH_4^+} = 5.70 \times 10^{-10}$

Subtracting the first reaction from the second and rearranging give

$NH_4^+ + OAc^- \rightleftharpoons NH_3 + HOAc \qquad K = K_{NH_4^+}/K_{HOAc}$

$$\frac{[NH_3][HOAc]}{[NH_4^+][OAc^-]} = \frac{5.70 \times 10^{-10}}{1.75 \times 10^{-5}} = 3.26 \times 10^{-5}$$

Chapter 11

11-33 (a) $K_1 = 1.12 \times 10^{-3}$ $K_2 = 3.91 \times 10^{-6}$ $K_1K_2 = 4.379 \times 10^{-9}$

At pH = 2.00

$[H_3O^+] = 1.00 \times 10^{-2}$ and $[H_3O^+]^2 = 1.00 \times 10^{-4}$

Substituting into Equation 11-17 gives

$$\alpha_0 = \frac{1.00 \times 10^{-4}}{1.00 \times 10^{-4} + 1.12 \times 10^{-5} + 4.379 \times 10^{-9}} = \frac{1.00 \times 10^{-4}}{1.112 \times 10^{-4}} = \underline{\underline{0.899}}$$

and into Equation 11-18 gives

$$\alpha_1 = \frac{(1.12 \times 10^{-3}) \times (1.00 \times 10^{-2})}{1.112 \times 10^{-4}} = \underline{\underline{0.101}}$$

and into Equation 11-19 gives

$$\alpha_2 = \frac{4.379 \times 10^{-9}}{1.112 \times 10^{-4}} = 3.94 \times 10^{-5}$$

At $[H_3O^+] = 1.00 \times 10^{-6}$ and $[H_3O^+]^2 = 1.00 \times 10^{-12}$ and $D = 5.500 \times 10^{-9}$

$\alpha_0 = [H_3O^+]^2/D = 1.00 \times 10^{-12}/5.500 \times 10^{-9} = \underline{\underline{1.82 \times 10^{-4}}}$

$\alpha_1 = [H_3O^+]K_1/D = 1.12 \times 10^{-9}/5.500 \times 10^{-9} = \underline{\underline{0.204}}$

$\alpha_2 = K_1K_2/D = 4.379 \times 10^{-9}/5.500 \times 10^{-9} = \underline{\underline{0.796}}$

At $[H_3O^+] = 1.00 \times 10^{-10}$ and $[H_3O^+]^2 = 1.00 \times 10^{-20}$ and $D = 4.379 \times 10^{-9}$

$\alpha_0 = 1.00 \times 10^{-20}/4.379 \times 10^{-9} = \underline{\underline{2.28 \times 10^{-12}}}$

$\alpha_1 = 1.12 \times 10^{-13}/4.379 \times 10^{-9} = \underline{\underline{2.56 \times 10^{-5}}}$

$\alpha_2 = 4.379 \times 10^{-9}/4.379 \times 10^{-9} = \underline{\underline{1.00}}$

(b) $K_1 = 7.11 \times 10^{-3}$ $K_2 = 6.32 \times 10^{-8}$ $K_3 = 4.5 \times 10^{-13}$

$K_1K_2 = 4.494 \times 10^{-10}$ $K_1K_2K_3 = 2.022 \times 10^{-22}$

173

Chapter 11

At $[H_3O^+] = 1.00 \times 10^{-2}$, the denominator D is given by

$$D = (1.00 \times 10^{-2})^3 + (7.11 \times 10^{-3})(1.00 \times 10^{-2})^2 +$$
$$(4.494 \times 10^{-10})(1.00 \times 10^{-2}) + 2.022 \times 10^{-22} = 1.711 \times 10^{-6}$$

$$\alpha_0 = [H_3O^+]^3/D = (1.00 \times 10^{-2})^3/1.711 \times 10^{-6} = 0.584$$

$$\alpha_1 = [H_3O^+]^2 K_1/D = 7.11 \times 10^{-3} \times (1.00 \times 10^{-2})^2/1.711 \times 10^{-6} = 0.416$$

$$\alpha_2 = [H_3O^+] K_1 K_2/D = 4.494 \times 10^{-10} \times 1.00 \times 10^{-2}/1.711 \times 10^{-6} = 2.63 \times 10^{-6}$$

$$\alpha_3 = K_1 K_2 K_3/D = 2.022 \times 10^{-22}/1.711 \times 10^{-6} = 1.18 \times 10^{-16}$$

The calculations at pH = 6.00 and 10.00 are performed in the same way. The results are given in the table that follows.

(c) and (d). These calculations are analogous to those in part (b). The results appear in the table that follows.

(e) and (f). These calculations are analogous to those in part (a). The results appear in the following table.

	pH	D	α_0	α_1	α_2	α_3
(a)	2.00	1.112×10^{-4}	0.899	0.101	3.94×10^{-5}	
	6.00	5.500×10^{-9}	1.82×10^{-4}	0.204	0.796	
	10.00	4.379×10^{-9}	2.28×10^{-12}	2.56×10^{-5}	1.000	
(b)	2.00	1.771×10^{-6}	0.584	0.416	2.63×10^{-6}	1.14×10^{-16}
	6.00	7.560×10^{-15}	1.32×10^{-4}	0.940	5.94×10^{-2}	2.67×10^{-8}
	10.00	4.521×10^{-20}	2.22×10^{-11}	1.58×10^{-3}	0.998	4.49×10^{-3}
(c)	2.00	1.075×10^{-6}	0.931	6.93×10^{-2}	1.20×10^{-4}	4.82×10^{-9}
	6.00	1.882×10^{-14}	5.31×10^{-5}	3.96×10^{-2}	0.685	0.275
	10.00	5.182×10^{-15}	1.93×10^{-16}	1.44×10^{-9}	2.49×10^{-4}	1.000
(d)	2.00	1.580×10^{-6}	0.633	0.367	4.04×10^{-6}	1.29×10^{-15}
	6.00	6.439×10^{-15}	1.55×10^{-4}	0.901	9.91×10^{-2}	3.17×10^{-7}
	10.00	6.590×10^{-20}	1.52×10^{-11}	8.80×10^{-4}	0.968	3.10×10^{-2}
(e)	2.00	4.000×10^{-4}	0.250	0.750	1.22×10^{-5}	
	6.00	3.486×10^{-8}	2.87×10^{-5}	0.861	0.139	
	10.00	4.863×10^{-9}	2.06×10^{-12}	6.17×10^{-4}	0.999	
(f)	2.00	6.630×10^{-4}	0.151	0.845	4.58×10^{-3}	
	6.00	3.091×10^{-6}	3.24×10^{-7}	1.81×10^{-2}	0.982	
	10.00	3.035×10^{-6}	3.30×10^{-15}	1.84×10^{-6}	1.000	

Chapter 11

11-34 Letting $H_3A = H_3AsO_4$ we write

$$K_1 = \frac{[H_3O^+][H_2A^-]}{[H_3A]} \quad (1) \qquad K_2 = \frac{[H_3O^+][HA^{2-}]}{[H_2A^-]} \qquad K_3 = \frac{[H_3O^+][A^{3-}]}{[HA^{2-}]}$$

$$K_1K_2 = \frac{[H_3O^+]^2[HA^{2-}]}{[H_3A]} \quad (2) \qquad K_1K_2K_3 = \frac{[H_3O^+]^3[A^{3-}]}{[H_3A]} \quad (3)$$

By definition

$$\alpha_0 = \frac{[H_3A]}{c_T} \qquad \alpha_1 = \frac{[H_2A^-]}{c_T} \qquad \alpha_2 = \frac{[HA^{2-}]}{c_T} \qquad \alpha_3 = \frac{[A^{3-}]}{c_T}$$

where

$$c_T = [H_3A] + [H_2A^-] + [HA^{2-}] + [A^{3-}] \qquad (4)$$

Substituting Equations (1), (2), and (3) into (4) yields

$$c_T = [H_3A] + \frac{K_1[H_3A]}{[H_3O^+]} + \frac{K_1K_2[H_3A]}{[H_3O^+]^2} + \frac{K_1K_2K_3[H_3A]}{[H_3O^+]^3}$$

$$\frac{c_T}{[H_3A]} = 1 + \frac{K_1}{[H_3O^+]} + \frac{K_1K_2}{[H_3O^+]^2} + \frac{K_1K_2K_3}{[H_3O^+]^3}$$

Multiplying the numerator and denominator of the right side of this equation by $[H_3O^+]^3$ gives

$$\frac{c_T}{[H_3A]} = \frac{[H_3O^+]^3 + K_1[H_3O^+]^2 + K_1K_2[H_3O^+] + K_1K_2K_3}{[H_3O^+]^3}$$

Letting $D = [H_3O^+]^3 + K_1[H_3O^+]^2 + K_1K_2[H_3O^+] + K_1K_2K_3 \qquad (5)$

gives

$$\frac{c_T}{[H_3A]} = \frac{D}{[H_3O^+]^3}$$

and inverting the two terms leads to α_0

$$\frac{[H_3A]}{c_T} = \frac{[H_3O^+]^3}{D} = \alpha_0 \qquad (6)$$

Substituting Equation (1) into (6) gives

$$\frac{[H_3O^+][H_2A^-]}{K_1 c_T} = \frac{[H_3O^+]^3}{D}$$

$$\frac{[H_2A^-]}{c_T} = \frac{K_2[H_3O^+]^2}{c_T} = \alpha_1$$

In the same way substituting Equation (2) into (6) gives upon rearrangement

$$\frac{[HA^{2-}]}{c_T} = \frac{K_1 K_2 [H_3O^+]}{D} = \alpha_2$$

Similarly, substituting Equation (3) into (6) yields

$$\frac{[A^{3-}]}{c_T} = \frac{K_1 K_2 K_3}{D} = \alpha_3$$

CHAPTER 12

12-1 Carbon dioxide is not strongly bonded by water molecules, and thus is readily volatilized from aqueous media. Gaseous HCl molecules, on the other hand, are fully dissociated into H_3O^+ and Cl^- when dissolved in water; neither of these species is volatile.

12-2 Nitric acid is seldom used as a standard because it is an oxidizing agent and thus will react with reducible species in titration mixtures.

12-3 Primary standard Na_2CO_3 can be obtained by heating primary standard grade $NaHCO_3$ for about an hour at 270° to 300°C. The reaction is

$$2NaHCO_3(s) \rightarrow Na_2CO_3(s) + H_2O(g) + CO_2(g)$$

12-4 Near the equivalence point in the titration of Na_2CO_3, the solution contains a buffer made up of a high concentration of H_2CO_3 and a small amount of Na_2CO_3. Boiling removes the H_2CO_3 as CO_2, which causes the pH of the solution to rise sharply (see Figure 12-1). Then the change in pH when titration is resumed is much greater than it would otherwise be. Thus, a sharper end point results.

12-5 For, let us say, a 40-mL titration of $KH(IO_3)_2$

$$40 \text{ mL NaOH} \times 0.010 \frac{\text{mmol}}{\text{mL}} \times \frac{1 \text{ mmol KH(IO}_3)_2}{\text{mmol NaOH}} \times \frac{0.390 \text{ g KH(IO}_3)_2}{\text{mmol}} =$$

$$0.16 \text{ g KH(IO}_3)_2$$

and for titration of HBz

$$40 \text{ mL NaOH} \times 0.010 \frac{\text{mmol}}{\text{mL}} \times \frac{1 \text{ mmol HBz}}{\text{mmol NaOH}} \times \frac{0.122 \text{ g HBz}}{\text{mmol}} = 0.045 \text{ g HBz}$$

The $KH(IO_3)_2$ is preferable because the relative weighing error would be less with a 0.16-g sample than with a 0.049-g sample. A second reason for preferring $KH(IO_3)_2$ is because it is a strong acid and HBz is not. The titration error would therefore be less.

12-6 If the base is to be used for titrations with an acid-range indicator the carbonate in the base will consume two analyte hydronium ions just as would the two hydroxides lost in the formation of Na_2CO_3.

Chapter 12

12-7 Unless a reducing agent is introduced into the H_2SO_4 prior to digestion, nitro, azo, and azoxy groups will be partially converted to N_2 or oxides of nitrogen, which are then lost by volatilization. Heterocyclic compounds containing nitrogen also yield low results in many instances because they tend to be incompletely decomposed under the usual digestion procedure.

12-8 (a)
$$2.00 \text{ L KOH} \times \frac{0.15 \text{ mol KOH}}{\text{L}} \times \frac{56.1 \text{ g KOH}}{\text{mol}} = \underline{\underline{17 \text{ g KOH}}}$$

Dissolve 17 g KOH and dilute to 2.0 L.

(b)
$$2.00 \text{ L} \times 0.015 \frac{\text{mol Ba(OH)}_2 \cdot 8H_2O}{\text{L}} \times \frac{315 \text{ g Ba(OH)}_2 \cdot 8H_2O}{\text{mol}} = \underline{\underline{9.46 \text{ g}}}$$

Dissolve 9.5 g $Ba(OH)_2 \cdot 8H_2O$ in H_2O and dilute to 2.0 L.

(c)
$$2.00 \text{ L HCl} \times 0.200 \frac{\text{mol HCl}}{\text{L HCl}} \times 36.46 \frac{\text{g HCl}}{\text{mol}} = 14.58 \text{ g HCl needed}$$

$$14.58 \text{ g HCl} \times \frac{1 \text{ mL reagent}}{1.058 \text{ g reagent}} \times \frac{100 \text{ g reagent}}{11.50 \text{ g HCl}} = \underline{\underline{119.8 \text{ mL reagent}}}$$

Dilute about 120 mL of the reagent to 2.00 L.

12-9 (a)
$$500 \text{ mL} \times 0.250 \frac{\text{mmol H}_2SO_4}{\text{mL}} \times 0.09808 \frac{\text{g H}_2SO_4}{\text{mmol}} = 12.26 \text{ g H}_2SO_4 \text{ needed}$$

$$12.60 \text{ g H}_2SO_4 \times \frac{1.00 \text{ mL reagent}}{1.1539 \text{ g reagent}} \times \frac{100 \text{ g reagent}}{21.8 \text{ g H}_2SO_4} = \underline{\underline{48.7 \text{ mL reagent}}}$$

Dilute about 49 mL of the reagent to 500 mL.

(b)
$$500 \text{ mL} \times 0.30 \frac{\text{mmol NaOH}}{\text{mL}} \times 0.0400 \frac{\text{g NaOH}}{\text{mmol NaOH}} = \underline{\underline{6.00 \text{ g NaOH}}}$$

Dissolve about 6.0 g NaOH and dilute to 500 mL.

(c)
$$500 \text{ mL} \times 0.08000 \frac{\text{mmol Na}_2CO_3}{\text{mL}} \times 0.10599 \frac{\text{g}}{\text{mmol Na}_2CO_3} = \underline{\underline{4.240 \text{ g}}}$$

Dissolve 4.24 g Na_2CO_3 and dilute to 500 mL.

Chapter 12

12-10 For the first data set,

$$c_1 = 0.7987 \text{ g KHP} \times \frac{1 \text{ mmol KHP}}{0.20422 \text{ g KHP}} \times \frac{1 \text{ mmol NaOH}}{\text{mmol KHP}} \times \frac{1}{38.29 \text{ mL NaOH}} = 0.10214 \frac{\text{mmol NaOH}}{\text{mL}}$$

The data shown below for c_i were obtained in the same way.

Sample	c_i, M	c_i^2
1	0.10214	0.01043258
2	0.10250	0.01050625
3	0.10305	0.01061930
4	0.10281	0.01056990
	$\Sigma c_i = 0.41050$	$\Sigma c_i^2 = 0.04212803$

(a) $\bar{c_i} = \dfrac{0.41050}{4} = \underline{\underline{0.1026 \text{ M}}}$

(b) $s = \sqrt{\dfrac{0.04212803 - (0.41050)^2/4}{4-1}} = \sqrt{\dfrac{0.000000466}{3}} = \underline{\underline{0.00039}}$

$\text{CV} = \dfrac{0.00039}{0.1026} \times 100\% = \underline{\underline{0.38\%}}$

12-11 For the first data set,

$$c_1 = 0.2068 \text{ g Na}_2\text{CO}_3 \times \frac{1 \text{ mmol Na}_2\text{CO}_3}{0.10599 \text{ g Na}_2\text{CO}_3} \times \frac{2 \text{ mmol HClO}_4}{1 \text{ mmol Na}_2\text{CO}_3} \times \frac{1}{36.31 \text{ mL HClO}_4} = 0.10747 \text{ M}$$

Chapter 12

Proceeding in the same way we find

Sample	c_i	c_i^2
1	0.10747	0.011549915
2	0.10733	0.011519623
3	0.10862	0.011798687
4	0.10742	0.011538536
	$\Sigma c_i = 0.43084$	$\Sigma c_i^2 = 0.046406401$

(a) $\bar{c}_i = 0.43084/4 = 0.10771 = \underline{\underline{0.1077}}$

(b) $s = \sqrt{\dfrac{0.046406401 - (0.43084)^2/4}{3}} = \underline{\underline{0.00061}}$

$CV = \dfrac{0.00061}{0.1077} \times 100\% = \underline{\underline{0.57\%}}$

(c) $Q = (0.10862 - 0.10747)/(0.10862 - 0.10733) = 0.89$

From Table 4-4, page 58,

$Q_{crit} = 0.829$ at the 95% confidence level

$Q_{crit} = 0.926$ at the 99% confidence level

Thus, reject at the 95% level and retain at the 99% level.

12-12 (a) With phenolphthalein, the CO_3^{2-} consumes but 1 mmol H_3O^+ per mmol of CO_3^{2-}.

Thus the effective amount of base is lowered by 11.2 mmol, and

$$c_{base} = \dfrac{1000 \text{ mL NaOH} \times 0.1500 \frac{\text{mmol}}{\text{mL}} - 11.2 \text{ mmol } CO_2 \times \frac{1 \text{ mmol NaOH}}{\text{mmol } CO_2}}{1000}$$

$= \underline{\underline{0.1388 \text{ M}}}$

(b) When bromocresol green is the indicator,

$CO_3^{2-} + 2H_3O^+ \rightarrow H_2CO_3 + 2H_2O$

Chapter 12

and the effective concentration of the base is unchanged. Thus,

$$c_{base} = \underline{\underline{0.1500 \text{ M}}}$$

12-13 As in part (a) of Problem 12-12,

$$500 \text{ mL NaOH} \times 0.1019 \frac{\text{mmol NaOH}}{\text{mL}} - 0.652 \text{ g CO}_2 \times \frac{1 \text{ mmol CO}_2}{0.04401 \text{ g}} \times \frac{1 \text{ mmol NaOH}}{\text{mmol CO}_2} =$$

$$36.135 \text{ mmol NaOH}$$

$$c_{NaOH} = \frac{36.135 \text{ mmol NaOH}}{500 \text{ mL}} = 0.07227 \text{ M}$$

$$\text{relative error in molarity} = \frac{0.07227 - 0.1019}{0.1019} \times 100\% = -29\%$$

The relative error in the determination of acetic acid will be the same as the relative error in the molarity or <u>-29%</u>.

12-14 (a)
$$0.6010 \text{ g AgCl} \times \frac{1 \text{ mmol AgCl}}{0.14332 \text{ g AgCl}} \times \frac{1 \text{ mmol HCl}}{\text{mmol AgCl}} \times \frac{1}{50.00 \text{ mL HCl}} =$$

$$\underline{\underline{0.08387 \frac{\text{mmol HCl}}{\text{mL}}}}$$

(b)
$$25.00 \text{ mL Ba(OH)}_2 \times 0.04010 \frac{\text{mmol Ba(OH)}_2}{\text{mL Ba(OH)}_2} \times \frac{2 \text{ mmol HCl}}{\text{mmol Ba(OH)}_2} \times \frac{1}{19.92 \text{ mL HCl}} =$$

$$\underline{\underline{0.1007 \frac{\text{mmol HCl}}{\text{mL HCl}}}}$$

(c)
$$0.2694 \text{ g Na}_2\text{CO}_3 \times \frac{1 \text{ mmol Na}_2\text{CO}_3}{0.10599 \text{ g Na}_2\text{CO}_3} \times \frac{2 \text{ mmol HCl}}{\text{mmol Na}_2\text{CO}_3} \times \frac{1}{38.77 \text{ mL HCl}} =$$

$$\underline{\underline{0.1311 \frac{\text{mmol HCl}}{\text{mL HCl}}}}$$

12-15 (a)
$$0.1684 \text{ g BaSO}_4 \times \frac{1 \text{ mmol BaSO}_4}{0.23339 \text{ g BaSO}_4} \times \frac{1 \text{ mmol Ba(OH)}_2}{\text{mmol BaSO}_4} \times \frac{1}{50.00 \text{ mL Ba(OH)}_2} =$$

$$\underline{\underline{0.01443 \text{ M Ba(OH)}_2}}$$

Chapter 12

(b) $$0.4815 \text{ g KHP} \times \frac{1 \text{ mmol KHP}}{0.20422 \text{ g KHP}} \times \frac{1 \text{ mmol Ba(OH)}_2}{2 \text{ mmol KHP}} \times \frac{1}{29.41 \text{ mL Ba(OH)}_2} =$$

$$\underline{\underline{0.04008 \text{ M Ba(OH)}_2}}$$

(c) $$\text{no. mmol HBz} = \frac{0.3614 \text{ g HBz}}{0.12212 \text{ g HBz}} = 2.9594$$

$$\text{no. mmol HCl} = 4.13 \text{ mL HCl} \times 0.05317 \frac{\text{mmol HCl}}{\text{mL HCl}} = \underline{0.21959}$$

$$\text{total no. mmol acid} = 3.17898$$

$$3.17898 \text{ mmol acid} \times \frac{1 \text{ mmol Ba(OH)}_2}{2 \text{ mmol acid}} \times \frac{1}{50.00 \text{ mL Ba(OH)}_2} = \underline{\underline{0.03179 \text{ M Ba(OH)}_2}}$$

12-16 (a) For 35 mL

$$35 \text{ mL HClO}_4 \times 0.150 \frac{\text{mmol HClO}_4}{\text{mL HClO}_4} \times \frac{1 \text{ mmol Na}_2\text{CO}_3}{2 \text{ mmol HClO}_4} \times \frac{0.10599 \text{ g Na}_2\text{CO}_3}{\text{mmol Na}_2\text{CO}_3} =$$

$$0.28 \text{ g Na}_2\text{CO}_3$$

Substituting 45 mL gives 0.36 g Na_2CO_3.

Thus, the range is $\underline{\underline{0.28 \text{ to } 0.36 \text{ g Na}_2\text{CO}_3}}$.

Proceeding in the same way, we obtain

(b) $\underline{\underline{0.18 \text{ to } 0.23 \text{ g Na}_2\text{C}_2\text{O}_4}}$

(c) 0.85 to 1.1 g of HBz

(d) 0.82 to 1.1 g $KH(IO_3)$

(e) 0.17 to 0.22 g TRIS

(f) 1.1 to 1.4 g $Na_2B_4O_7 \cdot 10 H_2O$

12-17 In Example 12-1 (page 250) we found that we should weigh out about 0.073 g TRIS, 0.032 g Na_2CO_3, and 0.11 g of $Na_2B_4O_7 \cdot 10H_2O$. In each case, the absolute standard deviation in molarity is

Chapter 12

(a) TRIS: $s_M = \dfrac{0.0002}{0.073} \times 0.0200 \text{ M} = \underline{\underline{0.00005 \text{ M}}}$

(b) Na_2CO_3: $s_M = \dfrac{0.0002}{0.032} \times 0.0200 \text{ M} = \underline{\underline{0.00013 \text{ M}}}$

(c) $Na_2B_4O_7 \cdot 10H_2O$: $s_M = \dfrac{0.0002}{0.23} \times 0.0200 \text{ M} = \underline{\underline{0.00002 \text{ M}}}$

12-18 (a) In each case,

no. mmol NaOH = 30 mL × 0.0400 mmol/mL = 1.20 mmol NaOH

For KHP:
$1.20 \text{ mmol NaOH} \times \dfrac{1 \text{ mmol KHP}}{\text{mmol NaOH}} \times \dfrac{0.2042 \text{ g KHP}}{\text{mmol KHP}} = \underline{\underline{0.245 \text{ g}}}$

For $KH(IO_3)_2$:
$1.20 \text{ mmol NaOH} \times \dfrac{1 \text{ mmol KH(IO}_3)_2}{\text{mmol NaOH}} \times \dfrac{0.38991 \text{ g KH(IO}_3)_2}{\text{mmol KH(IO}_3)_2} = \underline{\underline{0.468 \text{ g}}}$

For HBz:
$1.20 \text{ mmol NaOH} \times \dfrac{1 \text{ mmol HBz}}{\text{mmol NaOH}} \times \dfrac{0.12212 \text{ g HBz}}{\text{mmol HBz}} = \underline{\underline{0.147 \text{ g}}}$

(b) For KHP: $(s_M)_r = \dfrac{0.0002 \text{ g}}{0.245 \text{ g}} \times 100\% = \underline{\underline{0.082\%}}$

For $KH(IO_3)_2$: $(s_M)_r = \dfrac{0.0002 \text{ g}}{0.468 \text{ g}} \times 100\% = \underline{\underline{0.043\%}}$

For $Na_2B_4O_7 \cdot 10H_2O$: $(s_M)_r = \dfrac{0.0002 \text{ g}}{0.147 \text{ g}} \times 100\% = \underline{\underline{0.14\%}}$

12-19 $21.48 \text{ mL NaOH} \times 0.03776 \dfrac{\text{mmol NaOH}}{\text{mL NaOH}} \times \dfrac{1 \text{ mmol H}_2\text{T}}{2 \text{ mmol NaOH}} \times \dfrac{0.15009 \text{ g H}_2\text{T}}{\text{mmol H}_2\text{T}} \times \dfrac{100 \text{ mL}}{50 \text{ mL}} =$

$\underline{\underline{0.1217 \text{ g H}_2\text{T} / 100 \text{ mL}}}$

12-20 $\dfrac{(34.88 \times 0.09600) \text{ mmol NaOH} \times \dfrac{1 \text{ mmol HOAc}}{\text{mmol NaOH}} \times \dfrac{0.06005 \text{ g HOAc}}{\text{mmol}}}{25.00 \text{ mL sample} \times 50.0 \text{ mL} / 250 \text{ mL}} \times 100\% = \underline{\underline{4.02\% \text{ HOAc}}}$

Chapter 12

12-21 For each part, we may write

$$31.64 \text{ mL HCl} \times 0.1081 \frac{\text{mmol HCl}}{\text{mL HCl}} \times \frac{1}{0.7439 \text{ g sample}} = 4.5978 \frac{\text{mmol HCl}}{\text{g sample}}$$

(a)
$$0.45978 \frac{\text{mmol HCl}}{\text{g sample}} \times \frac{1 \text{ mmol Na}_2\text{B}_4\text{O}_7}{2 \text{ mmol HCl}} \times 0.20122 \frac{\text{g Na}_2\text{B}_4\text{O}_7}{\text{mmol Na}_2\text{B}_4\text{O}_7} \times 100\% =$$

$$\underline{\underline{46.25\% \text{ Na}_2\text{B}_4\text{O}_7}}$$

Proceeding in the same way

(b)
$$0.45978 \times \frac{1}{2} \times 0.38137 \times 100 = \underline{\underline{87.67\% \text{ Na}_2\text{B}_4\text{O}_7 \cdot 10\text{H}_2\text{O}}}$$

(c) $0.45978 \times 0.06962 \times 100 = \underline{\underline{32.01\% \text{ B}_2\text{O}_3}}$

(d)
$$0.45978 \frac{\text{mmol HCl}}{\text{g sample}} \times \frac{2 \text{ mmol B}}{\text{mmol HCl}} \times 0.010811 \frac{\text{g B}}{\text{mmol B}} \times 100\% = \underline{\underline{9.94\% \text{ B}}}$$

12-22
$$\frac{42.59 \text{ mL HCl} \times 0.1178 \frac{\text{mmol HCl}}{\text{mL HCl}} \times \frac{1 \text{ mmol OH}^-}{\text{mmol HCl}} \times \frac{1 \text{ mmol HgO}}{2 \text{ mmol OH}^-} \times 0.21659 \frac{\text{g HgO}}{\text{mmol HgO}}}{0.6334 \text{ sample}} \times 100\% =$$

$$\underline{\underline{85.78\% \text{ HgO}}}$$

12-23 no. mmol NaOH = no. mmol HCHO + 2 × no. mmol H_2SO_4

no. mmol HCHO = 50.0 × 0.0996 − 2 × 23.3 × 0.05250 = 2.5335

$$\frac{2.5335 \text{ mmol HCHO} \times 0.030026 \text{ g HCHO/mmol HCHO}}{0.3124 \text{ g sample}} \times 100\% = \underline{\underline{24.4\% \text{ HCHO}}}$$

12-24
$$\frac{(14.76 \times 0.0514) \text{ mmol NaOH} \times \frac{1 \text{ mmol NaBz}}{\text{mmol NaOH}} \times 0.1441 \frac{\text{g NaBz}}{\text{mmol NaBz}}}{106.3 \text{ g sample}} \times 100\% = \underline{\underline{0.103\% \text{ NaBz}}}$$

12-25 Letting RS_4 represent the compound,

1 mmol RS_4 ≡ 4 mmol SO_2 ≡ 4 mmol H_2SO_4 ≡ 8 mmol NaOH

Chapter 12

$$\frac{22.13 \text{ mL NaOH} \times 0.03736 \frac{\text{mmol NaOH}}{\text{mL}} \times \frac{1 \text{ mmol RS}_4}{8 \text{ mmol NaOH}} \times 0.29654 \frac{\text{g RS}_4}{\text{mmol}}}{0.4329 \text{ g sample}} \times 100\% = \underline{\underline{7.079\% \text{ RS}_4}}$$

12-26
$$\frac{40.38 \text{ mL} \times 0.2506 \frac{\text{mmol HCl}}{\text{mL HCl}} \times \frac{1 \text{ mmol NH}_3}{\text{mmol HCl}} \times 0.017031 \frac{\text{g NH}_3}{\text{mmol NH}_3}}{25.00 \text{ mL sample} \times 50.00 \text{ mL}/250.0 \text{ mL}} \times 100\% = \underline{\underline{3.447\% \text{ NH}_3 \text{ (w/v)}}}$$

12-27 no. mmol HCl = no. mmol NaOH + 2 × no. mmol CO_3^{2-}

no. mmol CO_3^{2-} = (50.00 × 0.1140 - 24.21 × 0.09802) / 2 = 1.6635

molar mass CO_3^{2-} = 1 × 12.01115 + 3 × 15.9994 = 60.01

$$\text{molar mass of the salt} = \frac{0.1401 \text{ g salt}}{1.6635 \text{ mmol salt}} \times \frac{10^3 \text{ mmol}}{\text{mol}} = 84.22 \text{ g/mol}$$

molar mass of the cation of the salt = 84.22 - 60.01 = 24.21

$\underline{\underline{\text{MgCO}_3}}$ with a molar mass of 84.31 seems a likely candidate.

12-28
$$\text{no. mmol NaA} = (28.62 \times 0.1084) \text{ mmol NaOH} \times \frac{1 \text{ mmol NaA}}{\text{mmol NaOH}} = 3.1024$$

$$\frac{0.2110 \text{ g NaA}}{3.1024 \text{ mmol NaA}} \times 10^3 \frac{\text{mmol NaA}}{\text{mol NaA}} = 68.01 \frac{\text{g NaA}}{\text{mol A}} = \mathcal{M}_{\text{NaA}}$$

molar mass HA = equivalent weight = $\mathcal{M}_{\text{NaA}} - \mathcal{M}_{\text{Na}} + \mathcal{M}_{\text{H}}$

= 68.01 − 22.99 + 1.008 = $\underline{\underline{46.03}}$

12-29 no. mmol Ba(OH)$_2$ = no. mmol CO$_2$ + no. mmol HCl/2

no. mmol CO$_2$ = 50.0 × 0.0116 - 23.6 × 0.0108/2 = 0.4526

$$\frac{0.4526 \text{ mmol CO}_2 \times 0.04401 \text{ g CO}_2/\text{mmol}}{3.00 \text{ L sample}} \times \frac{1 \text{ L CO}_2}{1.98 \text{ g CO}_2} \times 10^6 \text{ ppm} = \underline{\underline{3.35 \times 10^3 \text{ ppm}}}$$

12-30
$$11.1 \text{ mL NaOH} \times 0.00204 \frac{\text{mmol NaOH}}{\text{mL NaOH}} \times \frac{1 \text{ mmol SO}_2}{2 \text{ mmol NaOH}} \times 0.06406 \frac{\text{g SO}_2}{\text{mmol SO}_2} =$$

$$7.253 \times 10^{-4} \text{ g SO}_2$$

$$7.253 \times 10^{-4} \text{ g SO}_2 \times \frac{1 \text{ min}}{30.0 \text{ L sample}} \times \frac{1}{10.0 \text{ min}} \times \frac{1.00 \text{ L SO}_2}{2.85 \text{ g SO}_2} \times 10^6 \text{ ppm} =$$

$$\underline{\underline{0.848 \text{ ppm SO}_2}}$$

12-31 1 mmol P ≡ 1 mmol $(NH_4)_3PO_4 \cdot 12MoO_3$ ≡ 26 mmol NaOH

no. mmol NaOH = no. mmol HCl + 26 × no. mmol P

no. mmol P = (50.00 × 0.2000 − 14.17 × 0.1741)/26 = 0.28973

$$\frac{0.28973 \text{ mmol P} \times 0.030974 \text{ g P/mmol}}{0.1417 \text{ g sample}} \times 100\% = \underline{\underline{6.333\% \text{ P}}}$$

12-32 $C_6H_4(COOCH_3)_2 + 2OH^- \rightarrow 2CH_3OH + 2C_6H_4(COO^-)_2$

no. mmol NaOH = 2 × no. mmol analyte + no. mmol HCl

no. mmol analyte = (no. mmol NaOH − no. mmol HCl)/2

= (50.00 × 0.1031 − 24.27 × 0.1644)/2 = 0.58251

$$\frac{0.58251 \text{ mmol analyte} \times 0.19419 \text{ g analyte/mmol}}{0.8160 \text{ g sample}} \times 100\% = \underline{\underline{13.86\% \text{ analyte}}}$$

12-33 1 mmol RN_4 ≡ 4 mmol NH_3 ≡ 4 mmol HCl

$$\frac{(26.13 \times 0.01477) \text{ mmol HCl} \times \frac{1 \text{ mmol RN}_4}{4 \text{ mmol HCl}} \times 0.28537 \frac{\text{g RN}_4}{\text{mmol RN}_4}}{0.1247 \text{ g sample}} \times 100\% = \underline{\underline{22.08\% \text{ RN}_4}}$$

12-34 (100.0 × 0.1750 − 11.37 × 0.1080) = 16.272 mmol HCl consumed by analyte

$$\frac{16.272 \text{ mmol HCl} \times \frac{1 \text{ mmol CH}_5\text{N}_3}{3 \text{ mmol HCl}} \times \frac{59.07 \text{ mg CH}_5\text{N}_3}{\text{mmol CH}_5\text{N}_3}}{4 \text{ tablets}} = \underline{\underline{80.10 \text{ mg CH}_5\text{N}_3/\text{tablet}}}$$

Chapter 12

$$\text{no. tablets} = \frac{10 \text{ mg CH}_5\text{N}_3}{\text{kg}} \times 48 \text{ kg} \times \frac{1 \text{ tablet}}{80.10 \text{ mg CH}_5\text{N}_3} = 5.99 \text{ tablets} = \underline{\underline{6 \text{ tablets}}}$$

12-35
$$\%\text{N} = \frac{(24.61 \times 0.1180) \text{ mmol HCl} \times \frac{1 \text{ mmol N}}{\text{mmol HCl}} \times \frac{0.014007 \text{ g N}}{\text{mmol N}}}{1.047 \text{ g sample}} \times 100\% = \underline{\underline{3.885}}$$

12-36
$$3.885\% \text{ N} \times \frac{6.25\% \text{ protein}}{\% \text{ N}} = 24.28\% \text{ protein} \qquad \text{(see page 254)}$$

$$\frac{6.50 \text{ oz. tuna}}{\text{can}} \times \frac{28.3 \text{ g}}{\text{oz.}} \times \frac{24.28 \text{ g protein}}{100 \text{ g tuna}} = \underline{\underline{44.7 \text{ g protein/can}}}$$

12-37 In each part,

$$\frac{(50.00 \times 0.1062 - 11.89 \times 0.0925) \text{ mmol HCl}}{0.5843 \text{ g sample}} = 7.2506 \frac{\text{mmol HCl}}{\text{g sample}}$$

(a) $$7.2506 \frac{\text{mmol HCl}}{\text{g sample}} \times \frac{1 \text{ mmol N}}{\text{mmol HCl}} \times \frac{0.014007 \text{ g N}}{\text{mmol N}} \times 100\% = \underline{\underline{10.09\% \text{ N}}}$$

(b) $$7.2506 \frac{\text{mmol HCl}}{\text{g sample}} \times \frac{1 \text{ mmol urea}}{2 \text{ mmol HCl}} \times 0.06006 \frac{\text{g urea}}{\text{mmol urea}} = \underline{\underline{21.64\% \text{ urea}}}$$

(c) $$7.2506 \frac{\text{mmol HCl}}{\text{g sample}} \times \frac{1 \text{ mmol(NH}_4)_2\text{SO}_4}{2 \text{ mmol HCl}} \times 0.132141 \frac{\text{g (NH}_4)_2\text{SO}_4}{\text{mmol (NH}_4)_2\text{SO}_4} \times 100\% =$$

$$\underline{\underline{47.61\% \text{ (NH}_4)_2\text{SO}_4}}$$

(d) $$7.2506 \times \frac{1}{3} \times 0.14909 \times 100\% = \underline{\underline{35.81\%(\text{NH}_4)_3\text{PO}_4}}$$

12-38
$$\%\text{N} = \frac{(50.00 \times 0.05063 - 7.46 \times 0.04917) \text{ mmol HCl} \times \frac{1 \text{ mmol N}}{\text{mmol HCl}} \times \frac{0.014007 \text{ g N}}{\text{mmol N}}}{0.9092 \text{ g sample}} \times 100\% =$$

$$3.335\%$$

% protein = $3.335 \times 5.7 = \underline{\underline{19.0}}$ (see page 254)

12-39 In the first titration the analyte consumed

$$(30.00 \times 0.08421 - 10.17 \times 0.08802) = 1.63114 \text{ mmol HCl}$$

Chapter 12

and

$$1.63114 = \text{no. mmol } NH_4NO_3 + 2 \times \text{no. mmol } (NH_4)_2SO_4$$

The amounts of the two species in the entire sample are

$$\text{no. mmol } NH_4NO_3 + 2 \times \text{no. mmol } (NH_4)_2SO_4 = 1.63114 \times \frac{200 \text{ mL}}{50 \text{ mL}} \quad (1)$$

$$= 6.52455$$

In the second titration the analyte consumed

$$(30.00 \times 0.08421 - 14.16 \times 0.08802) = 1.27994 \text{ mmol HCl}$$

and

$$1.27994 = 2 \times \text{no. mmol } NH_4NO_3 + 2 \times \text{no. mmol } (NH_4)_2SO_4$$

The amounts of the two species in the entire sample are

$$2 \times \text{no. mmol } NH_4NO_3 + 2 \times \text{no. mmol } (NH_4)_2SO_4 = 1.27994 \times \frac{200 \text{ mL}}{25 \text{ mL}} \quad (2)$$

$$= 10.23949$$

Subtracting equation (1) from (2) gives

$$\text{no. mmol } NH_4NO_3 = 10.23949 - 6.52455 = 3.7149$$

$$\text{no. mmol } (NH_4)_2SO_4 = \frac{10.23949 - 2 \times 3.7149}{2} = 1.4085$$

$$\text{percent } NH_4NO_3 = \frac{3.7149 \text{ mmol } NH_4NO_3 \times 0.08004 \text{ g } NH_4NO_3/\text{mmol}}{1.219 \text{ g sample}} \times 100\%$$

$$= \underline{\underline{24.39\%}}$$

$$\text{percent } (NH_4)_2SO_4 = \frac{1.4085 \text{ mmol } (NH_4)_2SO_4 \times 0.13214 \text{ g } (NH_4)_2SO_4/\text{mmol}}{1.219 \text{ g sample}} \times 100\%$$

$$= \underline{\underline{15.23\%}}$$

Chapter 12

12-40 For first aliquot,

$$\text{no. mmol HCl} = \text{no. mmol NaOH} + \text{no. mmol KOH} + 2 \times \text{no. mmol K}_2\text{CO}_3$$

$$\text{no. mmol KOH} + 2 \times \text{no. mmol K}_2\text{CO}_3 = 40.00 \times 0.05304 - 4.74 \times 0.04983$$
$$= 1.8854$$

For second aliquot,

$$\text{no. mmol HCl} = \text{no. mmol KOH} = 28.56 \times 0.05304 = 1.5148$$

$$\text{no. mmol K}_2\text{CO}_3 = (1.8854 - 1.5148)/2 = 0.18530$$

$$\frac{1.5148 \text{ mmol KOH} \times 0.05611 \frac{\text{g KOH}}{\text{mmol}}}{1.217 \text{ g sample} \times 50.00 \text{ mL} / 500.0 \text{ mL}} \times 100\% = \underline{\underline{69.84\% \text{ KOH}}}$$

$$\frac{0.18530 \text{ mmol K}_2\text{CO}_3 \times 0.13821 \frac{\text{g K}_2\text{CO}_3}{\text{mmol}}}{1.217 \text{ g sample} \times 50.00 \text{ mL} / 500.0 \text{ mL}} \times 100\% = \underline{\underline{21.04\% \text{ K}_2\text{CO}_3}}$$

$$100\% - 69.84\% - 21.04\% = \underline{\underline{9.12\% \text{ H}_2\text{O}}}$$

12-41 For first aliquot,

$$\text{no. mmol HCl} = \text{no. mmol NaOH} + \text{no. mmol NaHCO}_3 + 2 \times \text{no. mmol Na}_2\text{CO}_3$$

$$\text{no. mmol NaHCO}_3 + 2 \times \text{no. mmol Na}_2\text{CO}_3 =$$
$$50.00 \times 0.01255 - 2.34 \times 0.01063 = 0.60263$$

For second aliquot,

$$\text{no. mmol NaHCO}_3 = \text{no. mmol NaOH} - \text{no. mmol HCl}$$
$$= 25.00 \times 0.01063 - 7.63 \times 0.01255 = 0.16999$$

$$\text{no. mmol Na}_2\text{CO}_3 = (0.60263 - 0.16999)/2 = 0.21632$$

$$\frac{0.16999 \text{ mmol NaHCO}_3 \times 0.08401 \text{ g NaHCO}_3/\text{mmol}}{0.5000 \text{ g sample} \times 25.00 \text{ mL}/250.0 \text{ mL}} \times 100\% = \underline{\underline{28.56\% \text{ NaHCO}_3}}$$

$$\frac{0.21632 \text{ mmol Na}_2\text{CO}_3 \times 0.10599 \text{ g Na}_2\text{CO}_3/\text{mmol}}{0.5000 \text{ g sample} \times 25.00 \text{ mL}/250.0 \text{ mL}} \times 100\% = \underline{\underline{45.86\% \text{ Na}_2\text{CO}_3}}$$

$$100\% - 28.56\% - 48.86\% = \underline{\underline{25.58\% \text{ H}_2\text{O}}}$$

12-42 (a)
$$20.0 \text{ mL} \times 0.05555 \frac{\text{mmol Na}_3\text{PO}_4}{\text{mL}} \times \frac{1 \text{ mmol HCl}}{\text{mmol Na}_3\text{PO}_4} \times \frac{1 \text{ mL HCl}}{0.06122 \text{ mmol HCl}} =$$

$$\underline{\underline{18.15 \text{ mL HCl}}}$$

(b)
$$25.00 \text{ mL} \times 0.05555 \frac{\text{mmol Na}_3\text{PO}_4}{\text{mL}} \times \frac{2 \text{ mmol HCl}}{\text{mmol Na}_3\text{PO}_4} \times \frac{1 \text{ mL HCl}}{0.06122 \text{ mmol HCl}} =$$

$$\underline{\underline{45.37 \text{ mL HCl}}}$$

(c)
$$40.00 \text{ mL} \times 0.02102 \frac{\text{mmol Na}_3\text{PO}_4}{\text{mL}} \times \frac{2 \text{ mmol HCl}}{\text{mmol Na}_3\text{PO}_4} = 1.6816 \text{ mmol HCl}$$

$$40.00 \text{ mL} \times 0.01655 \frac{\text{mmol Na}_2\text{HPO}_4}{\text{mL}} \times \frac{1 \text{ mmol HCl}}{\text{mmol Na}_2\text{HPO}_4} = 0.6620 \text{ mmol HCl}$$

$$(1.6816 + 0.6620) \text{ mmol HCl} \times 1 \text{ mL HCl}/0.06122 \text{ mmol} = \underline{\underline{38.28 \text{ mL HCl}}}$$

(d)
$$20.00 \text{ mL} \times 0.02102 \frac{\text{mmol Na}_3\text{PO}_4}{\text{mL}} \times \frac{1 \text{ mmol HCl}}{\text{mmol Na}_3\text{PO}_4} = 0.42040 \text{ mmol HCl}$$

$$20.00 \text{ mL} \times 0.01655 \frac{\text{mmol NaOH}}{\text{mL}} \times \frac{1 \text{ mmol HCl}}{\text{mmol NaOH}} = 0.3310 \text{ mmol HCl}$$

$$(0.4204 + 0.3310) \text{ mmol HCl} \times 1 \text{ mL HCl}/0.06122 \text{ mmol} = \underline{\underline{12.27 \text{ mL HCl}}}$$

12-43 (a) $\dfrac{(25.00 \times 0.03000 + 25.00 \times 0.01000) \text{ mmol NaOH}}{0.07731 \text{ mmol NaOH}/\text{mL}} = \underline{\underline{12.93 \text{ mL NaOH}}}$

(b) $\dfrac{(25.00 \times 0.03000 + 25.00 \times 0.01000 \times 2) \text{ mmol NaOH}}{0.07731 \text{ mmol NaOH}/\text{mL}} = \underline{\underline{16.17 \text{ mL NaOH}}}$

(c) $\dfrac{(30.00 \times 0.06407) \text{ mmol NaOH}}{0.07731 \text{ mmol NaOH}/\text{mL}} = \underline{\underline{24.86 \text{ mL NaOH}}}$

(d) $\dfrac{(25.00 \times 0.02000 \times 2 + 25.00 \times 0.0300) \text{ mmol NaOH}}{0.07731 \text{ mmol NaOH}/\text{mL}} = \underline{\underline{22.64 \text{ mL NaOH}}}$

Chapter 12

12-44 (a) (22.43×0.1202) mmol HCl $\times \dfrac{1 \text{ mmol NaOH}}{\text{mmol HCl}} \times \dfrac{40.00 \text{ mg NaOH}}{\text{mmol}} \times \dfrac{1}{25.00 \text{ mL sample}} =$

$\underline{\underline{4.314 \text{ mg NaOH/mL}}}$

(b) no. mmol Na_2CO_3 = $15.67 \times 0.1202 = 1.8835$

$2 \times$ no. mmol Na_2CO_3 + no. mmol $NaHCO_3$ = $42.13 \times 0.1202 = 5.0640$

no. mmol $NaHCO_3$ = $5.0640 - 2 \times 1.8835 = 1.2970$

1.8835 mmol $Na_2CO_3 \times 105.99 \dfrac{\text{mg } Na_2CO_3}{\text{mmol}} \times \dfrac{1}{25.00 \text{ mL sample}} =$

$\underline{\underline{7.985 \text{ mg } Na_2CO_3/\text{mL}}}$

1.2970 mmol $NaHCO_3 \times 84.01 \dfrac{\text{mg } NaHCO_3}{\text{mmol}} \times \dfrac{1}{25.00 \text{ mL sample}} =$

$\underline{\underline{4.358 \text{ mg } NaHCO_3/\text{mL}}}$

(c) no. mmol NaOH + no. mmol Na_2CO_3 = $29.64 \times 0.01202 = 3.5627$

no. mmol Na_2CO_3 = $(36.42 - 29.64) \times 0.1202 = 0.8150$

no. mmol NaOH = $3.5627 - 0.8150 = 2.7477$

0.8150 mmol $Na_2CO_3 \times 105.99 \dfrac{\text{mg } Na_2CO_3}{\text{mmol}} \times \dfrac{1}{25.00 \text{ mL sample}} =$

$\underline{\underline{3.455 \text{ mg } Na_2CO_3/\text{mL}}}$

2.7477 mmol NaOH $\times 40.00 \dfrac{\text{mg NaOH}}{\text{mmol}} \times \dfrac{1}{25.00 \text{ mL sample}} =$

$\underline{\underline{4.396 \text{ mg NaOH/mL}}}$

(d) 16.12 mL $\times 0.1202 \dfrac{\text{mmol HCl}}{\text{mL}} \times \dfrac{1 \text{ mmol } Na_2CO_3}{\text{mmol HCl}} \times \dfrac{105.99 \text{ mg } Na_2CO_3}{\text{mmol}} \times \dfrac{1}{25.00 \text{ mL sample}} =$

$\underline{\underline{8.215 \text{ mg } Na_2CO_3/\text{mL}}}$

(e) (33.33×0.1202) mmol HCl $\times \dfrac{1 \text{ mmol NaHCO}_3}{\text{mmol HCl}} \times \dfrac{84.01 \text{ mg NaHCO}_3}{\text{mmol}} \times \dfrac{1}{25.00 \text{ mL sample}} =$

$\underline{\underline{13.46 \text{ mg NaHCO}_3/\text{mL}}}$

12-45 (a) $\dfrac{(18.15 \times 0.08601) \text{ mmol HCl} \times \dfrac{1 \text{ mmol Na}_2\text{HAsO}_4}{\text{mmol HCl}} \times \dfrac{185.91 \text{ mg Na}_2\text{HAsO}_4}{\text{mmol}}}{25.00 \text{ mL sample}} =$

$\underline{\underline{11.61 \text{ mg Na}_2\text{HAsO}_4/\text{mL}}}$

(b) no. mmol NaOH + no. mmol Na$_2$AsO$_4$ = $21.00 \times 0.08601 = 1.8062$

no. mmol Na$_3$AsO$_4$ = $(28.15 - 21.00) \times 0.0861$ = $\underline{0.6150}$

no. mmol NaOH = 1.1912

0.6150 mmol Na$_3$AsO$_4 \times 207.89 \dfrac{\text{mg Na}_3\text{AsO}_4}{\text{mmol}} \times \dfrac{1}{25.00 \text{ mL}} = \underline{\underline{5.114 \text{ mg Na}_3\text{AsO}_4/\text{mL}}}$

1.1912 mmol NaOH $\times 40.00 \dfrac{\text{mg NaOH}}{\text{mmol}} \times \dfrac{1}{25.00 \text{ mL}} = \underline{\underline{1.906 \text{ mg Na}_3\text{AsO}_4/\text{mL}}}$

(c) (19.80×0.0861) mmol HCl $\times \dfrac{1 \text{ mmol Na}_3\text{AsO}_4}{\text{mmol HCl}} \times 207.89 \dfrac{\text{mg Na}_3\text{AsO}_4}{\text{mmol}} \times$

$\dfrac{1}{25.00 \text{ mL sample}} = \underline{\underline{14.16 \text{ mg Na}_3\text{AsO}_4/\text{mL}}}$

(d) (18.04×0.08601) mmol HCl $\times \dfrac{1 \text{ mmol NaOH}}{\text{mmol HCl}} \times 40.00 \dfrac{\text{mg NaOH}}{\text{mmol}} \times$

$\dfrac{1}{25.00 \text{ mL sample}} = \underline{\underline{2.483 \text{ mg NaOH/mL}}}$

(e) no. mmol Na$_3$AsO$_4$ = $16.00 \times 0.08601 = 1.3762$

no. mmol Na$_2$HAsO$_4$ + 2 × no. mmol Na$_3$AsO$_4$ = $37.37 \times 0.08601 = 3.2142$

no. mmol Na$_2$HAsO$_4$ = $3.2142 - 2 \times 1.3762 = 0.4618$

Chapter 12

$$0.4618 \text{ mmol Na}_2\text{HAsO}_4 \times 185.91 \frac{\text{mg Na}_2\text{HAsO}_4}{\text{mmol}} \times \frac{1}{25.00 \text{ mL sample}} =$$

$$\underline{\underline{3.434 \text{ mg Na}_2\text{HAsO}_4/\text{mL}}}$$

$$1.3762 \text{ mmol Na}_3\text{AsO}_4 \times 207.89 \frac{\text{mg Na}_3\text{AsO}_4}{\text{mmol}} \times \frac{1}{25.00 \text{ mL sample}} =$$

$$\underline{\underline{11.44 \text{ mg Na}_3\text{AsO}_4/\text{mL}}}$$

12-46 The equivalent weight of an acid is that weight of the pure material is that weight that contains one mole of titratable protons in a specified reaction. The equivalent weight of base is that weight of a pure compound that consumes one mole of protons in a specfied reaction.

12-47 (a) With this indicator only one of the two protons in the oxalic acid react. Therefore, the equivalent mass is the molar mass or 126.1 g. When phenolphthalein is the indicator, two of the protons are consumed. Therefore, the equiivalent mass of oxalic acid is one-half the molar mass or 63.1 g.

CHAPTER 13

13-1 The Fajans determination of chloride involves a direct titration, while a Volhard approach requires two standard solutions and a filtration step to eliminate AgCl.

13-2 The solubility of the silver salt of the analyte with respect to AgSCN in the acidic environment needed to keep the iron(III) indicator in solution determines whether a filtration step is required.

(a) The solubility of AgCl is unaffected by the acidity. Filtration is nevertheless required because AgCl is more soluble than AgSCN.

(b) The solubility of AgCN is less than that for AgSCN in a neutral or nearly neutral solution, but is appreciable in an acidic solution. Filtration is then required.

(c) Silver carbonate is more soluble than AgSCN; moreover, its solubility increases in acidic solution. Filtration is required on both counts.

13-3 In contrast to Ag_2CO_3 and AgCN, the solubility of AgI is unaffected by the acidity and additionally is less soluble than AgSCN. The filtration step is thus unnecessary, whereas it is with the other two compounds.

13-4 The ions that are preferentially adsorbed on the surface of an ionic solid are generally lattice ions. Thus, in a titration, one of the lattice ions is in excess and its charge determines the sign of the charge of the particles. After the equivalence point the ion of opposite charge is present in excess and determines the sign of the charge on the particle. Thus, in the equivalence-point region, the charge of the particles shifts from positive to negative, or the reverse.

13-5 Potassium is determined by precipitation with an excess of a standard solution of sodium tetraphenylboron. An excess of standard $AgNO_3$ is then added, which precipitates the excess tetraphenylboron ion. The excess $AgNO_3$ is then titrated with a standard solution of SCN^-. The reactions are

$$K^+ + B(C_6H_5)_4^- \rightleftharpoons KB(C_6H_5)_4(s) \qquad \text{[measured excess } B(C_6H_5)_4^-\text{]}$$

$$Ag^+ + B(C_6H_5)_4^- \rightleftharpoons AgB(C_6H_5)_4(s) \qquad \text{[measured excess } AgNO_3\text{]}$$

The excess $AgNO_3$ is then determined by a Volhard titration with KSCN.

Chapter 13

13-6 $Pb^{2+} + Cl^- + F^- \rightleftarrows PbClF(s)$ (neutral solution)

The PbClF is then filtered and dissolved in acid.

$PbClF(s) + H^+ \rightleftarrows Pb^{2+} + Cl^- + HF$ (acidic solution)

$Ag^+ + Cl^- \rightleftarrows AgCl(s)$ (excess standard $AgNO_3$)

The excess Ag^+ is then determined by a Volhard titration with standard KSCN.

13-7 $\dfrac{14.77 \text{ g}}{\text{L}} \times \dfrac{1 \text{ mol AgNO}_3}{169.873 \text{ g}} = 0.08695 \text{ M AgNO}_3$

(a) $0.2631 \text{ g} \times \dfrac{\text{mmol NaCl}}{0.05844 \text{ g}} \times \dfrac{1 \text{ mmol AgNO}_3}{\text{mmol NaCl}} \times \dfrac{1 \text{ mL AgNO}_3}{0.08695 \text{ mmol AgNO}_3} = \underline{\underline{51.78 \text{ mL AgNO}_3}}$

(b) $0.1799 \text{ g} \times \dfrac{1 \text{ mmol Na}_2\text{CrO}_4}{0.161973 \text{ g}} \times \dfrac{2 \text{ mmol AgNO}_3}{\text{mmol Na}_2\text{CrO}_4} \times \dfrac{1 \text{ mL}}{0.08695 \text{ mmol}} = \underline{\underline{25.55 \text{ mL AgNO}_3}}$

(c) $64.13 \text{ mg} \times \dfrac{\text{mmol Na}_3\text{AsO}_4}{207.88 \text{ mg}} \times \dfrac{3 \text{ mmol AgNO}_3}{\text{mmol Na}_3\text{AsO}_4} \times \dfrac{1 \text{ mL}}{0.08695 \text{ mmol}} = \underline{\underline{10.64 \text{ mL AgNO}_3}}$

(d) $381.1 \text{ mg} \times \dfrac{\text{mmol BaCl}_2 \cdot 2\text{H}_2\text{O}}{244.26 \text{ g}} \times \dfrac{2 \text{ mmol AgNO}_3}{\text{mmol BaCl}_2 \cdot 2\text{H}_2\text{O}} \times \dfrac{1 \text{ mL}}{0.08695 \text{ mmol}} =$

$\underline{\underline{35.89 \text{ mL AgNO}_3}}$

(e) $25.00 \text{ mL} \times \dfrac{0.05361 \text{ mmol Na}_3\text{PO}_4}{\text{mL}} \times \dfrac{3 \text{ mmol AgNO}_3}{\text{mmol Na}_3\text{PO}_4} \times \dfrac{1 \text{ mL}}{0.08695 \text{ mmol}} = \underline{\underline{46.24 \text{ mL AgNO}_3}}$

(f) $50.00 \times 0.01808 \dfrac{\text{mmol H}_2\text{S}}{\text{mL}} \times \dfrac{2 \text{ mmol AgNO}_3}{\text{mmol H}_2\text{S}} \times \dfrac{1 \text{ mL}}{0.08695 \text{ mmol}} = \underline{\underline{20.79 \text{ mL AgNO}_3}}$

13-8 (a) $\dfrac{0.2631 \text{ g NaCl}}{25.00 \text{ mL AgNO}_3} \times \dfrac{\text{mmol NaCl}}{0.05844 \text{ g}} \times \dfrac{\text{mmol AgNO}_3}{\text{mmol NaCl}} = \underline{\underline{0.1801 \text{ M AgNO}_3}}$

(b) $\dfrac{0.1799 \text{ g Na}_2\text{CrO}_4}{25.00 \text{ mL AgNO}_3} \times \dfrac{\text{mmol Na}_2\text{CrO}_4}{0.16197 \text{ g}} \times \dfrac{2 \text{ mmol AgNO}_3}{\text{mmol Na}_2\text{CrO}_4} = \underline{\underline{0.08886 \text{ M AgNO}_3}}$

(c) $\dfrac{64.13 \text{ mg Na}_3\text{AsO}_4}{25.00 \text{ mL AgNO}_3} \times \dfrac{\text{mmol Na}_3\text{AsO}_4}{207.888 \text{ g}} \times \dfrac{3 \text{ mmol AgNO}_3}{\text{mmol Na}_3\text{AsO}_4} = \underline{\underline{0.03702 \text{ M AgNO}_3}}$

(d) $\dfrac{38.11 \text{ mg BaCl}_2 \cdot 2\text{H}_2\text{O}}{25.00 \text{ mL AgNO}_3} \times \dfrac{\text{mmol BaCl}_2 \cdot 2\text{H}_2\text{O}}{244.26 \text{ mg}} \times \dfrac{2 \text{ mmol AgNO}_3}{\text{mmol BaCl}_2 \cdot 2\text{H}_2\text{O}} = \underline{\underline{0.01248 \text{ M AgNO}_3}}$

(e) $\dfrac{25.00 \text{ mL Na}_3\text{PO}_4}{25.00 \text{ mL AgNO}_3} \times \dfrac{0.05361 \text{ mmol Na}_3\text{PO}_4}{\text{mL}} \times \dfrac{3 \text{ mmol AgNO}_3}{\text{mmol Na}_3\text{PO}_4} = \underline{\underline{0.1608 \text{ M AgNO}_3}}$

(f) $\dfrac{50.00 \text{ mL H}_2\text{S}}{25.00 \text{ mL AgNO}_3} \times \dfrac{0.01808 \text{ mmol H}_2\text{S}}{\text{mL H}_2\text{S}} \times \dfrac{2 \text{ mmol AgNO}_3}{\text{mmol H}_2\text{S}} = \underline{\underline{0.07232 \text{ M AgNO}_3}}$

13-9 (a) An excess is assured if the calculation is based on a pure sample.

$0.2513 \text{ g NaCl} \times \dfrac{1 \text{ mmol NaCl}}{0.05844 \text{ g NaCl}} \times \dfrac{\text{mmol AgNO}_3}{\text{mmol NaCl}} \times \dfrac{1 \text{ mL AgNO}_3}{0.09621 \text{ mmol AgNO}_3} =$

$\underline{\underline{44.70 \text{ mL AgNO}_3}}$

(b) $0.3462 \text{ g ZnCl}_2 \times 0.7452 \dfrac{\text{g ZnCl}_2}{\text{g sample}} \times \dfrac{1 \text{ mmol ZnCl}_2}{0.1363 \text{ g ZnCl}_2} \times \dfrac{2 \text{ mmol AgNO}_3}{\text{mmol ZnCl}_2} \times$

$\dfrac{1 \text{ mL}}{0.09621 \text{ mmol AgNO}_3} = \underline{\underline{39.35 \text{ mL AgNO}_3}}$

(c) $25.00 \text{ mL AlCl}_3 \times \dfrac{0.01907 \text{ mmol AlCl}_3}{\text{mL AlCl}_3} \times \dfrac{3 \text{ mmol AgNO}_3}{\text{mmol AlCl}_3} \times \dfrac{1 \text{ mL AgNO}_3}{0.09621 \text{ mmol AgNO}_3} =$

$\underline{\underline{14.87 \text{ mL AgNO}_3}}$

13-10 (a) $\dfrac{45.32 \text{ mL AgNO}_3 \times 0.1046 \frac{\text{mmol AgNO}_3}{\text{mL}} \times \frac{1 \text{ mmol Cl}^-}{\text{mmol AgNO}_3} \times \frac{0.035453 \text{ g Cl}^-}{\text{mmol Cl}^-}}{0.7908 \text{ g sample}} \times 100\% =$

$\underline{\underline{21.25\% \text{ Cl}^-}}$

Chapter 13

(b) $$\frac{(45.32 \times 0.1046) \text{ mmol AgNO}_3 \times \frac{1 \text{ mmol BaCl}_2 \cdot 2\text{H}_2\text{O}}{2 \text{ mmol AgNO}_3} \times \frac{0.24426 \text{ g BaCl}_2 \cdot 2\text{H}_2\text{O}}{\text{mmol BaCl}_2 \cdot 2\text{H}_2\text{O}}}{0.7908 \text{ sample}} \times 100\% =$$

$$\underline{\underline{73.21\% \text{ BaCl}_2 \cdot 2\text{H}_2\text{O}}}$$

(c) $$\frac{(45.32 \times 0.1046) \text{ mmol AgNO}_3 \times \frac{1 \text{ mmol analyte}}{4 \text{ mmol AgNO}_3} \times 0.24328 \frac{\text{g analyte}}{\text{mmol analyte}}}{0.7908 \text{ g sample}} \times 100\% =$$

$$\underline{\underline{36.46\% \text{ analyte}}}$$

13-11 $$\frac{36.8 \text{ mL AgNO}_3 \times 0.1060 \frac{\text{mmol AgNO}_3}{\text{mL AgNO}_3} \times \frac{1 \text{ mmol Cl}}{\text{mmol AgNO}_3} \times \frac{0.03543 \text{ g Cl}}{\text{mmol Cl}}}{0.485 \text{ g sample}} \times 100\% = \underline{\underline{28.5\% \text{ Cl}}}$$

13-12 $$23.28 \text{ mL AgNO}_3 \times 0.03337 \frac{\text{mmol AgNO}_3}{\text{mL AgNO}_3} \times \frac{1 \text{ mmol Cl}}{1 \text{ mmol AgNO}_3} \times \frac{1 \text{ mmol C}_{12}\text{H}_8\text{Cl}_6}{6 \text{ mmol Cl}} =$$

$$0.12948 \text{ mmol C}_{12}\text{H}_8\text{Cl}_6$$

$$\frac{0.12948 \text{ mmol C}_{12}\text{H}_8\text{Cl}_6 \times 0.36492 \frac{\text{g C}_{12}\text{H}_8\text{Cl}_6}{\text{mmol}}}{0.1064 \text{ g sample}} \times 100\% = \underline{\underline{44.41\% \text{ C}_{12}\text{H}_8\text{Cl}_6}}$$

13-13 $$\frac{8.47 \text{ mL} \times 0.01310 \frac{\text{mmol AgNO}_3}{\text{mL}} \times \frac{1 \text{ mmol H}_2\text{S}}{2 \text{ mmol AgNO}_3} \times 0.03408 \frac{\text{g H}_2\text{S}}{\text{mmol}}}{100 \text{ mL H}_2\text{O} \times 1.00 \text{ g H}_2\text{O/mL}} \times 10^6 \text{ ppm} = \underline{\underline{18.9 \text{ ppm H}_2\text{S}}}$$

13-14 $$\frac{37.90 \text{ mL} \times 0.03981 \frac{\text{mmol Ag}^+}{\text{mL}} \times \frac{1 \text{ mmol K}}{\text{mmol Ag}^+} \times \frac{39.098 \text{ mg K}}{\text{mmol K}}}{2.000 \text{ L sample}} = \underline{\underline{29.50 \text{ mg K/L}}}$$

13-15 no. mmol AgNO$_3$ consumed by sample =
$(50.00 \times 0.0820 - 4.64 \times 0.0625 \times 250.0 / 50.00) = 2.650$

$$2.650 \text{ mmol AgNO}_3 \times \frac{1 \text{ mmol Ag}_3\text{PO}_4}{3 \text{ mmol AgNO}_3} \times \frac{1 \text{ mmol P}_2\text{O}_5}{2 \text{ mmol Ag}_3\text{PO}_4} = 0.44167 \text{ mmol P}_2\text{O}_5$$

$$\frac{0.44167 \text{ mmol P}_2\text{O}_5 \times 0.14194 \text{ g P}_2\text{O}_5/\text{mmol P}_2\text{O}_5}{4.258 \text{ sample}} \times 100\% = \underline{\underline{1.472\% \text{ P}_2\text{O}_5}}$$

13-16 $\dfrac{50.00 \text{ mL AgNO}_3}{22.98 \text{ mL NH}_4\text{SCN}} = 2.1758 \dfrac{\text{mL AgNO}_3}{\text{mL NH}_4\text{SCN}}$

$50.00 \text{ mL AgNO}_3 - \left(10.43 \text{ mL NH}_4\text{SCN} \times 2.1758 \dfrac{\text{mL AgNO}_3}{\text{mL NH}_4\text{SCN}}\right) =$

$\qquad\qquad\qquad\qquad\qquad\qquad 27.306 \text{ mL AgNO}_3 \text{ consumed by analyte}$

$27.306 \text{ mL AgNO}_3 \times 0.04521 \dfrac{\text{mmol AgNO}_3}{\text{mL AgNO}_3} \times \dfrac{1 \text{ mmol analyte}}{\text{mL AgNO}_3} \times$

$\qquad\qquad\qquad 94.50 \dfrac{\text{mg analyte}}{\text{mmol analyte}} = \underline{\underline{116.7 \text{ mg analyte}}}$

13-17 no. mmol AgNO$_3$ consumed by aliquot $= 50.00 \times 0.2221 - 3.36 \times 0.0397 = 10.972$

$\dfrac{10.972 \text{ mmol AgNO}_3 \times \frac{1 \text{ mmol KBH}_4}{8 \text{ mmol AgNO}_3} \times 0.053941 \frac{\text{g KBH}_4}{\text{mmol KBH}_4}}{3.213 \text{ g sample} \times 100 \text{ mL} / 500 \text{ mL}} \times 100\% = \underline{\underline{11.51\% \text{ KBH}_4}}$

13-18 From Problem 13-17, 10.972 mmol AgNO$_3$ were consumed by analyte, which would form 10.972 mmol Ag(s).

no. mmol KSCN $= 10.972 \text{ mmol Ag} \times \dfrac{50.00 \text{ mL}}{250.0 \text{ mL}} \times \dfrac{1 \text{ mmol KSCN}}{\text{mmol Ag}} = 2.1944$

vol KSCN $= \dfrac{2.1944 \text{ mmol KSCN}}{0.04642 \text{ mmol KSCN} / \text{mL KSCN}} = \underline{\underline{47.27 \text{ mL KSCN}}}$

13-19 Let V_{mL} be the volume AgNO$_3$ in milliliters and c be its molar concentration. Then,

$\% \text{ Cl}^- = \dfrac{V_{\text{mL}} \times c \frac{\text{mmol AgNO}_3}{\text{mL}} \times \frac{1 \text{ mmol Cl}}{\text{mmol AgNO}_3} \times 0.035453 \frac{\text{g Cl}}{\text{mmol Cl}}}{0.2500 \text{ g sample}} \times 100\%$

Substituting V_{mL} for %Cl in this equation and rearranging leads to

$c = 0.2500 / 3.5453 = \underline{\underline{0.07052 \text{ M}}}$

Chapter 13

13-20 $\mathcal{M}_{C_{10}H_5Cl_7} = 373.3 \text{ g/mol}$

The 37.33 found in the numerator of the equation is the product of three numbers: (1) the ratio of the number of millimoles of analyte to the number of millimoles of AgNO$_3$ consumed by the analyte; (2) the millimolar mass of the analyte (0.3733 g/mmol); and (3) 100%. That is,

$$37.33 = \frac{\text{no. mmol } C_{10}H_5Cl_7}{\text{no. mmol AgNO}_3} \times \frac{0.3733 \text{ g } C_{10}H_5Cl_7}{\text{mmol } C_{10}H_5Cl_7} \times 100\%$$

This equation rearranges to

$$\frac{\text{no. mmol AgNO}_3}{\text{no. mmol } C_{10}H_5Cl_7} = \frac{0.3733 \times 100}{37.33} = 1.00$$

Thus only one of the chlorines in the heptachor reacts with AgNO$_3$.

13-21
$$27.36 \text{ mL NaH}_2\text{PO}_4 \times 0.03369 \frac{\text{mmol NaH}_2\text{PO}_4}{\text{mL NaH}_2\text{PO}_4} \times \frac{1 \text{ mmol Bi}^{3+}}{\text{mmol NaH}_2\text{PO}_4} \times \frac{1 \text{ mmol 2Bi}_2\text{O}_3 \cdot 3\text{SiO}_2}{4 \text{ mmol Bi}^{3+}} =$$

$$0.23044 \text{ mmol 2Bi}_2\text{O}_3 \cdot 3\text{SiO}_2$$

$$\frac{0.23044 \text{ mmol 2Bi}_2\text{O}_3 \cdot 3\text{SiO}_2 \times \frac{1.112 \text{ g 2Bi}_2\text{O}_3 \cdot 3\text{SiO}_2}{\text{mmol}}}{0.6423 \text{ g sample}} \times 100\% = \underline{\underline{39.90\%}}$$

13-22 no. mmol AgNO$_3$ consumed by sample = $25.0 \times 0.0100 - 7.69 \times 0.0108 = 0.1669$

$$\frac{0.1669 \text{ mmol AgNO}_3 \times \frac{1 \text{ mmol } C_6H_8N_4O_2}{\text{mmol AgNO}_3} \times 0.1801 \frac{\text{g } C_6H_8N_4O_2}{\text{mmol}}}{2.95 \text{ g sample}} \times 100\% = \underline{\underline{1.02\% \, C_6H_8N_4O_2}}$$

13-23 no. mmol AgNO$_3$ consumd by sample = $20.00 \times 0.08181 - 2.81 \times 0.04124 = 1.52032$

$$1.52032 \text{ mmol AgNO}_3 \times \frac{1 \text{ mmol analyte}}{\text{mmol AgNO}_3} \times 0.20517 \frac{\text{g analyte}}{\text{mmol analyte}} = 0.31192 \text{ g saccharin}$$

$$0.31192 \times \frac{\text{g saccharin}}{20 \text{ tablets}} \times 10^3 \frac{\text{mg}}{\text{g}} = \underline{\underline{15.60 \text{ mg saccharin/tablet}}}$$

13-24 no. mmol CH$_2$O = no. mmol KCN − no. mmol AgNO$_3$ + no. mmol NH$_4$SCN

$\qquad\qquad = 30.00 \times 0.121 - 40.0 \times 0.100 + 16.1 \times 0.134$

$\qquad\qquad = 1.787$

$$\frac{1.787 \text{ mmol CH}_2\text{O} \times 0.03003 \text{ g CH}_2\text{O/mmol CH}_2\text{O}}{5.00 \text{ g sample} \times 25.0 \text{ mL}/500 \text{ mL}} \times 100\% = 21.5\% \text{ CH}_2\text{O}$$

13-25 no. mmol AgNO$_3$ consumed by analyte =
$\qquad\qquad 25.00 \times 0.02979 - 2.85 \times 0.05411 = 0.59054$

$$0.59054 \text{ mmol AgNO}_3 \times \frac{1 \text{ mmol CHI}_3}{3 \text{ mmol AgNO}_3} \times \frac{1 \text{ mmol C}_{19}\text{H}_{16}\text{O}_4}{1 \text{ mmol CHI}_3} =$$

$\qquad\qquad\qquad\qquad\qquad\qquad\qquad\qquad 0.19685 \text{ mmol C}_{19}\text{H}_{16}\text{O}_4$

$$\frac{0.19685 \text{ mmol C}_{19}\text{H}_{16}\text{O}_4 \times 0.30834 \frac{\text{g C}_{19}\text{H}_{16}\text{O}_4}{\text{mmol}}}{13.96 \text{ g sample}} \times 100\% = \underline{\underline{0.4348\% \text{ C}_{19}\text{H}_{16}\text{O}_4}}$$

13-26 no. mmol AgNO$_3$ consumed by forming Ag$_2$Se = $25.00 \times 0.0360 - 16.74 \times 0.01370$ =

$\qquad\qquad\qquad\qquad\qquad\qquad\qquad\qquad\qquad\qquad 0.6707$

$$0.6707 \text{ mmol AgNO}_3 \times \frac{2 \text{ mmol Ag}_2\text{Se}}{4 \text{ mmol AgNO}_3} \times \frac{3 \text{ mmol Se}}{2 \text{ mmol Ag}_2\text{Se}} = 0.5030 \text{ mmol Se}$$

$$\frac{0.5030 \text{ mmol Se} \times 78.96 \text{ mg Se/mmol}}{5.00 \text{ mL sample}} = \underline{\underline{7.94 \text{ mg Se/mL sample}}}$$

13-27
$\qquad$ no. mmol KCl = $41.36 \text{ mL AgNO}_3 \times 0.05818 \dfrac{\text{mmol AgNO}_3}{\text{mL}} \times \dfrac{1 \text{ mmol KCl}}{\text{mmol AgNO}_3}$

$\qquad\qquad\qquad = 2.40632$

$\qquad$ no. mmol K$^+$ = $49.98 \text{ mL AgNO}_3 \times 0.05818 \dfrac{\text{mmol AgNO}_3}{\text{mL}} \times \dfrac{1 \text{ mmol K}^+}{\text{mmol AgNO}_3}$

$\qquad\qquad\qquad = 2.90784$

Chapter 13

$$\left(2.90784 \text{ mmol K}^+ - 2.40632 \text{ mmol KCl} \times \frac{1 \text{ mmol K}^+}{1 \text{ mmol KCl}}\right) \frac{1 \text{ mmol K}_2\text{SO}_4}{2 \text{ mmol K}^+} =$$

$$0.25076 \text{ mmol K}_2\text{SO}_4$$

$$\frac{2.4063 \text{ mmol KCl} \times 0.07455 \frac{\text{g KCl}}{\text{mmol}}}{2.4414 \text{ sample} \times 50.00 \text{ mL}/250.00 \text{ mL}} \times 100\% = \underline{\underline{36.74\% \text{ KCl}}}$$

$$\frac{0.25076 \text{ mmol K}_2\text{SO}_4 \times 0.17426 \frac{\text{g K}_2\text{SO}_4}{\text{mmol}}}{2.4414 \text{ sample} \times 50.00 \text{ mL}/250.0 \text{ mL}} \times 100\% = \underline{\underline{8.949\% \text{ K}_2\text{SO}_4}}$$

13-28

$$\text{no. mmol Cl}^- = 13.97 \text{ mL} \times 0.08551 \text{ mmol AgNO}_3 \times \frac{1 \text{ mmol Cl}^-}{\text{mL AgNO}_3} = 1.19457$$

$$\text{no. mmol Cl}^- + \text{no. mmol ClO}_4^- =$$

$$40.12 \text{ mL} \times 0.08551 \frac{\text{mmol AgNO}_3}{\text{mL}} \times \frac{\text{mmol Cl}^- + \text{mmol ClO}_4^-}{\text{mmol AgNO}_3} = 3.43066$$

$$3.43066 \text{ (mmol Cl}^- + \text{mmol ClO}_4^-) - 1.19457 \text{ mmol Cl}^- = 2.23609 \text{ mmol ClO}_4^-$$

$$\frac{1.19457 \text{ mmol Cl}^- \times 0.035453 \text{ g Cl}^-/\text{mmol Cl}^-}{1.998 \text{ g sample} \times 50.00 \text{ mL}/250 \text{ mL}} \times 100\% = \underline{\underline{10.60\% \text{ Cl}^-}}$$

$$\frac{2.23609 \times 0.09945 \text{ g ClO}_4^-/\text{mmol ClO}_4^-}{1.998 \text{ g sample} \times 50.00 \text{ mL}/250.0 \text{ mL}} \times 100\% = \underline{\underline{55.65\% \text{ ClO}_4^-}}$$

13-29 (a) The equivalence point occurs at 50.00 mL NH$_4$SCN.

At 30.00 mL

$$c_{\text{AgNO}_3} \approx [\text{Ag}^+] = \frac{25.00 \times 0.0500 - 30.00 \times 0.02500}{25.00 + 30.00} = \underline{\underline{9.09 \times 10^{-3}}}$$

$$\text{pAg} = -\log(9.09 \times 10^{-3}) = \underline{\underline{2.04}}$$

$$[\text{SCN}^-] = K_{sp}/9.09 \times 10^{-3} = 1.1 \times 10^{-12}/9.09 \times 10^{-3} = \underline{\underline{1.2 \times 10^{-10}}}$$

Proceeding in the same way, we obtain the data for **40.00 mL** and **49.00 mL**. These data are displayed in the table at the end of the solution.

Chapter 13

At 50.00 mL

$$[Ag^+] = [SCN^-] = \sqrt{K_{sp}} = \sqrt{1.1 \times 10^{-12}} = \underline{\underline{1.05 \times 10^{-6}}}$$

$$pAg = -\log(1.05 \times 10^{-6}) = \underline{\underline{5.98}}$$

At 51.00 mL

$$c_{NH_4SCN} \approx [SCN^-] = \frac{51.00 \times 0.02500 - 25.00 \times 0.05000}{76.00} = \underline{\underline{3.3 \times 10^{-4}}}$$

$$[Ag^+] = 1.1 \times 10^{-12} / 3.3 \times 10^{-4} = \underline{\underline{3.3 \times 10^{-9}}}$$

$$pAg = -\log(3.3 \times 10^{-9}) = \underline{\underline{8.48}}$$

At 60.00 and 70.00 mL, pAg computed in the same way. The results are found in the table at the end of this solution.

(b) Proceeding as in part (a), we obtain the results shown in the table that follows.

(c) Proceeding as in part (a), we obtain the results shown in the table that follows.

(d) The equivalence point occurs at a volume of 70.00 mL of $Pb(NO_3)_2$.

$$K_{sp} = [Pb^{2+}][SO_4^{2-}] = 1.6 \times 10^{-8}$$

At 50.00 mL

$$[SO_4^{2-}] = \frac{35.00 \times 0.4000 - 50.00 \times 0.2000}{85.00} = \underline{\underline{4.71 \times 10^{-2}}}$$

$$[Pb^{2+}] = 1.6 \times 10^{-8} / (4.71 \times 10^{-2}) = \underline{\underline{3.4 \times 10^{-7}}}$$

$$pPb = -\log(3.4 \times 10^{-7}) = \underline{\underline{6.47}}$$

At 60.00 and 69.00 mL, the data are obtained in the same way and are shown in the table at the end of the problem.

At 70.00 mL

$$[Pb^{2+}] = [SO_4^{2-}] = \sqrt{K_{sp}} = \sqrt{1.6 \times 10^{-8}} = \underline{\underline{1.3 \times 10^{-4}}}$$

Chapter 13

$$pPb = -\log(1.3 \times 10^{-4}) = \underline{3.90}$$

At 71.00 mL

$$[Pb^{2+}] = c_{PbNO_3} = \frac{71.00 \times 0.2000 - 35.00 \times 0.4000}{106.0} = \underline{\underline{1.89 \times 10^{-3}}}$$

$$[SO_4^{2-}] = 1.6 \times 10^{-8} / 1.89 \times 10^{-3} = \underline{\underline{8.5 \times 10^{-6}}}$$

$$pPb = -\log(1.89 \times 10^{-3}) \, 10^{-3} = \underline{2.72}$$

At 80.00 and 90.00 mL, the results are obtained in the same way and are given in the table that follows.

(e) Proceeding as in part (a), we obtain the data in the table that follows.

(f) Proceeding as in part (d), we obtain the data in the table that follows.

(a)

V_{NH_4SCN} (mL)	[Ag$^+$]	[SCN$^-$]	pAg
30.00	9.09×10^{-3}	1.2×10^{-10}	2.04
40.00	3.85×10^{-3}	2.9×10^{-10}	2.42
49.00	3.38×10^{-4}	3.3×10^{-9}	3.47
50.00	1.05×10^{-6}	1.05×10^{-6}	5.98
51.00	3.3×10^{-9}	3.3×10^{-4}	8.48
60.00	3.7×10^{-10}	2.94×10^{-3}	9.43
70.00	2.1×10^{-10}	5.26×10^{-3}	9.68

(b)

V_{KI} (mL)	[Ag$^+$]	[I$^-$]	pAg
20.00	1.50×10^{-2}	5.5×10^{-15}	1.82
30.00	6.00×10^{-3}	1.4×10^{-14}	2.22
39.00	5.08×10^{-4}	1.6×10^{-13}	3.29
40.00	9.1×10^{-9}	9.1×10^{-9}	8.04
41.00	1.7×10^{-13}	4.92×10^{-4}	12.77
50.00	1.9×10^{-14}	4.29×10^{-3}	13.71
60.00	1.1×10^{-14}	7.50×10^{-3}	13.96

(c)

V_{NaCl} (mL)	[Ag$^+$]	[Cl$^-$]	pAg
10.00	3.75×10^{-2}	4.85×10^{-9}	1.43
20.00	1.50×10^{-2}	1.21×10^{-8}	1.82
29.00	1.27×10^{-3}	1.43×10^{-7}	2.90
30.00	1.35×10^{-5}	1.35×10^{-5}	4.87
31.00	1.48×10^{-7}	1.23×10^{-3}	6.83
40.00	1.70×10^{-8}	1.07×10^{-2}	7.77
50.00	9.71×10^{-9}	1.88×10^{-2}	8.01

Chapter 13

(d)

$V_{Pb(NO_3)_2}$ (mL)	$[Pb^{2+}]$	$[SO_4^{2-}]$	pPb
50.00	3.4×10^{-7}	4.71×10^{-2}	6.47
60.00	7.6×10^{-7}	2.11×10^{-2}	6.12
69.00	8.3×10^{-6}	1.92×10^{-3}	5.08
70.00	1.3×10^{-4}	1.3×10^{-4}	3.90
71.00	1.89×10^{-3}	8.5×10^{-6}	2.72
80.00	1.74×10^{-2}	9.2×10^{-7}	1.76
90.00	3.20×10^{-2}	5.0×10^{-7}	1.49

(e)

$V_{Na_2SO_4}$ (mL)	$[Ba^{2+}]$	$[SO_4^{2-}]$	pBa
0.00	2.50×10^{-2}	0.0	1.60
10.00	1.00×10^{-2}	1.1×10^{-8}	2.00
19.00	8.48×10^{-4}	1.3×10^{-7}	3.07
20.00	1.05×10^{-5}	1.05×10^{-5}	4.98
21.00	1.3×10^{-7}	8.20×10^{-4}	6.87
30.00	1.5×10^{-8}	7.14×10^{-3}	7.81
40.00	8.8×10^{-9}	1.25×10^{-2}	8.06

(f)

V_{TlNO_3} (mL)	$[Tl^+]$	$[I^-]$	pTl
5.00	4.5×10^{-7}	1.45×10^{-1}	6.35
15.00	1.1×10^{-6}	6.15×10^{-2}	5.98
24.00	1.2×10^{-5}	5.40×10^{-3}	4.92
25.00	2.55×10^{-4}	2.55×10^{-4}	3.59
26.00	5.26×10^{-3}	1.2×10^{-5}	2.28
35.00	4.71×10^{-2}	1.4×10^{-6}	1.33
45.00	8.42×10^{-2}	7.7×10^{-7}	1.07

13-30 **At 5.00 mL**

$$\text{original no. mmol KBr} = 50.00 \text{ mL} \times 0.0400 \frac{\text{mmol KBr}}{\text{mL}} = 2.00$$

$$\text{no. mmol KBr consumed} = 5.00 \text{ mL AgNO}_3 \times 0.0500 \frac{\text{mmol AgNO}_3}{\text{mL}} = 0.250$$

$$[Br^-] = c_{Br} = \frac{(2.00 - 0.250) \text{ mmol KBr}}{(50.00 + 5.00) \text{ mL soln}} = 3.182 \times 10^{-2} \text{ mmol/mL}$$

$$[Ag^+] = K_{sp}/[Br^-] = 5.0 \times 10^{-13}/3.18 \times 10^{-2} = 1.572 \times 10^{-11} \text{ M}$$

$$pAg = -\log(1.572 \times 10^{-11}) = \underline{10.80}$$

The calculations through 39.00 mL of $AgNO_3$ are performed in the same way. The results are listed in the table at the end of this problem.

At 40.00 mL. This is the equivalence point when

$$[Br^-] = [Ag^+]$$

$$[Ag^+] = \sqrt{K_{sp}} = \sqrt{5.0 \times 10^{-13}} = 7.07 \times 10^{-7} \text{ mmol/mL}$$

$$pAg = -\log(7.07 \times 10^{-7}) = \underline{6.15}$$

Chapter 13

At 45.00 mL

$$\text{no. mmol AgNO}_3 \text{ added} = 45.00 \text{ mL} \times \frac{0.0500 \text{ mmol AgNO}_3}{\text{mL AgNO}_3} = 2.250$$

$$\text{no. mmol AgNO}_3 \text{ consumed} = 50.00 \text{ mL} \times \frac{0.0400 \text{ mmol KBr}}{\text{mL KBr}} \times \frac{1 \text{ mmol AgNO}_3}{1 \text{ mmol KBr}}$$

$$= 2.000$$

$$c_{\text{AgNO}_3} = [\text{Ag}^+] = \frac{(2.250 - 2.000) \text{ mmol AgNO}_3}{(45.00 + 50.00) \text{ mL soln}} = 2.632 \times 10^{-3} \text{ M}$$

$$\text{pAg} = -\log(2.63 \times 10^{-3}) = \underline{2.58}$$

The calculation for **50.00 mL** is performed in the same way.

The results are in the following table.

Vol AgNO$_3$, mL	[Ag$^+$]	pAg
5.00	1.6×10^{-11}	10.80
15.00	2.6×10^{-11}	10.58
25.00	5.0×10^{-11}	10.30
30.00	8.0×10^{-11}	10.10
35.00	1.7×10^{-10}	9.77
39.00	8.9×10^{-10}	9.05
40.00	7.1×10^{-7}	6.15
45.00	2.6×10^{-3}	2.58
50.00	5.0×10^{-3}	2.30

CHAPTER 14

14-1 (a) A *chelate* is a cyclic complex consisting of metal ion and a reagent that contains two or more electron donor groups located in such a position that they can bond with the metal ion to form a heterocyclic ring structure.

(b) A *tetradentate chelating agent* is a molecule that contains four pairs of donor electrons located in such positions that they all can bond to a metal atom, thus forming two rings.

(c) A *ligand* is a species that contains one or more electron pair donor groups that tend to form bonds with metal ions.

(d) The *coordination number* is the number of covalent bonds that a cation tends to form with electron donor groups.

(e) A *conditional formation constant* is an equilibrium constant for the reaction between a metal ion and a complexing agent that applies only when the pH and/or the concentration of other complexing ions are carefully specified.

(f) *NTA* is the acronym for nitrilotriacetic acid, a tetradentate complexing agent that contains three carboxylate groups and one tertiary amine. As an electron donor, NYA has found applications in the titration of a variety of cations.

(g) *Water hardness* is the concentration of calcium carbonate that is equivalent to the total concentration of all of the multivalent metal carbonates in the water.

(h) In an *EDTA displacement titration*, an unmeasured excess of a solution containing the magnesium or zinc complex of EDTA is introduced into the solution of an analyte that forms a more stable complex than that of magnesium or zinc. The liberated zinc or magnesium ions are then titrated with a standard solution of EDTA. Displacement titrations are used for the determination of cations for which no good indicator exists.

14-2 Three general methods for performing EDTA titrations are (1) direct titration, (2) back titration, and (3) displacement titration. Method (1) is simple, rapid, and requires but one standard reagent. Method (2) is advantageous for those metals that react so slowly with EDTA as to make direct titration inconvenient. In addition, this procedure is useful for cations for which satisfactory indicators are not available. Finally, it is useful for

Chapter 14

analyzing samples that contain anions that form sparingly soluble precipitates with the analyte under the analytical conditions. Method (3) is particularly useful in situations where no satisfactory indicators are available for direct titration.

14-3 Multidentate ligands offer the advantage that they usually form more stable complexes than do unidentate ligands. Furthermore, they often form but a single complex with the cation, which simplifies their titration curves and makes end-point detection easier.

14-4 (a)
$$Ag^+ + S_2O_3^{2-} \rightleftharpoons AgS_2O_3^- \qquad K_1 = \frac{[AgS_2O_3^-]}{[Ag^+][S_2O_3^{2-}]}$$

$$AgS_2O_3^- + S_2O_3^{2-} \rightleftharpoons Ag(S_2O_3)_2^{3-} \qquad K_2 = \frac{[Ag(S_2O_3)_2^{3-}]}{[AgS_2O_3^-][S_2O_3^{2-}]}$$

(b)
$$Ni^{2+} + SCN^- \rightleftharpoons NiSCN^+ \qquad K_1 = \frac{[NiSCN^+]}{[Ni^{2+}][SCN^-]}$$

$$NiSCN^+ + SCN^- \rightleftharpoons Ni(SCN)_2 \qquad K_2 = \frac{[Ni(SCN)_2]}{[NiSCN^+][SCN^-]}$$

$$Ni(SCN)_2 + SCN^- \rightleftharpoons Ni(SCN)_3^- \qquad K_3 = \frac{[Ni(SCN)_3^-]}{[Ni(SCN)_2][SCN^-]}$$

14-5 The overall formation constant β_n is equal to the product of the individual stepwise constants. Thus the overall constant for formation of $Ni(SCN)_3^-$ in Example 14-4(b) is

$$\beta_3 = K_1 K_2 K_3 = \frac{[Ni(SCN)_3^-]}{[Ni^{2+}][SCN^-]^3}$$

which is the equilibrium constant for the reaction

$$Ni^{2+} + 3SCN^- \rightleftharpoons Ni(SCN)_3^-$$

and

$$\beta_2 = K_1 K_2 = \frac{Ni(SCN)_2}{[Ni^{2+}][SCN^-]^2}$$

where the overall constant β_2 is for the equation

Chapter 14

$$Ni^{2+} + 2SCN^- \rightleftarrows Ni(SCN)_2$$

14-6 Titrate the three ions in an aliquot of the sample that has been buffered to a pH of about 10. Buffer a second aliquot to a pH of about 4 and titrate the zinc and indium ions. Finally, titrate an aliquot that has been brought to a pH of about 1.5. Only the indium is complexed under this condition.

14-7 The MgY^{2-} is added to assure a sufficient analytical concentration of Mg^{2+} to provide a sharp end point with Eriochrome Black T indicator.

14-8
$$3.853 \text{ g reagent} \times \frac{99.7 \text{ g Na}_2\text{H}_2\text{Y} \cdot 2\text{H}_2\text{O}}{100 \text{ g reagent}} \times \frac{1 \text{ mol EDTA}}{372.24 \text{ g Na}_2\text{H}_2\text{Y} \cdot 2\text{H}_2\text{O}} \times \frac{1}{1.000 \text{ L}} =$$

$$\underline{\underline{0.01032 \text{ M EDTA}}}$$

14-9
$$50.00 \text{ mL} \times \frac{0.004517 \text{ mmol Mg}^{2+}}{\text{mL Mg}^{2+}} \times \frac{1 \text{ mmol EDTA}}{\text{mmol Mg}^{2+}} \times \frac{1}{32.22 \text{ mL EDTA}} = \underline{\underline{0.07010 \text{ M}}}$$

14-10 (a)
$$26.37 \text{ mL} \times 0.0741 \frac{\text{mmol Mg(NO}_3)_2}{\text{mL}} \times \frac{1 \text{ mmol EDTA}}{\text{mmol Mg(NO}_2)_2} \times \frac{1 \text{ mL EDTA}}{0.0500 \text{ mmol EDTA}} =$$

$$\underline{\underline{39.1 \text{ mL EDTA}}}$$

(b)
$$0.2145 \text{ g CaCO}_3 \times \frac{1 \text{ mmol CaCO}_3}{0.10009 \text{ g CaCO}_3} \times \frac{1 \text{ mmol EDTA}}{\text{mmol CaCO}_3} \times \frac{1 \text{ mL EDTA}}{0.0500 \text{ mmol EDTA}} =$$

$$\underline{\underline{42.9 \text{ mL EDTA}}}$$

(c) Letting A symbolize $CaHPO_4 \cdot 2H_2O$

$$(0.4397 \times 0.814) \text{ g A} \times \frac{1 \text{ mmol A}}{0.17209 \text{ g A}} \times \frac{1 \text{ mmol EDTA}}{\text{mmol A}} \times \frac{1 \text{ mL EDTA}}{0.0500 \text{ mmol EDTA}} =$$

$$\underline{\underline{41.6 \text{ mL EDTA}}}$$

(d) Proceeding in the same way, we obtain

$$\frac{0.2080}{0.3653} \text{ mmol A} \times \frac{4 \text{ mmol EDTA}}{1 \text{ mmol A}} \times \frac{1 \text{ mL EDTA}}{0.0500 \text{ mmol EDTA}} = \underline{\underline{45.6 \text{ mL EDTA}}}$$

Chapter 14

(e) Proceeding as in part (c), we write

$$(0.1557 \times 0.925) \text{ g A} \times \frac{\text{mmol A}}{0.1844 \text{ g}} \times \frac{2 \text{ mmol EDTA}}{1 \text{ mmol A}} \times \frac{1 \text{ mL EDTA}}{0.0500 \text{ mmol EDTA}} =$$

$$\underline{\underline{31.2 \text{ mL EDTA}}}$$

14-11

$$M_{CoSO_4} = 1.694 \frac{\text{mg CoSO}_4}{\text{mL CoSO}_4} \times 10^{-3} \frac{\text{g}}{\text{mg}} \times \frac{1 \text{ mmol CoSO}_4}{0.1550 \text{ g CoSO}_4} = 0.010929$$

In each part,

$$\text{no. mmol CoSO}_4 = 25.00 \text{ mL} \times 0.010929 \frac{\text{mmol CoSO}_4}{\text{mL CoSO}_4} = 0.27323$$

(a) $$0.27323 \text{ mmol CoSO}_4 \times \frac{1 \text{ mmol EDTA}}{\text{mmol CoSO}_4} \times \frac{1 \text{ mL EDTA}}{0.008640 \text{ mmol EDTA}} = \underline{\underline{31.62 \text{ mL EDTA}}}$$

(b) no. mmol excess EDTA = $50.00 \times 0.008640 - 0.27323 = 0.15877$

$$0.15877 \text{ mmol EDTA} \times \frac{1 \text{ mmol Zn}^{2+}}{\text{mmol EDTA}} \times \frac{1 \text{ mL Zn}^{2+}}{0.009450 \text{ mmol Zn}^{2+}} = \underline{\underline{16.80 \text{ mL Zn}^{2+}}}$$

(c) $$0.27323 \text{ mmol CoSO}_4 \times \frac{1 \text{ mmol Zn}^{2+}}{\text{mmol CoSO}_4} \times \frac{1 \text{ mmol EDTA}}{\text{mmol Zn}^{2+}} \times \frac{1 \text{ mmol EDTA}}{0.008640 \text{ mmol EDTA}} =$$

$$\underline{\underline{31.62 \text{ mL EDTA}}}$$

14-12

$$\frac{(21.27 \times 0.01645) \text{ mmol EDTA} \times \frac{1 \text{ mmol Zn}}{\text{mmol EDTA}} \times \frac{0.06539 \text{ g Zn}}{\text{mmol Zn}}}{0.7556 \text{ g sample}} \times 100\% = \underline{\underline{3.028\% \text{ Zn}}}$$

14-13

$$(15.00 \times 0.01768 - 4.30 \times 0.008120) \frac{\text{mmol Cr}}{3.00 \times 4.00 \text{ cm}^2} \times \frac{51.996 \text{ mg Cr}}{\text{mmol Cr}} = \underline{\underline{0.998 \frac{\text{mg Cr}}{\text{cm}^2}}}$$

14-14

$$\frac{(13.34 \times 0.03560) \text{ mmol EDTA}}{9.76 \text{ g sample}} \times \frac{1 \text{ mmol Tl}_2\text{SO}_4}{2 \text{ mmol EDTA}} \times \frac{0.5048 \text{ g Tl}_2\text{SO}_4}{\text{mmol Tl}_2\text{SO}_4} \times 100\% =$$

$$\underline{\underline{1.228\% \text{ Tl}_2\text{SO}_4}}$$

14-15 (a) $0.7682 \dfrac{\text{g MgCO}_3}{1000 \text{ mL}} \times \dfrac{1 \text{ mmol MgCO}_3}{0.084314 \text{ g MgCO}_3} \times 50.0 \text{ mL MgCO}_3 \times \dfrac{1 \text{ mmol EDTA}}{\text{mmol MgCO}_3} \times$

$\dfrac{1}{42.35 \text{ mL EDTA}} = 0.010757 = \underline{\underline{0.01076 \text{ M EDTA}}}$

(b) $\dfrac{\text{no. mmol CaCO}_3 + \text{no. mmol MgCO}_3}{\text{mL sample}} = \dfrac{18.81 \times 0.010757}{25.00} = 8.0943 \times 10^{-3}$

no. mmol CaCO$_3$/mL sample $= 31.54 \times 0.010757/50.00 = 6.7855 \times 10^{-3}$

no. mmol MgCO$_3$/mL sample $= (8.0943 - 6.7855) \times 10^{-3} = 1.3088 \times 10^{-3}$

$6.7855 \times 10^{-3} \dfrac{\text{mmol CaCO}_3}{\text{mL sample}} \times \dfrac{0.10009 \text{ g CaCO}_3}{\text{mmol}} \times \dfrac{1.00 \text{ mL sample}}{\text{g sample}} \times 10^6 \text{ ppm} =$

$\underline{\underline{679.2 \text{ ppm CaCO}_3}}$

(c) $1.3082 \times 10^{-3} \dfrac{\text{mmol MgCO}_3}{\text{mL sample}} \times \dfrac{0.08431 \text{ g MgCO}_3}{\text{mmol}} \times \dfrac{1.000 \text{ mL sample}}{\text{g sample}} \times 10^6 \text{ ppm} =$

$\underline{\underline{110.3 \text{ ppm MgCO}_3}}$

14-16 no. mmol Fe^{3+} $= (13.73 \times 0.01200) \text{ mmol EDTA} \times \dfrac{1 \text{ mmol Fe}^{2+}}{\text{mmol EDTA}} = 0.16476$

no. mmol Fe^{2+} $= (29.62 - 13.73) \text{ mL EDTA} \times 0.01200 \dfrac{\text{mmol EDTA}}{\text{mL EDTA}} \times \dfrac{1 \text{ mmol Fe}^{2+}}{\text{mmol EDTA}} =$

0.19068

$\dfrac{0.16476 \text{ mmol Fe}^{3+} \times 55.847 \text{ mg Fe}^{3+}/\text{mmol}}{50.00 \text{ mL} \times 10^{-3} \text{ L/mL}} = 184.0 \dfrac{\text{mg Fe}^{3+}}{\text{L}} = \underline{\underline{184.0 \text{ ppm Fe}^{3+}}}$

Similarly,

$0.19068 \times 55.847 / 0.05000 = \underline{\underline{213.0 \text{ ppm Fe}^{2+}}}$

14-17 no. mmol Mg^{2+} + no. mmol Ca^{2+} $= 26.81 \times 0.003474 = 9.3138 \times 10^{-2}$

no. mmol Ca^{2+} $= 11.63 \times 0.003474 = 4.0403 \times 10^{-2}$

Chapter 14

$$\text{no. mmol Mg}^{2+} = (9.3138 - 4.0403) \times 10^{-2} = 5.2735 \times 10^{-2}$$

$$4.0403 \times 10^{-2} \text{ mmol Ca}^{2+} \times \frac{40.08 \text{ mg Ca}^{2+}}{\text{mmol}} \times \frac{2000 \text{ mL}}{10 \text{ mL}} = \underline{\underline{323.9 \text{ mg Ca}^{2+}}}$$

$$5.2735 \times 10^{-2} \text{ mmol Mg}^{2+} \times \frac{24.305 \text{ mg Mg}^{2+}}{\text{mmol}} \times \frac{2000 \text{ mL}}{10 \text{ mL}} = \underline{\underline{256.3 \text{ mg Mg}^{2+}}}$$

Both within normal range.

14-18 no. mmol (Cd^{2+} + Pb^{2+}) = 28.89 × 0.06950 = 2.0079

no. mmol Pb^{2+} = 11.56 × 0.06950 = $\underline{0.8034}$

no. mmol Cd^{2+} = 1.2045

$$\frac{0.8034 \text{ mmol Pb}^{2+} \times 0.2072 \text{ g Pb/mmol Pb}}{1.509 \text{ g sample} \times 50.00 \text{ mL}/250.0 \text{ mL}} \times 100\% = \underline{\underline{55.16\% \text{ Pb}}}$$

Similarly,

$$\frac{1.2045 \times 0.11241}{1.509 \times 50.00/250.0} \times 100 = \underline{\underline{44.86\% \text{ Cd}}}$$

14-19 no. mmol (Ni^{2+} + Cu^{2+}) = 45.81 × 0.05285 = 2.4211

no. mmol Cu^{2+} = 22.85 × 0.07238 = $\underline{1.6539}$

no. mmol Ni^{2+} = 0.7672

$$\frac{1.6539 \text{ mmol Cu} \times 0.063546 \text{ g Cu/mmol Cu}}{0.6004 \text{ g sample} \times 25.00 \text{ mL}/100.0 \text{ mL}} \times 100\% = \underline{\underline{70.02\% \text{ Cu}}}$$

$$\frac{0.7672 \times 0.058693}{0.6004 \times 25.00/100.0} = \underline{\underline{30.00\% \text{ Ni}}}$$

14-20 $$\frac{(38.71 \times 0.01294) \text{ mmol EDTA} \times \frac{1 \text{ mmol ZnO}}{\text{mmol EDTA}} \times \frac{0.08139 \text{ g ZnO}}{\text{mmol ZnO}}}{1.022 \text{ g sample} \times 10.00 \text{ mL}/250.0 \text{ mL}} \times 100\% = \underline{\underline{99.7\% \text{ ZnO}}}$$

$$\frac{2.40 \text{ mL} \times 0.002727 \frac{\text{mmol ZnY}^{2-}}{\text{mL}} \times \frac{1 \text{ mmol Fe}_2\text{O}_3}{2 \text{ mmol ZnY}^{2-}} \times \frac{0.15969 \text{ g Fe}_2\text{O}_3}{\text{mmol Fe}_2\text{O}_3}}{1.022 \text{ g sample} \times 50.00 \text{ mL}/250.0 \text{ mL}} \times 100\% = \underline{\underline{0.256\% \text{ Fe}_2\text{O}_3}}$$

Chapter 14

14-21 1 mmol EDTA ≡ 1 mmol Ni^{2+} ≡ 2 mmol NaBr ≡ 2 mmol $NaBrO_3$

For the 10.00 mL aliquot,

$$\frac{\text{no. mmol NaBr} + \text{no. mmol NaBrO}_3}{\text{mL sample solution}} = \frac{21.94 \times 0.02089}{10.00} \times \frac{2}{1} = 9.1665 \times 10^{-2}$$

For the 25.00 mL aliquot,

$$\frac{\text{no. mmol NaBr}}{\text{mL sample soln}} = \frac{26.73 \times 0.02089}{25.00} \times \frac{2}{1} = 4.4671 \times 10^{-2}$$

no. mmol $NaBrO_3$ / mL = (9.1665 − 4.4671) × 10^{-2} = 4.6994 × 10^{-2}

$$\frac{4.4671 \times 10^{-2} \frac{\text{mmol NaBr}}{\text{mL}} \times 250.0 \text{ mL} \times 0.1029 \frac{\text{g NaBr}}{\text{mmol}}}{3.650 \text{ g sample}} \times 100\% = \underline{\underline{31.48\% \text{ NaBr}}}$$

$$\frac{4.6994 \times 10^{-2} \times 250.0 \times 0.1509}{3.650} \times 100\% = \underline{\underline{48.57\% \text{ NaBrO}_3}}$$

14-22 1 mmol Mg^{2+} ≡ 1 mmol EDTA ≡ 1/4 mmol $B(C_6H_5)_4^-$ ≡ 1/4 mmol K^+

$$(29.64 \times 0.05581) \text{ mmol Mg}^{2+} \times \frac{1 \text{ mmol K}^+}{4 \text{ mmol Mg}^{2+}} \times \frac{39.098 \text{ mg K}^+}{\text{mmol K}^+} \times \frac{1}{250 \times 10^{-3} \text{ L sample}} =$$

$$64.68 \text{ mg K}^+/\text{L} = \underline{\underline{64.68 \text{ ppm K}^+}}$$

14-23 no. mmol (Ni + Fe + Cr) in sample =

(50.00 × 0.05182 − 5.11 × 0.06241) × 250.0 / 50.00 = 11.360

no. mmol (Ni + Fe) in sample =

(36.28 × 0.05182) × 250.0 / 50.00 = 9.4001

no. mmol Cr in sample = 11.360 − 9.4001 = 1.9603

no. mmol Ni in sample = 25.91 × 0.05182 × 250 / 50.00 = 6.7133

no. mmol Fe in sample = 9.4001 − 6.7133 = 2.6868

% Cr = (1.9603 × 0.051996 / 0.6472) × 100 % = $\underline{\underline{15.75\%}}$

215

Chapter 14

$$\% \text{ Ni} = (6.7133 \times 0.05869 / 0.6472) \times 100\ \% = \underline{60.88\%}$$

$$\% \text{ Fe} = (2.6868 \times 0.055847 / 0.6472) \times 100\ \% = \underline{23.18\%}$$

14-24 no. mmol (Pb + Zn + Cu) in sample =

$$37.56 \times 0.002500 \times 500.0 / 10.00 = 4.6950$$

no. mmol (Pb + Zn) in sample =

$$27.67 \times 0.002500 \times 500.0 / 25.00 = 1.3835$$

no. mmol Pb in sample = $10.80 \times 0.002500 \times 500.0 / 100.00 = 0.1350$

no. mmol Zn in sample = $1.3835 - 0.1350 = 1.2485$

no. mmol Cu in sample = $4.6950 - 1.3835 = 3.3115$

$$\% \text{ Pb} = (0.1350 \times 0.2072 / 0.3284) \times 100\ \% = \underline{8.518\%}$$

$$\% \text{ Zn} = (1.2485 \times 0.06539 / 0.3284) \times 100\ \% = \underline{24.86\%}$$

$$\% \text{ Cu} = (3.3115 \times 0.06355 / 0.3284) \times 100\ \% = \underline{64.08\%}$$

$$\% \text{ Sn} = 100 - 8.518 - 24.86 - 64.08 = \underline{2.54\%}$$

14-25 $K'_{MY} = \alpha_4 K_{MY}$ (Equation 14 – 4)

For Fe(II), $K_{MY} = 2.1 \times 10^{14}$ (Table 14-1)

(a) At pH = 6.0, $\alpha_4 = 2.2 \times 10^{-5}$ (Table 14-2)

$$K'_{FeY} = 2.2 \times 10^{-5} \times 2.1 \times 10^{14} = \underline{4.6 \times 10^9}$$

(b) $5.4 \times 10^{-3} \times 2.1 \times 10^{14} = \underline{1.1 \times 10^{12}}$

(c) $3.5 \times 10^{-1} \times 2.1 \times 10^{14} = \underline{7.4 \times 10^{13}}$

14-26 Proceeding as in Solution 14-25, we obtain for K'_{BaY} at pH 7.0, 9.0, and 11.0:

(a) $K'_{BaY} = 4.8 \times 10^{-4} \times 5.8 \times 10^7 = \underline{2.8 \times 10^4}$

Chapter 14

(b) $K'_{BaY} = 5.2 \times 10^{-2} \times 5.8 \times 10^7 = \underline{\underline{3.0 \times 10^6}}$

(c) $K'_{BaY} = 0.85 \times 5.8 \times 10^7 = \underline{\underline{4.9 \times 10^7}}$

14-27 $K_{SrY} = 4.3 \times 10^8$ (Table 14-1) $\alpha_4 = 0.85$ (Table 14-2)

$$K'_{SrY} = 4.3 \times 10^8 \times 0.85 = 3.66 \times 10^8 = \frac{[SrY^{2-}]}{[Sr^{2+}]c_T} \quad \text{(Equation 14-4)}$$

0.00 mL

$[Sr^{2+}] = 0.01000$ and $pSr = -\log 0.01000 = \underline{\underline{2.00}}$

10.00 mL

$[Sr^{2+}] = (50.00 \times 0.01000 - 10.00 \times 0.02000)/60.00 = 5.00 \times 10^{-3}$

$pSr = -\log 5.00 \times 10^{-3} = \underline{\underline{2.30}}$

24.00 and 24.90 mL. Proceeding in the same way, we obtain

$[Sr^{2+}] = 2.70 \times 10^{-4}$ and $pSr = \underline{\underline{3.57}}$

$[Sr^{2+}] = 2.67 \times 10^{-5}$ and $pSr = \underline{\underline{4.57}}$

25.00 mL

Here, $[Sr^{2+}] = c_T$ and

$[SrY^{2-}] = (50.00 \times 0.01000)/75.00 - [Sr^{2+}] \approx 6.67 \times 10^{-3}$

$K'_{SrY} = 3.66 \times 10^8 = 6.67 \times 10^{-3}/(c_T[Sr^{2+}]) = 6.67 \times 10^{-3}/[Sr^{2+}]^2$

$[Sr^{2+}] = \sqrt{6.67 \times 10^{-3}/3.66 \times 10^8} = 4.27 \times 10^{-6}$

$pSr = -\log 4.27 \times 10^{-6} = \underline{\underline{5.37}}$

25.10 mL

$c_T = (25.10 \times 0.02000 - 50.00 \times 0.01000)/75.10 = 2.66 \times 10^{-5}$

$[Sr^{2+}] = (50.00 \times 0.01000)/75.10 = 6.66 \times 10^{-3}$

Chapter 14

$$\frac{6.66 \times 10^{-3}}{[Sr^{2+}] \times 2.66 \times 10^{-5}} = 3.66 \times 10^{8}$$

$[Sr^{2+}] = 6.66 \times 10^{-3} / (2.66 \times 10^{-5} \times 3.66 \times 10^{8}) = 6.84 \times 10^{-7}$

$pSr = -\log 6.84 \times 10^{-7} = \underline{6.16}$

26.00 and 30.00. Proceeding in the same way, we obtain

$[Sr^{2+}] = 6.84 \times 10^{-8}$ and $pSr = \underline{7.16}$

$[Sr^{2+}] = 1.37 \times 10^{-8}$ and $pSr = \underline{7.86}$

14-28 $K_{Fe(II)Y} = 2.1 \times 10^{14}$ and $\alpha_4 = 4.8 \times 10^{-4}$

$K'_{Fe(II)Y} = 2.1 \times 10^{14} \times 4.8 \times 10^{-4} = 1.01 \times 10^{11}$

Proceeding in the same way as in Problem 14-27, we obtain the following data.

Vol 0.03000 M EDTA, mL	$[Fe^{2+}]$	pFe(II)	Vol 0.03000 M EDTA, mL	$[Fe^{2+}]$	pFe(II)
0.00	1.50×10^{-2}	1.82	25.00	3.15×10^{-7}	6.50
10.00	7.50×10^{-3}	2.12	25.10	2.48×10^{-9}	8.61
24.00	4.05×10^{-4}	3.39	26.00	2.48×10^{-10}	9.61
24.90	4.00×10^{-5}	4.40	30.00	4.96×10^{-11}	10.30

CHAPTER 15

15-1 (a) *Oxidation* is a process in which a species loses one or more electrons.

(b) An *oxidizing agent* is an electron acceptor.

(c) A *salt bridge* is a device that provides electrical contact but prevents mixing of dissimilar solutions in an electrochemical cell.

(d) A *liquid junction* is the interface between dissimilar liquids. A potential develops across this interface.

(e) The *Nernst equation* relates the potential to the concentrations (strictly, activities) of the participants in an electrochemical reaction.

15-2 (a) The *electrode potential* is the potential of an electrochemical cell in which a standard hydrogen electrode acts as anode and the half-cell of interest is the cathode.

(b) The *formal potential* of a half-reaction is the potential of the system (measured against the SHE) when the concentration of each solute participating in the half-reaction has a concentration of exactly one molar and the concentrations of all other constituents of the solution are carefully specified.

(c) The *standard electrode potential* for a half-reaction is the potential of a *cell* consisting of a cathode at which that half-reaction is occurring and a standard hydrogen electrode behaving as the anode. The activities of all of the participants in the half-reaction are specified as having a value of unity. The additional specification that the standard hydrogen electrode is the anode implies that the standard potential for a half-reaction is always a *reduction potential*.

(d) A *liquid junction potential* is the potential that develops across the interface between two dissimilar solutions.

(e) An *oxidation potential* is the potential of an electrochemical cell in which the cathode is a standard hydrogen electrode and the half-cell of interest acts as anode.

15-3 (a) *Reduction* is the process whereby a substance acquires electrons; a *reducing agent* is a supplier of electrons.

(b) A *galvanic cell* is one in which a spontaneous electrochemical reaction occurs and is thus a source of energy. The reaction in an *electrolytic cell* is forced in a nonspontaneous direction through application of an external source of electrical energy.

Chapter 15

(c) The *anode* of an electrochemical cell is the electrode at which oxidation occurs. The *cathode* is the electrode at which reduction occurs.

(d) In a *reversible cell*, alteration of the direction of the current simply causes a reversal in the electrochemical process. In an *irreversible cell*, reversal of the current results in a different reaction at one or both of the electrodes.

(e) The *standard electrode potential* is the potential of an electrochemical cell in which the standard hydrogen electrode acts as an anode and all participants in the cathode process have unit activity. The formal potential differs in that the molar *concentrations* of the reactants and products are unity and the concentration of other species in the solution are carefully specified.

15-4 The first standard potential is for a solution that is saturated with I_2, which has an $I_2(aq)$ activity significantly less than one. The second potential is for a *hypothetical* half-cell in which the $I_2(aq)$ activity is unity. Such a half-cell, if it existed, would have a greater potential since the driving force for the reduction would be greater at the higher I_2 concentration. The second half-cell potential, although hypothetical, is nevertheless useful for calculating electrode potentials for solutions that are undersaturated in I_2.

15-5 It is necessary to bubble hydrogen through the electrolyte in a hydrogen electrode in order to keep the solution saturated with the gas. Only under these circumstances is the hydrogen activity constant so that the electrode potential is constant and reproducible.

15-6 The potential in the presence of base would be more negative because the nickel ion activity in this solution would be far less than 1 M. Consequently the driving force for the reduction if Ni(II) to the metallic state would also be far less, and the electrode potential would be significantly more negative. (In fact the standard electrode potential for the reaction $Ni(OH)_2(s) + 2e^- \rightleftarrows Ni(s) + 2OH^-$ has a value of -0.72 V.)

15-7 (a) $2Fe^{3+} + Sn^{2+} \rightarrow 2Fe^{2+} + Sn^{4+}$

(b) $Cr(s) + 3Ag^+ \rightarrow Cr^{3+} + 3Ag(s)$

(c) $2NO_3^- + Cu(s) + 4H^+ \rightarrow 2NO_2(g) + 2H_2O + Cu^{2+}$

(d) $2MnO_4^- + 5H_2SO_3 \rightarrow 2Mn^{2+} + 5SO_4^{2-} + 4H^+ + 3H_2O$

(e) $Ti^{3+} + Fe(CN)_6^{3-} + H_2O \rightarrow TiO^{2+} + Fe(CN)_6^{4-} + 2H^+$

Chapter 15

(f) $H_2O_2 + 2Ce^{4+} \rightarrow O_2(g) + 2Ce^{3+} + 2H^+$

(g) $2Ag(s) + 2I^- + Sn^{4+} \rightarrow 2AgI(s) + Sn^{2+}$

(h) $UO_2^{2+} + Zn(s) + 4H^+ \rightarrow U^{4+} + Zn^{2+} + 2H_2O$

(i) $5HNO_2 + 2MnO_4^- + H^+ \rightarrow 5NO_3^- + 2Mn^{2+} + 3H_2O$

(j) $H_2NNH_2 + IO_3^- + 2H^+ + 2Cl^- \rightarrow N_2(g) + ICl_2^- + 3H_2O$

15-8 (a) Oxidizing Agent Fe^{3+}; $Fe^{3+} + e^- \rightleftarrows Fe^{2+}$

Reducing Agent Sn^{2+}; $Sn^{2+} \rightleftarrows Sn^{4+} + 2e^-$

(b) Oxidizing Agent Ag^+; $Ag^+ + e^- \rightleftarrows Ag(s)$

Reducing Agent Cr; $Cr(s) \rightleftarrows Cr^{3+} + 3e^-$

(c) Oxidizing Agent NO_3^-; $NO_3^- + 2H^+ + e^- \rightleftarrows NO_2(g) + H_2O$

Reducing Agent Cu; $Cu(s) \rightleftarrows Cu^{2+} + 2e^-$

(d) Oxidizing Agent MnO_4^-; $MnO_4^- + 8H^+ + 5e^- \rightleftarrows Mn^{2+} + 4H_2O$

Reducing Agent H_2SO_3; $H_2SO_3 + H_2O \rightleftarrows SO_4^{2-} + 4H^+ + 2e^-$

(e) Oxidizing Agent $Fe(CN)_6^{3-}$; $Fe(CN)_6^{3-} + e^- \rightleftarrows Fe(CN)_6^{4-}$

Reducing Agent Ti^{3+}; $Ti^{3+} + H_2O \rightleftarrows TiO^{2+} + 2H^+ + e^-$

(f) Oxidizing Agent Ce^{4+}; $Cu^{4+} + e^- \rightleftarrows Ce^{3+}$

Reducing Agent H_2O_2; $H_2O_2 \rightleftarrows O_2(g) + 2H^+ + 2e^-$

(g) Oxidizing Agent Sn^{4+}; $Sn^{4+} + 2e^- \rightleftarrows Sn^{2+}$

Reducing Agent Ag; $Ag(s) + I^- \rightleftarrows AgI(s) + e^-$

Chapter 15

    (h) Oxidizing Agent UO_2^{2+}; $UO_2^{2+} + 4H^+ + 2e^- \rightleftarrows U^{4+} + 2H_2O$

        Reducing Agent Zn; $Zn(s) \rightleftarrows Zn^{2+} + 2e^-$

    (i) Oxidizing Agent MnO_4^-; $MnO_4^- + 8H^+ + 5e^- \rightleftarrows Mn^{2+} + 4H_2O$

        Reducing Agent HNO_2; $HNO_2 + H_2O \rightleftarrows NO_3^- + 3H^+ + 2e^-$

    (j) Oxidizing Agent IO_3^-; $IO_3^- + 2Cl^- + 6H^+ + 4e^- \rightleftarrows ICl_2^- + 3H_2O$

        Reducing Agent H_2NNH_2; $N_2H_4 \rightleftarrows N_2(g) + 4H^+ + 4e^-$

15-9 (a) $MnO_4^- + 5VO^{2+} + 11H_2O \rightarrow Mn^{2+} + 5V(OH)_4^+ + 2H^+$

    (b) $I_2 + H_2S(g) \rightarrow 2I^- + S(s) + 2H^+$

    (c) $Cr_2O_7^{2-} + 3U^{4+} + 2H^+ \rightarrow 2Cr^{3+} + 3UO_2^{2+} + H_2O$

    (d) $2Cl^- + MnO_2(s) + 4H^+ \rightarrow Cl_2(g) + Mn^{2+} + 2H_2O$

    (e) $IO_3^- + 5I^- + 6H^+ \rightarrow 3I_2 + H_2O$

    (f) $IO_3^- + 2I^- + 6Cl^- + 6H^+ \rightarrow 3ICl_2^- + 3H_2O$

    (g) $HPO_3^{2-} + 2MnO_4^- + 3OH^- \rightarrow PO_4^{3-} + 2MnO_4^{2-} + 2H_2O$

    (h) $SCN^- + BrO_3^- + H_2O \rightarrow SO_4^{2-} + HCN + Br^- + H^+$

    (i) $V^{2+} + 2V(OH)_4^+ + 2H^+ \rightarrow 3VO^{2+} + 5H_2O$

    (j) $2MnO_4^- + 3Mn^{2+} + 4OH^- \rightarrow 5MnO_2(s) + 2H_2O$

15-10 (a) Oxidizing Agent MnO_4^-; $MnO_4^- + 8H^+ + 5e^- \rightleftarrows Mn^{2+} + 4H_2O$

        Reducing Agent VO^{2+}; $VO^{2+} + 3H_2O \rightleftarrows V(OH)_4^+ + 2H^+ + e^-$

(b) Oxidizing Agent I_2 ; $\quad I_2(aq) + 2e^- \rightleftharpoons 2I^-$

Reducing Agent H_2S ; $\quad H_2S(g) \rightleftharpoons S(s) + 2H^+ + 2e^-$

(c) Oxidizing Agent $Cr_2O_7^{2-}$; $\quad Cr_2O_7^{2-} + 14H^+ + 6e^- \rightleftharpoons 2Cr^{3+} + 7H_2O$

Reducing Agent U^{4+} ; $\quad U^{4+} + 2H_2O \rightleftharpoons UO_2^{2+} + 4H^+ + 2e^-$

(d) Oxidizing Agent MnO_2 ; $\quad MnO_2(s) + 4H^+ + 2e^- \rightleftharpoons 2H_2O + Mn^{2+}$

Reducing Agent Cl^- ; $\quad 2Cl^- \rightleftharpoons Cl_2 + 2e^-$

(e) Oxidizing Agent IO_3^- ; $\quad IO_3^- + 6H^+ + 5e^- \rightleftharpoons \frac{1}{2}I_2 + 3H_2O$

Reducing Agent I^- ; $\quad I^- \rightleftharpoons \frac{1}{2}I_2 + e^-$

(f) Oxidizing Agent IO_3^- ; $\quad IO_3^- + 2Cl^- + 6H^+ + 4e^- \rightleftharpoons ICl_2^- + 3H_2O$

Reducing Agent I^- ; $\quad I^- + 2Cl^- \rightleftharpoons ICl_2^- + 2e^-$

(g) Oxidizing Agent MnO_4^- ; $\quad MnO_4^- + e^- \rightleftharpoons MnO_4^{2-}$

Reducing Agent HPO_3^{2-} ; $\quad HPO_3^{2-} + 3OH^- \rightleftharpoons PO_4^{3-} + 2H_2O + 2e^-$

(h) Oxidizing Agent BrO_3^- ; $\quad BrO_3^- + 6H^+ + 6e^- \rightleftharpoons Br^- + 3H_2O$

Reducing Agent SCN^- ; $\quad SCN^- + 4H_2O \rightleftharpoons SO_4^{2-} + HCN + 7H^+ + 6e^-$

(i) Oxidizing Agent $V(OH)_4^+$; $\quad V(OH)_4^+ + 2H^+ + e^- \rightleftharpoons VO^{2+} + 3H_2O$

Reducing Agent V^{2+} ; $\quad V^{2+} + H_2O \rightleftharpoons VO^{2+} + 2H^+ + 2e^-$

(j) Oxidizing Agent MnO_4^- ; $\quad MnO_4^- + 4H^+ + 3e^- \rightleftharpoons MnO_2(s) + 2H_2O$

Reducing Agent Mn^{2+} ; $\quad Mn^{2+} + 2H_2O \rightleftharpoons MnO_2(s) + 4H^+ + 2e^-$

Chapter 15

15-11 (a)

(a)	(b), (c)	E^0
$AgBr(s) + e^- \rightleftarrows Ag(s) + Br^-$	$S_2O_8^{2-} + 2e^- \rightleftarrows 2SO_4^{2-}$	2.01
$V^{2+} \rightleftarrows V^{3+} + e^-$	$Tl^{3+} + 2e^- \rightleftarrows Tl^+$	1.25
$Tl^{3+} + 2e^- \rightleftarrows Tl^+$	$Fe(CN)_6^{3-} + e^- \rightleftarrows Fe(CN)_6^{4-}$	0.36
$Fe(CN)_6^{4-} \rightleftarrows Fe(CN)_6^{3-} + e^-$	$AgBr(s) + e^- \rightleftarrows Ag(s) + Br^-$	0.073
$V^{3+} + e^- \rightleftarrows V^{2+}$	$V^{3+} + e^- \rightleftarrows V^{2+}$	-0.256
$Zn \rightleftarrows Zn^{2+} + 2e^-$	$Zn^{2+} + 2e^- \rightleftarrows Zn(s)$	-0.763
$Fe(CN)_6^{3-} + e^- \rightleftarrows Fe(CN)_6^{4-}$		
$Ag(s) + Br^- \rightleftarrows AgBr(s) + e^-$		
$S_2O_8^{2-} + 2e^- \rightleftarrows 2SO_4^{2-}$		
$Tl^+ \rightleftarrows Tl^{3+} + 2e^-$		

Chapter 15

15-12 (a)

(a)	(b), (c)	E^0
$Sn(s) \rightleftarrows Sn^{2+} + 2e^-$	$Ag^+ + e^- \rightleftarrows Ag$	0.799
$2H^+ + 2e^- \rightleftarrows H_2(g)$	$Fe^{3+} + e^- \rightleftarrows Fe^{2+}$	0.771
$Ag^+ + e^- \rightleftarrows Ag(s)$	$Sn^{4+} + 2e^- \rightleftarrows Sn^{2+}$	0.154
$Fe^{2+} \rightleftarrows Fe^{3+} + e^-$	$2H^+ + 2e^- \rightleftarrows H_2(g)$	0.00
$Sn^{4+} + 2e^- \rightleftarrows Sn^{2+}$	$Sn^{2+} + 2e^- \rightleftarrows Sn(s)$	-0.136
$H_2(g) \rightleftarrows 2H^+ + 2e^-$	$Co^{2+} + 2e^- \rightleftarrows Co(s)$	-0.277
$Fe^{3+} + e^- \rightleftarrows Fe^{2+}$		
$Sn^{2+} \rightleftarrows Sn^{4+} + 2e^-$		
$Sn^{2+} + 2e^- \rightleftarrows Sn(s)$		
$Co(s) \rightleftarrows Co^{2+} + 2e^-$		

15-13 (a) $E_{Cu} = 0.337 - \dfrac{0.0592}{2} \log \dfrac{1}{0.0440} = \underline{\underline{0.297 \text{ V}}}$

(b) $K_{sp} CuCl = 1.9 \times 10^{-7} = [Cu^+][Cl^-]$

$E_{Cu} = 0.521 - \dfrac{0.0592}{1} \log \dfrac{1}{[Cu^+]} = 0.521 - \dfrac{0.0591}{1} \log \dfrac{[Cl^-]}{K_{sp}}$

$= 0.521 - \dfrac{0.0592}{1} \log \dfrac{0.0750}{1.9 \times 10^{-7}} = 0.521 - \dfrac{0.0592}{1} \log 3.95 \times 10^5$

$= 0.521 - 0.331 = \underline{\underline{0.190 \text{ V}}}$

(c) $K_{sp} Cu(OH)_2 = 4.8 \times 10^{-20}$ $[Cu^{2+}] = \dfrac{4.8 \times 10^{-20}}{[OH^-]^2}$

225

Chapter 15

$$E_{Cu} = 0.337 - \frac{0.0592}{2} \log \frac{[OH^-]^2}{K_{sp}} = 0.337 - \frac{0.0592}{2} \log \frac{(0.0400)^2}{4.8 \times 10^{-20}}$$

$$= 0.337 - \frac{0.0592}{2} \log 3.33 \times 10^{16} = 0.337 - 0.489 = \underline{\underline{-0.152 \text{ V}}}$$

(d)
$$\beta_4 = 5.62 \times 10^{11} = \frac{[Cu(NH_3)_4^{2+}]}{[Cu^{2+}][NH_3]^4}$$

$$\frac{1}{[Cu^{2+}]} = \frac{(5.62 \times 10^{11})(0.128)^4}{0.0250} = 6.034 \times 10^9$$

$$E_{Cu} = 0.337 - \frac{0.0592}{2} \log 6.034 \times 10^9 = 0.337 - 0.289 = \underline{\underline{0.047 \text{ V}}}$$

(e) Substituting into Equation 14-4 gives

$$\frac{[CuY^{2-}]}{[Cu^{2+}]c_T} = \alpha_4 K_{CuY} = 3.6 \times 10^{-9} \times 6.3 \times 10^{18} = 2.27 \times 10^{10}$$

$$[CuY^{2-}] \cong 4.00 \times 10^{-3}$$

$$c_T = 2.90 \times 10^{-2} - 4.00 \times 10^{-3} = 2.50 \times 10^{-2}$$

$$\frac{4.00 \times 10^{-3}}{[Cu^{2+}] 2.50 \times 10^{-2}} = 2.27 \times 10^{10}$$

$$\frac{1}{[Cu^{2+}]} = \frac{2.27 \times 10^{10} \times 2.50 \times 10^{-2}}{4.00 \times 10^{-3}} = 1.42 \times 10^{11}$$

$$E = 0.337 - \frac{0.0592}{2} \log \frac{1}{[Cu^{2+}]} = 0.337 - \frac{0.0592}{2} \log 1.42 \times 10^{11}$$

$$= 0.337 - 0.330 = \underline{\underline{0.007 \text{ V}}}$$

15-14 (a) $E_{Zn} = -0.763 - \frac{0.0592}{2} \log \frac{1}{0.0600} = -0.763 - 0.036 = \underline{\underline{-0.799 \text{ V}}}$

(b) $K_{sp} \text{Zn(OH)}_2 = 3.0 \times 10^{-16} = [Zn^{2+}][OH^-]^2$

$$E_{Zn} = -0.763 - \frac{0.0592}{2} \log \frac{1}{[Zn^{2+}]} = -0.763 - \frac{0.0592}{2} \log \frac{[OH^-]^2}{3.0 \times 10^{-16}}$$

$$= -0.763 - \frac{0.0592}{2} \log \frac{1}{3.0 \times 10^{-16}} - \frac{0.0592}{2} \log (0.0100)^2$$

$$= -0.763 - 0.459 + 0.118 = \underline{\underline{-1.104 \text{ V}}}$$

(c)
$$Zn^{2+} + 4NH_3 \rightleftarrows Zn(NH_3)_4^{2+} \qquad \beta_4 = \frac{[Zn(NH_3)_4^{2+}]}{[Zn^{2+}][NH_3]^4} = 7.76 \times 10^8$$

$$7.76 \times 10^8 = \frac{0.0100}{[Zn^{2+}](0.250)^4}$$

$$\frac{1}{[Zn^{2+}]} = 3.03 \times 10^8$$

$$E = -0.763 - \frac{0.0592}{2} \log 3.03 \times 10^8 = \underline{\underline{-1.014 \text{ V}}}$$

(d) Proceeding as in Solution 15-13(e)

$$K'_{ZnY} = \alpha_4 K_{ZnY} = 5.2 \times 10^{-2} \times 3.2 \times 10^{16} = 1.66 \times 10^{15}$$

$$1.66 \times 10^{15} = \frac{[ZnY^{2-}]}{[Zn^{2+}]c_T}$$

$$[ZnY^{2-}] = 5.00 \times 10^{-3}$$

$$c_T = 4.45 \times 10^{-2} - 5.00 \times 10^{-3} = 3.95 \times 10^{-2}$$

$$1.66 \times 10^{15} = \frac{5.00 \times 10^{-3}}{[Zn^{2+}] 3.95 \times 10^{-2}} \quad \text{and} \quad \frac{1}{[Zn^{2+}]} = 1.31 \times 10^{16}$$

$$E = -0.763 - \frac{0.0592}{2} \log 1.31 \times 10^{16} = \underline{\underline{-1.240 \text{ V}}}$$

Chapter 15

15-15 $2H^+ + 2e^- \rightleftharpoons H_2(g)$

$$E = E^0 - \frac{0.0592}{2}\log\frac{p_{H_2}}{a_{H^+}^2} = 0.00 - \frac{0.0592}{2}\log\frac{1.00}{[H^+]^2 \times \gamma_{H^+}^2}$$

The ionic strength of the solution μ is given by

$$\mu = \frac{1}{2}(0.0100 \times 1^2 + 0.0100 \times 1^2) = 0.0100$$

From Table 8-1, page 154

$$\gamma_{H^+} = 0.914$$

$$E = 0.00 - \frac{0.0592}{2}\log\frac{1.00}{(0.0100 \times 0.914)^2}$$

$$= 0.00 - 0.121 = \underline{\underline{-0.121 \text{ V}}}$$

15-16 $PtCl_4^{2-} + 2e^- \rightleftharpoons Pt(s) + 4Cl^-$ $\quad E^0 = 0.73 \text{ V}$

(a) $E_{Pt} = 0.73 - \frac{0.0592}{2}\log\frac{(0.1492)^4}{0.0263} = 0.73 - (-0.051) = \underline{\underline{0.78 \text{ V}}}$

(b) $E_{Pt} = 0.154 - \frac{0.0592}{2}\log\frac{2.50 \times 10^{-3}}{7.50 \times 10^{-2}} = 0.154 - (-0.044) = \underline{\underline{0.198 \text{ V}}}$

(c) $E_{Pt} = 0.000 - \frac{0.0592}{2}\log\frac{1.00}{(1.00 \times 10^{-6})^2} = \underline{\underline{-0.355 \text{ V}}}$

(d) $VO^{2+} + 2H^+ + e^- \rightleftharpoons V^{3+} + H_2O$ $\quad E^0 = 0.359 \text{ V}$

$$E_{Pt} = 0.359 - \frac{0.0592}{1}\log\frac{0.0586 \times 2}{(0.0353)(0.100)^2} = 0.359 - 0.149 = \underline{\underline{0.210 \text{ V}}}$$

Chapter 15

(e) $2Fe^{3+} + Sn^{2+} \rightarrow 2Fe^{2+} + Sn^{4+}$

no. mmol Sn^{2+} taken $= 25.00 \times 0.0918 = 2.295$

no. mmol Fe^{3+} taken $= 25.00 \times 0.1568 = 3.920$

no. mmol Sn^{4+} formed $= 3.920 \text{ mmol } Fe^{2+} \times \dfrac{1 \text{ mmol } Sn^{4+}}{2 \text{ mmol } Fe^{3+}} = 1.960$

no. mmol Sn^{2+} remaining $= 2.295 - 1.960 = 0.335$

$E_{Pt} = 0.154 - \dfrac{0.0592}{2} \log \dfrac{0.335/50.0}{1.96/50.0} = 0.154 - (-0.023) = \underline{\underline{0.177 \text{ V}}}$

(f) $V(OH)_4^+ + V^{3+} \rightarrow 2VO^{2+} + 2H_2O$

no. mmol $V(OH)_4^+$ taken $= 25.00 \times 0.0832 = 2.080$

no. mmol V^{3+} taken $= 2 \times 50.00 \times 0.01087 = 1.087$

no. mmol VO^{2+} formed $= 2 \times 1.087 = 2.174$

no. mmol $V(OH)_4^+$ remaining $= 2.080 - 1.087 = 0.993$

$E_{Pt} = 1.00 - 0.0592 \log \dfrac{2.174/75.00}{(0.993/75.00)(0.1000)^2} = 1.00 - 0.139 = \underline{\underline{0.86 \text{ V}}}$

15-17 (a) $E_{Pt} = 0.36 - \dfrac{0.0592}{1} \log \dfrac{0.0813}{0.00566} = 0.36 - 0.068 \approx \underline{\underline{0.29 \text{ V}}}$

(b) $E_{Pt} = 0.771 - \dfrac{0.0592}{1} \log \dfrac{0.0400}{2 \times 0.00845} = 0.771 - 0.022 = \underline{\underline{0.749 \text{ V}}}$

(c) pH $= 5.55 \quad [H_3O^+] = 2.82 \times 10^{-6}$

$E_{Pt} = 0.000 - \dfrac{0.0592}{2} \log \dfrac{1.00}{(2.82 \times 10^{-6})^2} = 0.000 - 0.329 = \underline{\underline{-0.329 \text{ V}}}$

Chapter 15

(d) $E_{Pt} = 1.00 - \dfrac{0.0592}{1} \log \dfrac{0.0789}{(0.1996)(0.0800)^2} = 1.00 - 0.106 = \underline{\underline{0.894 \text{ V}}}$

(e) no. mmol Ce^{4+} taken $= 50.00 \times 0.0607 = 3.035$

no. mmol Fe^{2+} taken $= 50.00 \times 0.100 = 5.00$

no. mmol Fe^{3+} formed $= 3.035$

no. mmol Fe^{2+} remaining $= 5.00 - 3.035 = 1.965$

$E_{Pt} = 0.68 - 0.0592 \log \dfrac{1.965/100.0}{3.035/100.0} = 0.68 - (-0.011) = \underline{\underline{0.69 \text{ V}}}$

(f) $V(OH)_4^+ + V^{3+} \rightarrow 2VO^{2+} + 2H_2O$

no. mmol V^{3+} taken $= 25.00 \times 2 \times 0.0832 = 4.160$

no. mmol $V(OH)_4^+$ taken $= 50.00 \times 0.00628 = 0.314$

no. mmol VO^{2+} formed $= 2 \times 0.314 = 0.628$

no. mmol V^{3+} remaining $= 4.160 - 0.314 = 3.846$

$E_{Pt} = 0.359 - 0.0592 \log \dfrac{3.846/75.00}{(0.628/75.00)(0.100)^2} = 0.359 - 0.165 = \underline{\underline{0.194 \text{ V}}}$

15-18 (a) $E_{Ni} = -0.250 - \dfrac{0.0592}{2} \log \dfrac{1}{0.0943} = -0.250 - 0.030 = \underline{\underline{-0.280}} \quad \text{anode}$

(b) $E_{Ag} = -0.151 - \dfrac{0.0592}{1} \log 0.0922 = -0.151 + 0.061 = \underline{\underline{-0.090 \text{ V}}} \quad \text{anode}$

(c) $E_{O_2} = 1.229 - \dfrac{0.0592}{4} \log \dfrac{1}{(780/760)(1.50 \times 10^{-4})^4} = 1.229 - 0.226 =$

$\underline{\underline{1.003 \text{ V}}} \quad \text{cathode}$

(d) $E_{Pt} = 0.154 - \dfrac{0.0592}{2} \log \dfrac{0.0944}{0.350} = 0.154 + 0.017 = \underline{\underline{0.171 \text{ V}}} \quad \text{cathode}$

(e) $E_{Ag} = 0.017 - \dfrac{0.0592}{1} \log \dfrac{(0.1439)^2}{0.00753} = 0.017 - 0.026 = \underline{\underline{-0.009 \text{ V}}} \quad \text{anode}$

15-19 (a) $E_{Cu} = 0.337 - \dfrac{0.0592}{2} \log \dfrac{1}{0.0897} = 0.337 - 0.031 = \underline{\underline{0.306 \text{ V}}} \quad \text{cathode}$

(b) $E_{Pt} = -0.185 - \dfrac{0.0592}{1} \log 0.1214 = -0.185 + 0.054 = \underline{\underline{-0.131 \text{ V}}} \quad \text{anode}$

(c) $E_{Pt} = 0.00 - \dfrac{0.0592}{2} \log \dfrac{0.984}{(1.00 \times 10^{-4})^2} = 0.000 - 0.237 = \underline{\underline{-0.237 \text{ V}}} \quad \text{anode}$

(d) $E_{Pt} = 0.771 - \dfrac{0.0592}{1} \log \dfrac{0.1628}{0.0906} = 0.771 - 0.015 = \underline{\underline{0.756 \text{ V}}} \quad \text{cathode}$

(e) $E_{Ag} = -0.31 - \dfrac{0.0592}{1} \log \dfrac{(0.0699)^2}{0.0827} = -0.31 + 0.073 = \underline{\underline{-0.24 \text{ V}}} \quad \text{anode}$

15-20 $2Ag^+ + 2e^- \rightleftarrows 2Ag(s) \qquad E^0 = 0.799 \text{ V}$

$[Ag^+]^2[SO_3^{2-}] = 1.5 \times 10^{-14} = K_{sp}$

$E = 0.799 - \dfrac{0.0592}{2} \log \dfrac{1}{[Ag^{2+}]^2} = 0.799 - \dfrac{0.0592}{2} \log \dfrac{[SO_3^{2-}]}{K_{sp}}$

When $[SO_3^{2-}] = 1.00$, $E = E^0$ for $Ag_2SO_3(s) + 2e^- \rightleftarrows 2Ag(s) + SO_3^{2-}$

Thus,

$E^0 = 0.799 - 0.0296 \log (1.00/K_{sp}) = 0.799 - 0.0296 \log (1/1.5 \times 10^{-14})$

$= 0.799 - 0.409 = \underline{\underline{0.390 \text{ V}}}$

15-21 Proceeding as in Problem 15-20,

$2Ni^{2+} + 4e^- \rightleftarrows 2Ni(s) \qquad E^0 = -0.250 \qquad [Ni^{2+}]^2[P_2O_7^{4-}] = 1.7 \times 10^{-13}$

$E = -0.250 - \dfrac{0.0592}{4} \log \dfrac{1}{[Ni^{2+}]^2} = -0.250 - \dfrac{0.0592}{4} \log \dfrac{[P_2O_7^{4-}]}{1.7 \times 10^{-13}}$

Chapter 15

When $[P_2O_7^{4-}] = 1.000$, $E = E^0$ for $Ni_2P_2O_7(s) + 4e^- \rightleftharpoons 2Ni(s) + P_2O_7^{4-}$

Thus,

$$E^0 = -0.250 - \frac{0.0592}{4} \log \frac{1.00}{1.7 \times 10^{-13}} = \underline{\underline{-0.439 \text{ V}}}$$

15-22 $2Tl^+ + 2e^- \rightleftharpoons 2Tl \qquad E^0 = -0.336 \text{ V}$

$[Tl^+]^2 [S^{2-}] = 6 \times 10^{-22}$

$$E = -0.336 - \frac{0.0592}{2} \log \frac{1}{[Tl^+]^2} = -0.336 - \frac{0.0592}{2} \log \frac{[S^{2-}]}{K_{sp}}$$

When $[S^{2-}] = 1.00$,

$$E = E^0_{Tl_2S} = -0.336 - \frac{0.0592}{2} \log \frac{1}{6 \times 10^{-22}} = \underline{\underline{-0.964 \text{ V}}}$$

15-23 $3Pb^{2+} + 6e^- \rightleftharpoons 3Pb$; $E^0 = -0.126 \text{ V}$; and $[Pb^{2+}]^3[AsO_4^{2-}]^2 = 4.1 \times 10^{-36}$

$$E = -0.126 - \frac{0.0592}{6} \log \frac{1}{[Pb^{2+}]^3} = -0.126 - \frac{0.0592}{6} \log \frac{[AsO_4^{2-}]^2}{4.1 \times 10^{-36}}$$

When $[AsO_4^{2-}] = 1.00$,

$$E = E^0_{Pb_3(AsO_4)_2} = -0.126 - \frac{0.0592}{6} \log \frac{1}{4.1 \times 10^{-36}} = \underline{\underline{-0.475 \text{ V}}}$$

15-24
$$E = -0.763 - 0.0296 \log \frac{1}{[Zn^{2+}]}$$

$$\frac{[ZnY^{2-}]}{[Y^{4-}][Zn^{2+}]} = 3.2 \times 10^{16} \quad \text{or} \quad [Zn^{2+}] = \frac{[ZnY^{2-}]}{[Y^{4-}] 3.2 \times 10^{16}}$$

$$E = -0.763 - 0.0296 \log \frac{[Y^{4-}] 3.2 \times 10^{16}}{[ZnY^{2-}]}$$

When $[Y^{4-}] = [ZnY^{2-}] = 1.00$, $E = E^0_{ZnY^{2-}}$

$$E^0_{ZnY^{2-}} = -0.763 - 0.0296 \log(1.00 \times 3.2 \times 10^{16}/1.00) = \underline{\underline{-1.25 \text{ V}}}$$

15-25 $[Fe^{3+}] = \dfrac{[FeY^-]}{1.3 \times 10^{25}[Y^{4-}]}$ and $[Fe^{2+}] = \dfrac{[FeY^{2-}]}{2.1 \times 10^{14}[Y^{4-}]}$

$$E = 0.771 - 0.0592 \log \frac{[Fe^{2+}]}{[Fe^{3+}]}$$

$$= 0.771 - 0.0592 \log \frac{[FeY^{2-}] \, 1.3 \times 10^{25}[Y^{4-}]}{2.1 \times 10^{14}[Y^{4-}][FeY^-]}$$

$E = E^0_{FeY^-}$ when $[FeY^{2-}]$, $[FeY^-]$, and $[Y^{4-}] = 1.000$

$$E^0 = 0.771 - 0.0592 \log(1.3 \times 10^{25}/2.1 \times 10^{14})$$

$$= 0.771 - 0.639 = \underline{\underline{0.13 \text{ V}}}$$

15-26 Proceeding as in Solution 15-25, we write

$$E = 0.153 - 0.0592 \log [Cu^+]/[Cu^{2+}]$$

$$[Cu^+] = \frac{[Cu(NH_3)_2^+]}{[NH_3]^2 \, 7.2 \times 10^{10}} \quad \text{and} \quad [Cu^{2+}] = \frac{[Cu(NH_3)_4^{2+}]}{[NH_3]^4 \, 5.62 \times 10^{11}}$$

Substituting into the first equation and letting

$[Cu(NH_3)_2^+]$, $[Cu(NH_3)_4^{2+}]$, and $[NH_3] = 1.00$

$$E^0 = 0.153 - 0.0592 \log(5.62 \times 10^{11}/7.2 \times 10^{10}) = \underline{\underline{0.100 \text{ V}}}$$

CHAPTER 16

16-1 The electrode potential of a system is the electrode potential of all half-cell processes at equilibrium in the system.

16-2 (a) *Equilibrium* is the state that a system assumes after each addition of reagent. *Equivalence* refers to a particular equilibrium state when a stoichiometric amount of titrant has been added.

(b) A *specific indicator* exhibits its color change as a result of reactions with a particular solute species. A true oxidation/reduction indicator owes its color change to changes in the electrode potential of the system.

16-3 The electrode potentials for all half-cell processes in an oxidation/reduction system have the same numerical value when the system is at equilibrium.

16-4 For points before equivalence, potential data are computed from the analyte standard potential and the analytical concentrations of the analyte and its reaction product. Post-equivalence point data are based upon the standard potential for the titrant and its analytical concentrations. The equivalence point potential is computed from the two standard potentials and the stoichiometric relation between the analyte and titrant.

16-5 In contrast to all other points on the titration curve, the concentrations of all of the participants in one of the half-reactions or the other cannot be derived from stoichiometric calculations.

16-6 An asymmetric titration curve will be encountered whenever the titrant and the analyte react in a ratio that is not 1:1.

16-7 (a)
$$E_{cathode} = -0.403 - \frac{0.0592}{2}\log\frac{1}{0.0511} = -0.441 \text{ V}$$

$$E_{anode} = -0.126 - \frac{0.0592}{2}\log\frac{1}{0.1393} = -0.151 \text{ V}$$

$$E_{cell} = E_{cathode} - E_{anode} = -0.441 - (-0.151) = \underline{\underline{-0.290 \text{ V}}}$$

Because E_{cell} is negative, the cell is <u>electrolytic</u>.

(b)
$$E_{cathode} = 1.25 - \frac{0.0592}{2}\log\frac{0.0620}{9.06 \times 10^{-3}} = 1.23 \text{ V}$$

$$E_{anode} = -0.763 - \frac{0.0592}{2}\log\frac{1}{0.0364} = -0.806 \text{ V}$$

$$E_{cell} = E_{cathode} - E_{anode} = 1.23 - (-0.806) = \underline{\underline{2.03 \text{ V}}} \quad \text{galvanic}$$

(c)
$$E_{cathode} = -0.250 - \frac{0.0592}{2}\log\frac{1}{0.0214} = -0.299 \text{ V}$$

$$E_{anode} = 0.000 - \frac{0.0592}{2}\log\frac{765/760}{(1.00\times 10^{-4})^2} = -0.237 \text{ V}$$

$$E_{cell} = E_{cathode} - E_{anode} = -0.299 - (-0.237) = \underline{\underline{-0.062 \text{ V}}} \quad \text{electrolytic}$$

(d)
$$E_{cathode} = 0.854 - \frac{0.0592}{2}\log\frac{1}{4.59\times 10^{-3}} = 0.785 \text{ V}$$

$$[Pb^{2+}][I^-]^2 = 7.9\times 10^{-9} \quad \text{and} \quad [Pb^{2+}] = 7.9\times 10^{-9}/[I^-]^2$$

$$E_{anode} = -0.126 - \frac{0.0592}{2}\log\frac{(0.0120)^2}{7.9\times 10^{-9}} = -0.252 \text{ V}$$

$$E_{cell} = E_{cathode} - E_{anode} = 0.785 - (-0.252) = \underline{\underline{1.037 \text{ V}}} \quad \text{galvanic}$$

(e)
$$\frac{[H_3O^+][NH_3]}{[NH_4^+]} = 5.70\times 10^{-10} = \frac{[H_3O^+]0.438}{0.379}$$

$$[H_3O^+] = 5.70\times 10^{-10}\times 0.379/0.438 = 4.93\times 10^{-10}$$

$$E_{cathode} = 0.000 \text{ V}$$

$$E_{anode} = 0.000 - \frac{0.0592}{2}\log\frac{1.00}{(4.93\times 10^{-10})^2} = -0.551 \text{ V}$$

$$E_{cell} = E_{cathode} - E_{anode} = 0.000 - (-0.551) = \underline{\underline{0.551 \text{ V}}} \quad \text{galvanic}$$

(f)
$$E_{cathode} = 0.359 - 0.0592\log\frac{0.0784}{0.1340\,(0.0538)^2} = 0.223 \text{ V}$$

Chapter 16

$$E_{anode} = 0.099 - 0.0592 \log \frac{0.00918}{0.0790\,(1.47 \times 10^{-2})^2} = -0.063 \text{ V}$$

$$E_{cell} = E_{cathode} - E_{anode} = 0.223 - (-0.063) = \underline{\underline{0.286 \text{ V}}} \quad \text{galvanic}$$

16-8 (a)
$$E_{cathode} = -0.277 - \frac{0.0592}{2} \log \frac{1}{6.78 \times 10^{-3}} = -0.341 \text{ V}$$

$$E_{anode} = -0.763 - \frac{0.0592}{2} \log \frac{1}{0.0955} = -0.793 \text{ V}$$

$$E_{cell} = E_{cathode} - E_{anode} = -0.341 - (-0.793) = \underline{\underline{0.452 \text{ V}}} \quad \text{galvanic}$$

(b)
$$E_{cathode} = 0.854 - \frac{0.0592}{2} \log \frac{1}{0.0671} = 0.819 \text{ V}$$

$$E_{anode} = 0.771 - 0.0592 \log \frac{0.0681}{0.1310} = 0.788 \text{ V}$$

$$E_{cell} = E_{cathode} - E_{anode} = 0.819 - 0.788 = \underline{\underline{0.031 \text{ V}}} \quad \text{galvanic}$$

(c)
$$E_{cathode} = 1.229 - \frac{0.0592}{4} \log \frac{1}{1.12\,(0.0794)^4} = 1.165 \text{ V}$$

$$E_{anode} = 0.799 - 0.0592 \log \frac{1}{0.1544} = 0.751 \text{ V}$$

$$E_{cell} = E_{cathode} - E_{anode} = 1.165 - 0.751 = \underline{\underline{0.414 \text{ V}}} \quad \text{galvanic}$$

(d) $E_{cathode} = -0.151 - 0.0592 \log 0.1350 = -0.100 \text{ V}$

$$E_{anode} = 0.337 - \frac{0.0592}{2} \log \frac{1}{0.0601} = 0.301 \text{ V}$$

$$E_{cell} = E_{cathode} - E_{anode} = -0.100 - 0.301 = \underline{\underline{-0.401 \text{ V}}} \quad \text{electrolytic}$$

(e) $$\frac{[H_3O^+][HCOO^-]}{[HCOOH]} = 1.80 \times 10^{-4} = \frac{[H_3O^+] \times 0.0764}{0.1302}$$

$$[H_3O^+] = 1.80 \times 10^{-4} \times 0.1302/0.0764 = 3.07 \times 10^{-4}$$

$$E_{\text{cathode}} = 0.000 - \frac{0.0592}{2}\log\frac{1.00}{(3.07\times 10^{-4})^2} = -0.208 \text{ V}$$

$$E_{\text{anode}} = 0.000 \text{ V}$$

$$E_{\text{cell}} = -0.208 - 0.000 = \underline{\underline{-0.208 \text{ V}}} \quad \text{electrolytic}$$

(f) $$E_{\text{cathode}} = 0.771 - 0.0592\log\frac{0.1134}{0.003876} = 0.684 \text{ V}$$

$$E_{\text{anode}} = 0.334 - \frac{0.0592}{2}\log\frac{6.37\times 10^{-2}}{7.93\times 10^{-3}(1.16\times 10^{-3})^4} = -0.040 \text{ V}$$

$$E_{\text{cell}} = E_{\text{cathode}} - E_{\text{anode}} = 0.684 - (-0.040) = \underline{\underline{0.724 \text{ V}}} \quad \text{galvanic}$$

16-9 (a) $$E_{\text{Pb}^{2+}} = -0.126 - \frac{0.0592}{2}\log\frac{1}{0.0848} = -0.158 \text{ V}$$

$$E_{\text{Zn}^{2+}} = -0.763 - \frac{0.0592}{2}\log\frac{1}{0.1364} = -0.789 \text{ V}$$

$$E_{\text{cell}} = E_{\text{cathode}} - E_{\text{anode}} = -0.158 - (-0.789) = \underline{\underline{0.631 \text{ V}}}$$

(b) $$E_{\text{Fe}^{3+}} = 0.771 - 0.0592\log\frac{0.0760}{0.0301} = 0.747 \text{ V}$$

$$E_{\text{Fe(CN)}_6^{3-}} = 0.36 - 0.0592\log\frac{0.00309}{0.1564} = 0.461 \text{ V}$$

$$E_{\text{cell}} = E_{\text{cathode}} - E_{\text{anode}} = 0.747 - 0.461 = \underline{\underline{0.286 \text{ V}}}$$

(c) $E_{\text{SHE}} = 0.000 \text{ V}$

$$E_{\text{TiO}_2^{2+}} = 0.099 - 0.0592\log\frac{0.02723}{1.46\times 10^{-3}(10^{-3})^2} = -0.331 \text{ V}$$

$$E_{\text{cell}} = E_{\text{cathode}} - E_{\text{anode}} = 0.000 - (-0.331) = \underline{\underline{0.331 \text{ V}}}$$

(d) $$E_{\text{Co}^{2+}} = -0.277 - \frac{0.0592}{2}\log\frac{1}{0.0767} = -0.310 \text{ V}$$

Chapter 16

$$E_{O_2} = 1.229 - \frac{0.0592}{4} \log \frac{1}{1.00\,(0.200)^4} = 1.188 \text{ V}$$

$$E_{cell} = E_{cathode} - E_{anode} = -0.310 - 1.188 = \underline{\underline{-1.498 \text{ V}}}$$

(e) $E_{AgBr} = 0.073 - 0.0592 \log 0.0791 = 0.138 \text{ V}$

$$E_{Ag^+} = 0.799 - 0.0592 \log \frac{1}{0.2058} = 0.758 \text{ V}$$

$$E_{cell} = E_{cathode} - E_{anode} = 0.138 - 0.758 = \underline{\underline{-0.620 \text{ V}}}$$

(f) $$E_{I_3^-} = 0.536 - \frac{0.0592}{2} \log \frac{(0.1523)^3}{0.0364} = 0.566 \text{ V}$$

$$E_{Fe^{3+}} = 0.771 - 0.0592 \log \frac{0.1037}{0.02105} = 0.730 \text{ V}$$

$$E_{cell} = E_{cathode} - E_{anode} = 0.566 - 0.730 = \underline{\underline{-0.164 \text{ V}}}$$

16-10 (a) $Zn|Zn^{2+}(0.1364 \text{ M})||Pb^{2+}(0.0848 \text{ M})|Pb$

(b) $Pt|Fe(CN)_6^{4-}(0.00309 \text{ M}),Fe(CN)_6^{3-}(0.1564 \text{ M})||Fe^{3+}(0.0301 \text{ M}),Fe^{2+}(0.0760 \text{ M})|Pt$

(c) $Pt|TiO^{2+}(1.46 \times 10^{-3} \text{ M}),Ti^{3+}(0.02723 \text{ M}),H^+(1.00 \times 10^{-3} \text{ M})||SHE$

(d) $Pt|O_2(1.00 \text{ atm}),HCl(0.200 \text{ M})||Co^{2+}(0.0767 \text{ M})|Co$

(e) $Ag|Ag^+(0.2058 \text{ M})||KBr(0.0791 \text{ M}),AgBr(sat'd)|Ag$

(f) $Pt|Fe^{3+}(0.02105 \text{ M}),Fe^{2+}(0.1037 \text{ M})||I^-(0.1523 \text{ M}),I_3^-(0.0364 \text{ M})|Pt$

16-11 Note that in these calculations, it is necessary to round the answers to either one or two significant figures because the final step involves taking an antilogarithm of a large number (see page 41, text).

(a) $Fe^{3+} + V^{2+} = Fe^{2+} + V^{3+}$ $E^0_{Fe^{3+}} = 0.771$ $E_{V^{3+}} = -0.256$

$$0.771 - \frac{0.0592}{1} \log \frac{[Fe^{2+}]}{[Fe^{3+}]} = -0.256 - \frac{0.0592}{1} \log \frac{[V^{2+}]}{[V^{3+}]}$$

Chapter 16

$$\frac{0.771 - (-0.256)}{0.0592} = \log\frac{[V^{3+}][Fe^{2+}]}{[V^{2+}][Fe^{3+}]} = \log K_{eq} = 17.348$$

$$K_{eq} = 2.23 \times 10^{17} = \underline{\underline{2.2 \times 10^{17}}} = \frac{[Fe^{2+}][V^{3+}]}{[Fe^{3+}][V^{2+}]}$$

(b) $Fe(CN)_6^{3-} + Cr^{2+} \rightleftarrows Fe(CN)_6^{4-} + Cr^{3+}$ $E^0_{Fe(CN)_6^{3-}} = 0.36$ $E^0_{Cr^{3+}} = -0.408$

$$0.36 - \frac{0.0592}{1}\log\frac{[Fe(CN)_6^{4-}]}{[Fe(CN)_6^{3-}]} = -0.408 - \frac{0.0592}{1}\log\frac{[Cr^{2+}]}{[Cr^{3+}]}$$

$$\frac{0.36 - (-0.408)}{0.0592} = \log\frac{[Cr^{3+}][Fe(CN)_6^{4-}]}{[Cr^{2+}][Fe(CN)_6^{3-}]} = \log K_{eq} = 12.973$$

$$K_{eq} = 9.4 \times 10^{12} = \underline{\underline{9 \times 10^{12}}} = \frac{[Cr^{3+}][Fe(CN)_6^{4+}]}{[Cr^{2+}][Fe(CN)_6^{3+}]}$$

(c) $2V(OH)_4^+ + U^{4+} \rightleftarrows 2VO^{2+} + UO_2^{2+} + 4H_2O$ $E^0_V = 1.00$ $E^0_U = 0.334$

$$1.00 - \frac{0.0592}{2}\log\frac{[VO^{2+}]^2}{[V(OH)_4^+]^2[H^+]^4} = 0.334 - \frac{0.0592}{2}\log\frac{[U^{4+}]}{[UO_2^{2+}][H^+]^4}$$

$$\frac{2(1.00 - 0.334)}{0.0592} = \log\frac{[VO^{2+}]^2[UO_2^{2+}][H^+]^4}{[V(OH)_4^{2+}]^2[U^{4+}][H^+]^4} = \log K_{eq} = 22.50$$

$$K_{eq} = \underline{\underline{3.2 \times 10^{22}}} = \frac{[VO^{2+}]^2[UO_2^{2+}]}{[V(OH)_4^+]^2[U^{4+}]}$$

(d) $Tl^{3+} + 2Fe^{2+} = Tl^+ + 2Fe^{3+}$ $E^0_{Tl} = 1.25$ V $E^0_{Fe} = 0.771$ V

$$1.25 - \frac{0.0592}{2}\log\frac{[Tl^+]}{[Tl^{3+}]} = 0.771 - \frac{0.0592}{2}\log\frac{[Fe^{2+}]^2}{[Fe^{3+}]^2}$$

$$\frac{2(1.25 - 0.771)}{0.0592} = \log\frac{[Tl^+][Fe^{3+}]^2}{[Tl^{3+}][Fe^{2+}]^2} = \log K_{eq} = 16.18$$

Chapter 16

$$K_{eq} = 1.5 \times 10^{16} = \underline{\underline{2 \times 10^{16}}} = \frac{[Tl^+][Fe^{3+}]^2}{[Tl^{3+}][Fe^{2+}]^2}$$

(e) $2Ce^{4+} + H_3AsO_3 + H_2O \rightleftarrows 2Ce^{3+} + H_3AsO_4 + 2H^+$

In 1 M $HClO_4$, $\quad Ce^{4+} + e^- \rightleftarrows Ce^{3+} \quad E^{0'} = 1.70$ V

$$H_3AsO_4 + 2H^+ + 2e^- \rightleftarrows H_3AsO_3 + H_2O \quad E^{0'} = 0.577 \text{ V}$$

$$1.70 - \frac{0.0592}{2}\log\frac{[Ce^{3+}]^2}{[Ce^{4+}]^2} = 0.577 - \frac{0.0592}{2}\log\frac{[H_3AsO_3]}{[H_3AsO_4][H^+]^2}$$

$$\frac{2(1.70 - 0.577)}{0.0592} = \log\frac{[Ce^{3+}]^2[H_3AsO_4]}{[Ce^{4+}]^2[H_3AsO_3][H^+]^2} = \log K_{eq}$$

$\log K_{eq} = 37.94 \quad$ and $\quad K_{eq} = 8.9 \times 10^{37} = \underline{\underline{9 \times 10^{37}}}$

(f) $2V(OH)_4^+ + H_2SO_3 \rightleftarrows 2VO^{2+} + SO_4^{2-} + 5H_2O \quad E_V^0 = 1.00 \quad E_S^0 = 0.172$

$$1.00 - \frac{0.0592}{2}\log\frac{[VO^{2+}]^2}{[V(OH)_4^+]^2[H^+]^4} = 0.172 - \frac{0.0592}{2}\log\frac{[H_2SO_3]}{[SO_4^{2-}][H^+]^4}$$

$$\frac{2(1.00 - 0.172)}{0.0592} = \log\frac{[VO^{2+}]^2[SO_4^{2-}][H^+]^4}{[V(OH)_4^+]^2[H^+]^4[H_2SO_3]} = \log K_{eq} = 27.97$$

$$K_{eq} = 9.4 \times 10^{27} = \underline{\underline{9 \times 10^{27}}} = \frac{[VO^{2+}]^2[SO_4^{2-}]}{[V(OH)_4^+]^2[H_2SO_3]}$$

(g) $VO^{2+} + V^{2+} + 2H^+ = 2V^{3+} + H_2O \quad E_{VO}^0 = 0.359 \quad E_{V^{2+}}^0 = -0.256$

$$0.359 - \frac{0.0592}{1}\log\frac{[V^{3+}]}{[VO^{2+}][H^+]^2} = -0.256 - \frac{0.0592}{1}\log\frac{[V^{2+}]}{[V^{3+}]}$$

$$\frac{0.359 - (-0.256)}{0.0592} = \log\frac{[V^{3+}][V^{3+}]}{[VO^{2+}][H^+]^2[V^{2+}]} = \log K_{eq} = 10.389$$

$$K_{eq} = 2.446 \times 10^{10} = \underline{\underline{2.4 \times 10^{10}}} = \frac{[V^{3+}]^2}{[VO^{2+}][V^{2+}][H^+]^2}$$

(h) $TiO^{2+} + Ti^{2+} + 2H^+ \rightleftarrows 2Ti^{3+} + H_2O$ $E^0_{TiO} = 0.099\ V$ $E^0_{Ti} = -0.369\ V$

$$0.099 - \frac{0.0592}{1} \log \frac{[Ti^{3+}]}{[TiO^{2+}][H^+]^2} = -0.369 - \frac{0.0592}{1} \log \frac{[Ti^{2+}]}{[Ti^{3+}]}$$

$$\frac{0.099 - (-0.369)}{0.0592} = \log \frac{[Ti^{3+}]^2}{[TiO^{2+}][H^+]^2[Ti^{2+}]} = \log K_{eq} = 7.9054$$

$$K_{eq} = \underline{\underline{8.0 \times 10^7}} = \frac{[Ti^{3+}]^2}{[TiO^{2+}][Ti^{2+}][H^+]^2}$$

16-12 At equivalence, $[Fe^{2+}] = [V^{3+}]$ $[Fe^{3+}] = [V^{2+}]$

(a)
$$E_{eq} = E^0_{Fe} - \frac{0.0592}{1} \log \frac{[Fe^{2+}]}{[Fe^{3+}]} = 0.771 - 0.0592 \log \frac{[Fe^{2+}]}{[Fe^{3+}]}$$

$$E_{eq} = E^0_V - \frac{0.0592}{1} \log \frac{[V^{2+}]}{[V^{3+}]} = -0.256 - 0.0592 \log \frac{[V^{2+}]}{[V^{3+}]}$$

$$2E_{eq} = E^0_{Fe} + E^0_V - 0.0592 \log \frac{\cancel{[Fe^{2+}][V^{2+}]}}{\cancel{[Fe^{3+}][V^{3+}]}}$$

$$E_{eq} = \frac{0.771 - 0.256}{2} = \underline{\underline{0.258\ V}}$$

(b) At equivalence, $[Fe(CN)_6^{3-}] = [Cr^{2+}]$ $[Fe(CN)_6^{4-}] = [Cr^{3+}]$

$$E_{eq} = 0.36 - \frac{0.0592}{1} \log \frac{[Fe(CN)_6^{4-}]}{[Fe(CN)_6^{3-}]}$$

$$E_{eq} = -0.408 - \frac{0.0592}{1} \log \frac{[Cr^{2+}]}{[Cr^{3+}]}$$

$$2E_{eq} = 0.36 - 0.408 - 0.0592 \log \frac{\cancel{[Fe(CN)_6^{4-}][Cr^{3+}]}}{\cancel{[Fe(CN)_6^{3-}][Cr^{3+}]}}$$

Chapter 16

$$E_{eq} = \frac{0.36 - 0.408}{2} = -0.024 \text{ V}$$

(c) At equivalence, $[VO^{2+}] = 2[UO_2^{2+}]$ and $[V(OH)_4^+] = 2[U^{4+}]$

$$E_{eq} = 1.00 - \frac{0.0592}{1} \log \frac{[VO^{2+}]}{[V(OH)_4^+][H^+]^2}$$

$$2E_{eq} = 2 \times 0.344 - 0.0592 \log \frac{[U^{4+}]}{[UO_2^{2+}][H^+]^4}$$

$$3E_{eq} = 1.688 - 0.0592 \log \frac{[VO^{2+}][U^{4+}]}{[V(OH)_4^+][UO_2^{2+}][H^+]^6}$$

$$= 1.688 - 0.0592 \log \frac{\cancel{2[UO_2^{2+}]}\cancel{[U^{4+}]}}{\cancel{2[U^{4+}]}\cancel{[UO_2^{2+}]}[H^+]^6}$$

$$3E_{eq} = 1.688 - 0.0592 \log \frac{1}{(0.100)^6} = 1.688 - 0.355 = 1.333$$

$$E_{eq} = 1.333/3 = \underline{\underline{0.444 \text{ V}}}$$

(d) At equivalence, $[Fe^{2+}] = 2[Tl^{3+}]$ and $[Fe^{3+}] = 2[Tl^+]$

$$E_{eq} = 0.771 - 0.0592 \log \frac{[Fe^{2+}]}{[Fe^{3+}]}$$

$$2E_{eq} = 2 \times 1.25 - 0.0592 \log \frac{[Tl^+]}{[Tl^{3+}]}$$

$$3E_{eq} = 3.27 - 0.0592 \log \frac{[Fe^{2+}][Tl^+]}{[Fe^{3+}][Tl^{3+}]} = 3.27 - 0.0592 \log \frac{\cancel{2[Tl^{3+}]}\cancel{[Tl^+]}}{\cancel{2[Tl^+]}\cancel{[Tl^{3+}]}}$$

$$E_{eq} = \frac{3.27 - 0}{3} = \underline{\underline{1.09 \text{ V}}}$$

(e) At equivalence,
$$[Ce^{3+}] = 2[H_3AsO_4] \quad [Ce^{4+}] = 2[H_3AsO_3] \quad [H^+] = 1.00$$

242

Chapter 16

$$E_{eq} = 1.70 - 0.0592 \log \frac{[Ce^{3+}]}{[Ce^{4+}]}$$

$$2E_{eq} = 2 \times 0.577 - 0.0592 \log \frac{[H_3AsO_3]}{[H_3AsO_4][H^+]^2}$$

$$3E_{eq} = 2.854 - 0.0592 \log \frac{[Ce^{3+}][H_3AsO_3]}{[Ce^{4+}][H_3AsO_4][H^+]^2}$$

$$E_{eq} = 0.951 - \frac{0.0592}{3} \log \frac{\cancel{2[H_3AsO_4]}\cancel{[H_3AsO_3]}}{\cancel{2[H_3AsO_3]}\cancel{[H_3AsO_4]}(1.00)^2} = \underline{\underline{0.951 \text{ V}}}$$

(f) At equivalence, $[V(OH)_4^+] = 2[H_2SO_3]$ $\quad$ $[VO^{2+}] = 2[SO_4^{2-}]$

$$E_{eq} = 1.00 - 0.0592 \log \frac{[VO^{2+}]}{[V(OH)_4^+][H^+]^2}$$

$$2E_{eq} = 2 \times 0.172 - 0.0592 \log \frac{[H_2SO_3]}{[SO_4^{2-}][H^+]^4}$$

$$3E_{eq} = 1.344 - 0.0592 \log \frac{[VO^{2+}][H_2SO_3]}{[V(OH)_4^+][SO_4^{2-}][H^+]^6}$$

$$= 1.344 - 0.0592 \log \frac{\cancel{2[SO_4^{2-}]}\cancel{[H_2SO_3]}}{\cancel{2[H_2SO_3]}\cancel{[SO_4^{2-}]}[H^+]^6}$$

$$E_{eq} = 0.448 - \frac{0.0592}{3} \log \frac{1}{(0.100)^6} = 0.448 - 0.118 = \underline{\underline{0.330 \text{ V}}}$$

(g) At equivalence, $[VO^{2+}] = [V^{2+}]$

$$E_{eq} = 0.359 - 0.0592 \log \frac{[V^{3+}]}{[VO^{2+}][H^+]^2}$$

$$E_{eq} = -0.256 - 0.0592 \log \frac{[V^{2+}]}{[V^{3+}]}$$

Chapter 16

$$2E_{eq} = 0.103 - 0.0592 \log \frac{\cancel{[V^{3+}][V^{2+}]}}{\cancel{[VO^{2+}][V^{3+}]}[H^+]^2}$$

$$E_{eq} = 0.0515 - \frac{0.0592}{2} \log \frac{1}{1 \times 10^{-2}} = \underline{\underline{-0.008 \text{ V}}}$$

(h) At equivalence, $[Ti^{2+}] = [TiO^{2+}]$

$$E_{eq} = 0.099 - 0.0592 \log \frac{[Ti^{3+}]}{[TiO^{2+}][H^+]^2}$$

$$E_{eq} = -0.369 - 0.0592 \log \frac{[Ti^{2+}]}{[Ti^{3+}]}$$

$$2E_{eq} = 0.099 - 0.369 - 0.0592 \log \frac{\cancel{[Ti^{3+}][Ti^{2+}]}}{\cancel{[TiO^{2+}][Ti^{3+}]}[H^+]^2}$$

$$E_{eq} = -\frac{0.270}{2} - \frac{0.0592}{2} \log \frac{1}{1 \times 10^{-2}} = -0.135 - 0.0592 = \underline{\underline{-0.194 \text{ V}}}$$

16-13

	E_{eq}, V	Indicator
(a)	0.258	Phenosafranine
(b)	-0.024	None
(c)	0.444	Indigo tetrasulfonate or Methylene blue
(d)	1.09	1,10-Phenanthroline iron(II) complex
(e)	0.951	Erioglaucin A
(f)	0.330	Indigo tetrasulfonate
(g)	-0.008	None
(h)	-0.194	None

Chapter 16

16-14 (a) $2V^{2+} + Sn^{4+} \rightleftarrows 2V^{3+} + Sn^{2+}$

10.00 mL

$[V^{3+}] = (10.00 \times 0.0500 \times 2)/60.00 = 1.67 \times 10^{-2}$

$[V^{2+}] = (50.00 \times 0.1000 - 10.00 \times 0.05000 \times 2)/60.00$

$= 6.67 \times 10^{-2}$

$E = -0.256 - 0.0592 \log(6.67 \times 10^{-2}/1.67 \times 10^{-2}) = \underline{\underline{-0.292 \text{ V}}}$

The remaining preequivalence point data are treated in the same way. The answers appear in the table that follows.

50.00 mL

Proceeding as in Problem 16-12, we write

$E = -0.256 - 0.0592 \log [V^{2+}]/[V^{3+}]$

$2E = 2 \times 0.154 - 0.0592 \log [Sn^{2+}]/[Sn^{4+}]$

$3E = -0.256 + 0.308 - 0.0592 \log \dfrac{[V^{2+}][Sn^{2+}]}{[V^{3+}][Sn^{4+}]}$

At equivalence, $[V^{2+}] = 2[Sn^{4+}]$ and $[V^{3+}] = 2[Sn^{2+}]$. Thus,

$E_{eq} = (-0.256 + 0.308)/3 - (0.0592/3) \log 1.00 = \underline{\underline{0.017 \text{ V}}}$

50.10 mL

$c_{Sn^{2+}} = 50.00 \times 0.1000/(2 \times 100.0) = 0.0250 \approx [Sn^{2+}]$

$c_{Sn^{4+}} = \dfrac{50.10 \times 0.0500 - 50.00 \times 0.1000/2}{100.0} = 5.00 \times 10^{-5}$

$E = 0.154 - \dfrac{0.0592}{2} \log \dfrac{[Sn^{2+}]}{[Sn^{4+}]} = 0.154 - 0.0296 \log \dfrac{0.0250}{5.00 \times 10^{-5}}$

$= \underline{\underline{0.074 \text{ V}}}$

Chapter 16

The remaining data, obtained in the same way, are given in the table that follows.

(b) $Fe(CN)_6^- + Cr^{2+} \rightleftharpoons Fe(CN)_6^{4-} + Cr^{3+}$

The preequivalence point data are obtained by substituting concentrations into the equation

$$E = 0.36 - 0.0592 \log [Fe(CN)_6^{4-}]/[Fe(CN)_6^{3-}]$$

The post equivalence point data are obtained with the Nernst expression for the Cr^{2+}/Cr^{3+} system. That is,

$$E = -0.0408 - 0.0592 \log ([Cr^{2+}]/[Cr^{3+}])$$

The equivalence point potential is found following the procedure in Problem 16-12. The results are found in the table that follows.

(c) The data for this titration, which are given in the table that follows, are obtained in the same way as those for parts (a) and (b).

(d) The data for this titration, which are given in the table that follows, are obtained in the same way as those for parts (a) and (b).

(e) $2MnO_4^- + 5U^{4+} + 2H_2O \rightleftharpoons 2Mn^{2+} + 5UO_2^{2+} + 4H^+$

10.00 mL

10.00 mL $KMnO_4 \times 0.02000$ mmol $KMnO_4$/mL = 0.2000 mmol $KMnO_4$

50.00 mL $U^{4+} \times 0.05000$ mmol U^{4+}/mL = 2.500 mmol U^{4+}

$$c_{UO_2^{2+}} = [UO_2^{2+}] = \frac{0.2000 \text{ mmol KMnO}_4 \times 5 \text{ mmol UO}_2^{2+}/2 \text{ mmol KMnO}_4}{60.00 \text{ mL solution}}$$

$$= 0.5000 \text{ mmol } UO_2^+/60.00 \text{mL} = 8.33 \times 10^{-3} \text{ M } UO_2^+$$

$$c_{U^{4+}} = [U^{4+}] = \frac{2.5000 \text{ mmol } U^{4+} - 0.5000 \text{ mmol } UO_2^{2+}}{60.00 \text{ mL solution}}$$

$$= 2.000 \text{ mmol } U^{4+}/60.00 \text{ mL} = 3.33 \times 10^{-2} \text{ M}$$

$$E = 0.334 - 0.0296 \log [3.33 \times 10^{-2} / 8.33 \times 10^{-3} (1.00)^4]$$

$$= \underline{\underline{0.316 \text{ V}}}$$

Additional preequivalence point data, obtained in the same way, are given in the table that follows.

50.00 mL

$$5E_{eq} = 5 \times 1.51 - 0.0592 \log \frac{[Mn^{2+}]}{[MnO_4^-][H^+]^8}$$

$$2E_{eq} = 2 \times 0.334 - 0.0592 \log \frac{[U^{4+}]}{[UO_2^{2+}][H^+]^4}$$

Adding the two equations gives

$$7E_{eq} = 8.218 - 0.0592 \log \frac{[Mn^{2+}][U^{4+}]}{[MnO_4^-][UO_2^{2+}][H^+]^{12}}$$

At equivalence, $[MnO_4^-] = \frac{2}{5}[U^{4+}]$ and $[Mn^{2+}] = \frac{2}{5}[UO_2^{2+}]$

Substituting these equalities and $[H^+] = 1.00$ into this gives

$$E_{eq} = \frac{8.218}{7} - \frac{0.0592}{7} \log 1.00 = \underline{\underline{1.17 \text{ V}}}$$

50.10 mL

total no. mmol $KMnO_4$ taken $= 50.10 \times 0.02000 = 1.0020$

no. mmol Mn^{2+} formed $= 50.00 \times 0.05000 \times 2/5 = 1.0000$

no. mmol $KMnO_4$ remaining $= 0.0020$

$$E = 1.51 - \frac{0.0592}{5} \log \frac{1.00/100.1}{(0.0020/100.1)(1.00)^8} = \underline{\underline{1.48 \text{ V}}}$$

The remaining post equivalence point data are derived in the same way and are given in the table that follows.

		E, V			
Vol, mL	(a)	(b)	(c)	(d)	(e)
10.00	-0.292	0.40	0.32	0.807	0.316
25.00	-0.256	0.36	0.36	0.771	0.334
49.00	-0.156	0.26	0.46	0.671	0.384
49.90	-0.097	0.20	0.52	0.612	0.414
50.00	0.017	-0.02	0.95	0.36	1.17
50.10	0.074	-0.248	1.17	0.234	1.48
51.00	0.104	-0.308	1.20	0.204	1.49
60.00	0.133	-0.367	1.23	0.175	1.50

CHAPTER 17

17-1 (a) $2Mn^{2+} + 5S_2O_8^{2-} + 8H_2O \rightarrow 10SO_4^{2-} + 2MnO_4^- + 16H^+$

(b) $NaBiO_3(s) + 2Ce^{3+} + 4H^+ \rightarrow BiO^+ + 2Ce^{4+} + 2H_2O + Na^+$

(c) $H_2O_2 + U^{4+} \rightarrow UO_2^{2+} + 2H^+$

(d) $V(OH)_4^+ + Ag(s) + Cl^- + 2H^+ \rightarrow VO^{2+} + AgCl(s) + 3H_2O$

(e) $2MnO_4^- + 5H_2O_2 + 6H^+ \rightarrow 5O_2 + 2Mn^{2+} + 8H_2O$

(f) $ClO_3^- + 6I^- + 6H^+ \rightarrow 3I_2 + Cl^- + 3H_2O$

17-2 (a) $2Fe^{3+} + SO_2(g) + 2H_2O \rightarrow 2Fe^{2+} + SO_4^{2-} + 4H^+$

(b) $2H_2MoO_4 + 3Zn(s) + 12H^+ \rightarrow 3Zn^{2+} + 2MoO^{3+} + 8H_2O$

(c) $2MnO_4^- + 5HNO_2 + H^+ \rightarrow 2Mn^{2+} + 5NO_3^- + 3H_2O$

(d) $BrO_3^- + 5Br^- + C_6H_5NH_2 + 3H^+ \rightarrow Br_3C_6H_2NH_2 + 3Br^- + 3H_2O$

(e) $2HAsO_3^{2-} + O_2 \rightarrow 2HAsO_4^{2-}$

(f) $2I^- + 2HNO_2 + 2H^+ \rightarrow I_2 + 2NO + 2H_2O$

17-3 Only in the presence of Cl⁻ ion is Ag a sufficiently good reducing agent to be very useful for prereductions. In the presence of Cl⁻ ion, the half-reaction occurring in the Walden reductor is

$$Ag(s) + Cl^- \rightarrow AgCl(s) + e^-$$

The excess HCl increases the tendency of this reaction to occur by the common ion effect.

17-4 Amalgamation of the zinc prevents loss of reagent by reaction of the zinc with hydronium ions.

17-5 $UO_2^{2+} + 2Ag(s) + 4H^+ + 2Cl^- \rightleftharpoons U^{4+} + 2AgCl(s) + H_2O$

Chapter 17

17-6 $2TiO^{2+} + Zn(s) + 4H^+ \rightarrow 2Ti^{3+} + Zn^{2+} + 2H_2O$

17-7 Standard solutions of reductants find somewhat limited use because of their susceptibility to air oxidation.

17-8 Standard KMnO$_4$ solutions are seldom used to titrate solutions containing HCl because of the tendency of MnO_4^- to oxidize Cl$^-$ to Cl$_2$, thus causing an over-consumption of MnO_4^-.

17-9 Cerium(IV) precipitates as a basic oxide in alkaline solution.

17-10 $2MnO_4^- + 3Mn^{2+} + 2H_2O \rightarrow 5MnO_2(s) + 4H^+$

17-11 Freshly prepared solutions of permanganate are inevitably contaminated with small amounts of solid manganese dioxide, which catalyzes the further decomposition of permanganate ion. By removing the dioxide at the outset, a much more stable standard reagent is produced.

17-12 Standard permanganate and thiosulfate solutions are generally stored in the dark because their decomposition reactions are catalyzed by light.

17-13 $4MnO_4^- + 2H_2O \rightarrow \underset{\text{brown}}{4MnO_2(s)} + 3O_2 + 4OH^-$

17-14 Solutions of K$_2$Cr$_2$O$_7$ are used extensively for back-titrating solutions of Fe^{2+} when the latter is being used as a standard reductant for the determination of oxidizing agents.

17-15 Iodine is not sufficiently soluble in water to produce a useful standard reagent. It is quite soluble in solutions that contain an excess of iodide, however, as a consequence of the formation of the triiodide complex. The rate at which iodine dissolves in iodide solutions increases as the concentration of iodide ion becomes greater. For this reason, iodine is always dissolved in a very concentrated solution of potassium iodide and diluted only after solution is complete.

17-16 The solution becomes stronger because of air oxidation of the excess I$^-$ present, which increases the I_3^- concentration. The reaction is

$$6I^- + O_2(g) + 4H^+ \rightarrow 2I_3^- + 2H_2O$$

17-17 $S_2O_3^{2-} + H^+ \rightarrow HSO_3^- + S(s)$

Chapter 17

17-18 When a measured volume of a standard solution of KIO$_3$ is introduced into an acidic solution containing an excess of iodide ion, a known amount of iodine is produced as a consequence of the reaction

$$IO_3^- + 5I^- + 6H^+ \rightarrow 3I_2 + 3H_2O$$

17-19 $BrO_3^- + \underset{\text{excess}}{6I^-} + 6H^+ \rightarrow Br^- + 3I_2 + 3H_2O$

$$I_2 + 2S_2O_3^{2-} \rightarrow 2I^- + S_4O_6^{2-}$$

17-20 $Cr_2O_7^{2-} + \underset{\text{excess}}{6I^-} + 14H^+ \rightarrow 2Cr^{3+} + 3I_2 + 7H_2O$

$$I_2 + 2S_2O_3^{2-} \rightarrow 2I^- + S_4O_6^{2-}$$

17-21 $2I_2 + N_2H_4 \rightarrow N_2 + 4H^+ + 4I^-$

17-22 Starch is decomposed in the presence of high concentrations of iodine to give products that do not behave satisfactorily as indicators. This reaction is prevented by delaying the addition of the starch until the iodine concentration is very small.

17-23
$$0.2464 \text{ g} \times \frac{1 \text{ mmol Fe}}{0.055847 \text{ g Fe}} = 4.4121 \text{ mmol Fe}$$

(a) $\dfrac{4.4121 \text{ mmol Fe}}{39.31 \text{ mL}} \times \dfrac{\text{mmol Ce}^{4+}}{\text{mmol Fe}} = \underline{\underline{0.1122 \text{ M Ce}^{4+}}}$

(b) $\dfrac{4.4121 \text{ mmol Fe}}{39.31 \text{ mL}} \times \dfrac{\text{mmol Cr}_2O_7^{2-}}{6 \text{ mmol Fe}} = \underline{\underline{0.01871 \text{ M Cr}_2O_7^{2-}}}$

(c) $\dfrac{4.4121 \text{ mmol Fe}}{39.31 \text{ mL}} \times \dfrac{\text{mmol MnO}_4^-}{5 \text{ mmol Fe}} = \underline{\underline{0.02245 \text{ M MnO}_4^-}}$

(d) $\dfrac{4.4121 \text{ mmol Fe}}{39.31 \text{ mL}} \times \dfrac{\text{mmol V(OH)}_4^+}{\text{mmol Fe}} = \underline{\underline{0.1122 \text{ M V(OH)}_4^+}}$

(e) $\dfrac{4.4121 \text{ mmol Fe}}{39.31 \text{ mL}} \times \dfrac{\text{mmol IO}_3^-}{4 \text{ mmol Fe}} = \underline{\underline{0.02806 \text{ M IO}_3^-}}$

Chapter 17

17-24 $250 \text{ mL} \times 0.03500 \dfrac{\text{mmol K}_2\text{Cr}_2\text{O}_7}{\text{mL}} \times \dfrac{0.294185 \text{ g}}{\text{mmol K}_2\text{Cr}_2\text{O}_7} = 2.57412 \text{ g}$

Dissolve 2.574 g $K_2Cr_2O_7$ in sufficient water to give 250.0 mL of solution.

17-25 $2000 \text{ mL} \times \dfrac{0.02500 \text{ mmol KBrO}_3}{\text{mL}} \times \dfrac{0.167001 \text{ g}}{\text{mmol KBrO}_3} = 8.35005 \text{ g}$

Dissolve 8.350 g of $KBrO_3$ in sufficient water to give 2.000 L of solution.

17-26 $1.5 \text{ L} \times \dfrac{0.1 \text{ mol KMnO}_4}{\text{L}} \times \dfrac{158.034 \text{ g}}{\text{mol KMnO}_4} = 23.705 \text{ g}$

Dissolve about 24 g of $KMnO_4$ in 1.5 L of water.

17-27 $2.0 \text{ L} \times \dfrac{0.05 \text{ mol I}_2}{\text{L}} \times \dfrac{253.809 \text{ g}}{\text{mol I}_2} = 25.38 \text{ g}$

Dissolve about 26 g I_2 in a concentrated solution of KI; then dilute to about 2 L.

17-28 $\dfrac{0.1467 \text{ g Na}_2\text{Ox}}{28.85 \text{ mL KMnO}_4} \times \dfrac{1 \text{ mmol Na}_2\text{Ox}}{0.133999 \text{ g Na}_2\text{Ox}} \times \dfrac{2 \text{ mmol KMnO}_4}{5 \text{ mmol Na}_2\text{Ox}} = \underline{\underline{0.01518 \text{ M KMnO}_4}}$

17-29 $\dfrac{0.1809 \text{ g Fe}}{31.33 \text{ mL Ce}} \times \dfrac{1 \text{ mmol Fe}}{0.055847 \text{ g Fe}} \times \dfrac{1 \text{ mmol Ce}}{\text{mmol Fe}} = \underline{\underline{0.1034 \text{ M Ce}^{3+}}}$

17-30 $\text{Cr}_2\text{O}_7^{2-} + 6\text{I}^- + 14\text{H}^+ = 2\text{Cr}^{3+} + 3\text{I}_2 + 7\text{H}_2\text{O}$

$1 \text{ mmol Cr}_2\text{O}_7^{2-} \equiv 3 \text{ mmol I}_2 \equiv 6 \text{ mmol S}_2\text{O}_3^{2-}$

$\dfrac{0.1518 \text{ g K}_2\text{Cr}_2\text{O}_7}{46.13 \text{ mL Na}_2\text{S}_2\text{O}_3} \times \dfrac{1 \text{ mmol K}_2\text{Cr}_2\text{O}_7}{0.294185 \text{ g K}_2\text{Cr}_2\text{O}_7} \times \dfrac{6 \text{ mmol S}_2\text{O}_3^{2-}}{\text{mmol K}_2\text{Cr}_2\text{O}_7} = \underline{\underline{0.06711 \text{ M Na}_2\text{S}_2\text{O}_3}}$

17-31 $\text{BrO}_3^- + 6\text{I}^- + 6\text{H}^+ = \text{Br}^- + 3\text{I}_2 + 3\text{H}_2\text{O}$

$1 \text{ mmol BrO}_3^- \equiv 3 \text{ mmol I}_2 \equiv 6 \text{ mmol S}_2\text{O}_3^{2-}$

$\dfrac{0.1017 \text{ g KBrO}_3}{39.75 \text{ mL Na}_2\text{S}_2\text{O}_3} \times \dfrac{1 \text{ mmol KBrO}_3}{0.167001 \text{ g KBrO}_3} \times \dfrac{6 \text{ mmol Na}_2\text{S}_2\text{O}_3}{\text{mmol KBrO}_3} = \underline{\underline{0.09192 \text{ M Na}_2\text{S}_2\text{O}_3}}$

17-32 1 mmol I_2 ≡ 1 mmol Sb ≡ 2 mmol Sb_2S_3

$$\frac{41.67 \text{ mL} \times 0.03134 \frac{\text{mmol } I_2}{\text{mL } I_2} \times \frac{1 \text{ mmol Sb}}{\text{mmol } I_2} \times 0.12176 \frac{\text{g Sb}}{\text{mmol Sb}}}{1.080 \text{ g sample}} \times 100\% = \underline{\underline{14.72\% \text{ Sb}}}$$

$$\frac{41.67 \times 0.03134 \times \frac{1}{2} \times 0.33971}{1.080} \times 100\% = \underline{\underline{20.54\% \text{ Sb}_2S_3}}$$

17-33 1 mmol MnO_2 ≡ 1 mmol I_2 ≡ 2 mmol $Na_2S_2O_3$

$$\frac{32.30 \text{ mmol Na}_2S_2O_3}{0.1344 \text{ g sample}} \times 0.07220 \frac{\text{mmol Na}_2S_2O_3}{\text{mL Na}_2S_2O_3} \times \frac{1 \text{ mmol MnO}_2}{2 \text{ mmol Na}_2S_2O_3} \times$$

$$\frac{0.086937 \text{ g MnO}_2}{\text{mmol MnO}_2} \times 100\% = \underline{\underline{75.42\% \text{ MnO}_2}}$$

17-34 Letting A = $CS(NH_2)_2$, 4 mol $KBrO_3$ = 3 mol A

$$\frac{14.1 \text{ mL KBrO}_3}{0.0715 \text{ g sample}} \times 0.00833 \times \frac{\text{mmol KBrO}_3}{\text{mL KBrO}_3} \times \frac{3 \text{ mmol A}}{4 \text{ mmol KBrO}_3} \times$$

$$\frac{0.076122 \text{ g A}}{\text{mmol A}} \times 100\% = \underline{\underline{9.38\% \text{ A}}}$$

17-35 1 mmol $KMnO_4$ ≡ 5 mmol Fe ≡ $\frac{5}{2}$ mmol Fe_2O_3

no. mmol $KMnO_4$ = 39.21 mL $KMnO_4 \times \frac{0.02086 \text{ mmol KMnO}_4}{\text{mL KMnO}_4}$ = 0.81792

(a) $\frac{0.81792 \text{ mmol KMnO}_4}{0.7120 \text{ g sample}} \times \frac{5 \text{ mmol Fe}}{\text{mmol KMnO}_4} \times \frac{0.055847 \text{ g Fe}}{\text{mmol Fe}} \times 100\% = \underline{\underline{32.08\% \text{ Fe}}}$

(b) $\frac{0.81792 \text{ mmol KMnO}_4}{0.7120 \text{ g sample}} \times \frac{5 \text{ mmol Fe}_2O_3}{2 \text{ mmol KMnO}_4} \times \frac{0.159692 \text{ g Fe}_2O_3}{\text{mmol Fe}_2O_3} \times 100\% =$

$\underline{\underline{45.86\% \text{ Fe}_2O_3}}$

Chapter 17

17-36 $3Sn^{2+} + Cr_2O_7^{2-} + 14H^+ \rightleftarrows 3Sn^{4+} + 2Cr^{3+} + 7H_2O$

1 mmol $Cr_2O_7^{2-}$ ≡ 3 mmol Sn ≡ 3 mmol SnO_2

$$29.77 \text{ mL K}_2\text{Cr}_2\text{O}_7 \times 0.01735 \frac{\text{mmol K}_2\text{Cr}_2\text{O}_7}{\text{mL K}_2\text{Cr}_2\text{O}_7} \times \frac{3 \text{ mmol Sn}}{1 \text{ mmol K}_2\text{Cr}_2\text{O}_7} = 1.54953 \text{ mmol Sn}$$

$$= 1.54953 \text{ mmol SnO}_2$$

(a) $\dfrac{1.54953 \text{ mmol Sn}}{0.4352 \text{ g sample}} \times \dfrac{0.11871 \text{ g Sn}}{\text{mmol Sn}} \times 100\% = \underline{\underline{42.27\% \text{ Sn}}}$

(b) $\dfrac{1.54953 \text{ mmol SnO}_2}{0.4352 \text{ g sample}} \times \dfrac{0.15071 \text{ g SnO}_2}{\text{mmol SnO}_2} \times 100\% = \underline{\underline{53.66\% \text{ SnO}_2}}$

17-37 1 mmol $K_2Cr_2O_7$ ≡ 6 mmol Fe ≡ 3 mmol H_2NOH

$$\frac{23.61 \text{ mL K}_2\text{Cr}_2\text{O}_7}{50.00 \text{ mL sample}} \times 0.02170 \frac{\text{mmol K}_2\text{Cr}_2\text{O}_7}{\text{mL K}_2\text{Cr}_2\text{O}_7} \times \frac{3 \text{ mmol H}_2\text{NOH}}{\text{mmol K}_2\text{Cr}_2\text{O}_7} = \underline{\underline{0.03074 \text{ M}}}$$

17-38 2 mmol $KMnO_4$ ≡ 5 mmol $Na_2C_2O_4$ ≡ 5 mmol ZnO

$$\frac{37.81 \text{ mL KMnO}_4}{0.9280 \text{ g sample}} \times 0.01508 \frac{\text{mmol KMnO}_4}{\text{mL KMnO}_4} \times \frac{5 \text{ mmol ZnO}}{2 \text{ mmol KMnO}_4} \times$$

$$\frac{0.08139 \text{ g ZnO}}{\text{mmol ZnO}} \times 100\% = \underline{\underline{12.50\% \text{ ZnO}}}$$

17-39

no. mmol Fe^{2+} = $50.00 \text{ mL Fe}^{2+} \times 0.09601 \dfrac{\text{mmol Fe}^{2+}}{\text{mmol Fe}^{2+}} = 4.8005$

no. mmol Fe^{2+} titrated by Ce^{4+} = $12.99 \text{ mL Ce}^{4+} \times 0.08362 \dfrac{\text{mmol Ce}^{4+}}{\text{mL}} \times \dfrac{1 \text{ mmol Fe}^{2+}}{\text{mmol Ce}^{4+}}$

= 1.08622

no. mmol Fe^{2+} consumed by analyte = 4.8005 − 1.08622 = 3.71428

$$\frac{3.71428 \text{ mmol Fe}^{2+}}{0.1342 \text{ g sample}} \times \frac{1 \text{ mmol KClO}_3}{6 \text{ mmol Fe}^{2+}} \times \frac{0.122549 \text{ g KClO}_3}{\text{mmol KClO}_3} \times 100\% = \underline{\underline{56.53\% \text{ KClO}_3}}$$

17-40 1 mmol I_2 ≡ 2 mmol $Na_2S_2O_3$ ≡ 1 mmol A   (A = analyte)

total no. mmol I_2 = 15.00 mL I_2 × 0.02095 $\dfrac{\text{mmol } I_2}{\text{mL } I_2}$ = 0.31425

no. mmol I_2 consumed by $Na_2S_2O_3$ = 6.09 mL $Na_2S_2O_3$ × 0.03465 $\dfrac{\text{mmol}}{\text{mL}}$ × $\dfrac{1 \text{ mmol } I_2}{2 \text{ mmol } Na_2S_2O_3}$

= 0.10551

no. mmol I_2 consumed by analyte = 0.31425 − 0.10551 = 0.20874

$\dfrac{0.20874 \text{ mmol } I_2}{25.00 \text{ mL sample}} \times \dfrac{1 \text{ mmol A}}{\text{mmol } I_2} \times \dfrac{323.4 \text{ mg A}}{\text{mmol A}} \times \dfrac{1000 \text{ mL sample}}{\text{L sample}} = \underline{\underline{2700 \text{ mg A / L sample}}}$

17-41 $H_3AsO_3 + I_2 + H_2O \rightleftarrows H_3AsO_4 + 2I^- + 2H^+$

2 mmol I_2 ≡ 2 mmol H_2AsO_3 ≡ 1 mmol As_2O_3

$\dfrac{23.77 \text{ mL } I_2}{8.13 \text{ g sample}} \times 0.02425 \dfrac{\text{mmol } I_2}{\text{mL } I_2} \times \dfrac{1 \text{ mmol } As_2O_3}{2 \text{ mmol } I_2} \times \dfrac{0.197841 \text{ g } As_2O_3}{\text{mmol } As_2O_3} \times 100\% =$

$\underline{\underline{0.701\% \; As_2O_3}}$

17-42 $Cr_2O_7^{2-} + 3U^{4+} + 2H^+ \rightleftarrows 3UO_2^{2+} + 2Cr^{3+} + H_2O$

1 mmol $K_2Cr_2O_7$ ≡ 3 mmol U^{4+} ≡ 1 mmol NaCl

$\dfrac{19.9 \text{ mL } K_2Cr_2O_7}{0.800 \, (25.0/500.0) \text{ g sample}} \times 0.01000 \dfrac{\text{mmol } K_2Cr_2O_7}{\text{mL } K_2Cr_2O_7} \times \dfrac{1 \text{ mmol NaCl}}{\text{mmol } K_2Cr_2O_7} \times$

$0.058442 \dfrac{\text{g NaCl}}{\text{mmol NaCl}} \times 100\% = \underline{\underline{29.1\% \text{ NaCl}}}$

17-43

no. mmol I_2 taken = 50.0 mL × 0.01194 $\dfrac{\text{mmol } I_2}{\text{mL } I_2}$ = 0.5970

Chapter 17

$$\text{no. mmol I}_2 \text{ consumed by Na}_2\text{S}_2\text{O}_3 = 16.77 \text{ mL} \times 0.01325 \frac{\text{mmol}}{\text{mL}} \times \frac{1 \text{ mmol I}_2}{2 \text{ mmol Na}_2\text{S}_2\text{O}_3}$$

$$= 0.1111$$

no. mmol I_2 consumed by analyte = $0.5970 - 0.1111$ = 0.4859

$$\frac{0.4859 \text{ mmol I}_2}{1.657 \text{ g sample}} \times \frac{2 \text{ mmol C}_2\text{H}_5\text{SH}}{\text{mmol I}_2} \times 0.06214 \frac{\text{g C}_2\text{H}_5\text{SH}}{\text{mmol C}_2\text{H}_5\text{SH}} \times 100\% = \underline{\underline{3.64\% \text{ C}_2\text{H}_5\text{SH}}}$$

17-44

$$\text{no. mmol K}_2\text{Cr}_2\text{O}_7 = 50.00 \text{ mL} \times 0.03114 \frac{\text{mmol K}_2\text{Cr}_2\text{O}_7}{\text{mL}} = 1.55700$$

$$\text{no. mmol K}_2\text{Cr}_2\text{O}_7 \text{ consumed by Fe}^{2+} = 10.05 \text{ mL Fe}^{2+} \times 0.1135 \frac{\text{mmol}}{\text{mL}} \times$$

$$\frac{1 \text{ mmol K}_2\text{Cr}_2\text{O}_7}{6 \text{ mmol Fe}^{2+}} = 0.19011$$

no. mmol $K_2Cr_2O_7$ for analyte = $1.55700 - 0.19011$ = 1.36689

$$\frac{1.36689 \text{ mmol K}_2\text{Cr}_2\text{O}_7}{4.971 \text{ g sample}} \times \frac{3 \text{ mmol TeO}_2}{\text{mmol K}_2\text{Cr}_2\text{O}_7} \times \frac{0.1596 \text{ g TeO}_2}{\text{mmol TeO}_2} \times 100\% = \underline{\underline{13.17\% \text{ TeO}_2}}$$

17-45 1 mmol KI ≡ 1 mmol IO_3^- ≡ 3 mmol I_2 ≡ 6 mmol $Na_2S_2O_3$

$$\frac{20.66 \text{ mL Na}_2\text{S}_2\text{O}_3}{1.204 \text{ g sample}} \times 0.05551 \frac{\text{mmol Na}_2\text{S}_2\text{O}_3}{\text{mL}} \times \frac{\text{mmol KI}}{6 \text{ mmol Na}_2\text{S}_2\text{O}_3} \times$$

$$0.16600 \frac{\text{g KI}}{\text{mmol KI}} \times 100\% = \underline{\underline{2.635\% \text{ KI}}}$$

17-46

$$\text{no. mmol Fe} = (13.72 \times 0.01920) \text{ mmol KMnO}_4 \times \frac{5 \text{ mmol Fe}}{\text{mmol KMnO}_4} \times \frac{500.0 \text{ mL}}{50.00 \text{ mL}}$$

$$= 13.171 \text{ mmol Fe}$$

Similarly,

no. mmol (Fe + Cr) = $36.43 \times 0.01920 \times 5 \times 500/100$ = 17.486

no. mmol Cr = $17.486 - 13.171$ = 4.315

256

$$\frac{13.171 \text{ mmol Fe} \times 0.055847 \text{ g Fe/mmol}}{1.065 \text{ g sample}} \times 100\% = \underline{\underline{69.07\% \text{ Fe}}}$$

$$\frac{4.315 \text{ mmol Cr} \times 0.051996 \text{ g Cr/mmol}}{1.065 \text{ g sample}} \times 100\% = \underline{\underline{21.07\% \text{ Cr}}}$$

17-47 In Walden reductor, $V(OH)_4^+ + 2H^+ + e^- \rightarrow VO^{2+} + 3H_2O$

In Jones reductor, $V(OH)_4^+ + 4H^+ + 3e^- \rightarrow V^{2+} + 4H_2O$

In first titration,

no. mmol Ce^{4+} = 17.74×0.1000 = no. mmol Fe + no. mmol V

In second titration,

no. mmol Ce^{4+} = 44.67×0.1000 = no. mmol Fe + $3 \times$ no. mmol V

Subtracting the first relationship from the second gives

$(44.67 - 17.74)\, 0.1000 = 2.6930 = 2 \times$ no. mmol V

no. mmol V = 2.6930/2 = 1.3465;   no. mmol V_2O_5 = 1.3465/2 = 0.67325

no. mmol Fe = 1.7740 − 1.3465 = 0.4275

no. mmol Fe_2O_3 = 0.4275/2 = 0.21375

$$\frac{0.67325 \text{ mmol } V_2O_5 \times 0.18188 \text{ g } V_2O_5/\text{mmol}}{2.559 \text{ g sample} \times 50.00 \text{ mL}/500.0 \text{ mL}} \times 100\% = \underline{\underline{47.85\% \, V_2O_5}}$$

$$\frac{0.21375 \text{ mmol } Fe_2O_3 \times 0.15969 \text{ g } Fe_2O_3/\text{mmol}}{2.559 \text{ g sample} \times 50.00 \text{ mL}/500.0 \text{ mL}} \times 100\% = \underline{\underline{13.34\% \, Fe_2O_3}}$$

17-48 6 mmol Fe^{2+} ≡ 1 mmol $Cr_2O_7^{2-}$ ≡ 4 mmol Tl^+

$$(40.60 \times 0.1004) \text{ mmol Fe} \times \frac{4 \text{ mmol Tl}}{6 \text{ mmol Fe}} \times 0.20438 \frac{\text{g Tl}}{\text{mmol Tl}} = \underline{\underline{0.5554 \text{ g Tl}}}$$

17-49 1 mmol IO_3^- ≡ 2 mmol H_2SO_3 ≡ 2 mmol SO_2

In $2.50 \frac{L}{\text{min}} \times 64.0 \text{ min} = 160.0$ L of sample, there are

Chapter 17

$$4.98 \text{ mL IO}_3^- \times 0.003125 \frac{\text{mmol IO}_3^-}{\text{mL IO}_3^-} \times \frac{2 \text{ mmol SO}_2}{\text{mmol IO}_3^-} \times \frac{0.064065 \text{ g SO}_2}{\text{mmol SO}_2} = 0.0019940 \text{ g SO}_2$$

$$\frac{0.001994 \text{ g SO}_2}{160.0 \text{ L} \times 1.2 \text{ g/L}} \times 10^6 \text{ ppm} = 10.39 \approx \underline{\underline{10.4 \text{ ppm SO}_2}}$$

17-50 $1 \text{ mmol I}_2 \equiv 5 \text{ mmol CO} \equiv 2 \text{ mmol S}_2\text{O}_3^{2-}$

$$7.76 \text{ mL} \times 0.00221 \frac{\text{mmol S}_2\text{O}_3}{\text{mL}} \times \frac{5 \text{ mmol CO}}{2 \text{ mmol S}_2\text{O}_3^{2-}} \times \frac{0.02801 \text{ g}}{\text{mmol CO}} = 0.0012009 \text{ g CO}$$

$$\frac{0.0012009 \text{ g CO}}{24.7 \text{ L} \times 1.2 \text{ g/L}} \times 10^6 \text{ ppm} = \underline{\underline{40.5 \text{ ppm}}}$$

Work place is in compliance.

17-51 $1 \text{ mmol I}_2 \equiv 1 \text{ mmol H}_2\text{S} \equiv 2 \text{ mmol Na}_2\text{S}_2\text{O}_3$

$$\text{no. mmol I}_2 \text{ taken} = 10.00 \text{ mL} \times 0.01070 \frac{\text{mmol I}_2}{\text{mL}} = 0.1070$$

$$\text{no. mmol I}_2 \text{ in excess} = 12.85 \text{ mL} \times 0.01344 \frac{\text{mmol Na}_2\text{S}_2\text{O}_3}{\text{mL}} \times \frac{1 \text{ mmol I}_2}{2 \text{ mmol Na}_2\text{S}_2\text{O}_3}$$

$$= 0.08635$$

$$\text{no. mmol I}_2 = \text{mmol H}_2\text{S} = 0.1070 - 0.08635 = 0.02065$$

$$\frac{0.02065 \text{ mmol H}_2\text{S}}{30.00 \text{ L sample} \times 1.2 \text{ g sample/L}} \times 0.034082 \frac{\text{g H}_2\text{S}}{\text{mmol}} \times 10^6 \text{ ppm} = \underline{\underline{19.5 \text{ ppm H}_2\text{S}}}$$

17-52 (a) $\text{AgI} + 2\text{S}_2\text{O}_3^{2-} \rightarrow \text{Ag(S}_2\text{O}_3)_2^{3-} + \text{I}^-$

$3\text{Br}_2 + \text{I}^- + 3\text{H}_2\text{O} \rightarrow \text{IO}_3^- + 6\text{Br}^- + 6\text{H}^+$

$5\text{I}^- + \text{IO}_3^- + 6\text{H}^+ \rightarrow 3\text{I}_2 + 3\text{H}_2\text{O}$

$3\text{I}_2 + 6\text{S}_2\text{O}_3^{2-} \rightarrow 6\text{I}^- + 3\text{S}_4\text{O}_6^{2-}$

(b) $1 \text{ mmol IO}_3^- \equiv 3 \text{ mmol I}_2 \equiv 1 \text{ mmol AgI} \equiv 6 \text{ mmol S}_2\text{O}_3^{2-}$

$$\frac{(13.7 \times 0.0352) \text{ mmol Na}_2\text{S}_2\text{O}_3 \times \frac{1 \text{ mmol AgI}}{6 \text{ mmol Na}_2\text{S}_2\text{O}_3} \times \frac{234.77 \text{ mg AgI}}{\text{mmol AgI}}}{4.00 \text{ cm}^2} = \underline{\underline{4.72 \text{ mg AgI/cm}^2}}$$

CHAPTER 18

18-1 (a) An indicator electrode is an electrode used in potentiometry that responds to variations in the activity of an analyte ion or molecule.

(b) A reference electrode is an electrode whose potential is known, constant, and independent of the type of solution in which it is immersed.

(c) An electrode of the first kind is a metal electrode that is used to determine the concentration of the cation of that metal in a solution.

(d) An electrode of the second kind is a metal electrode that is used to determine the concentration of an anion that forms a precipitate or a stable complex with the cation of the electrode metal.

18-2 (a) A liquid-junction potential is the potential that develops across the interface between two solutions having different electrolyte compositions.

(b) A boundary potential is the potential that develops across an ion-sensitive membrane when the two sides of the membrane are bathed in solutions having different concentrations of the ion to which the membrane is sensitive.

(c) The asymmetry potential is a potential that develops across an ion-sensitive membrane when the concentrations of that ion are the same on either side of the membrane. This potential arises from dissimilarities between the inner and outer surface of the membrane.

18-3 (a) An electrode of the first kind for Hg(II) would take the form

$$\|Hg^{2+}(x\ M)|Hg$$

$$E_{Hg} = E^0_{Hg} - \frac{0.0592}{2}\log\frac{1}{[Hg^{2+}]} = E^0_{Hg} + \frac{0.0592}{2}pHg$$

(b) An electrode of the second kind for EDTA would take the form

$$\|HgY^{2-}(y\ M),Y^{4-}(x\ M)|Hg$$

where a small and fixed amount of HgY^{2-} is introduced into the analyte solution. Here the potential of the mercury electrode is given by

$$E_{Hg} = K - \frac{0.0592}{2}\log[Y^{4-}] = K + \frac{0.0592}{2}pY$$

where
$$K = E^0_{HgY^{2-}} - \frac{0.0592}{2}\log\frac{1}{a_{HgY^{2-}}} \approx 0.21 - \frac{0.0592}{2}\log\frac{1}{[HgY^{2-}]}$$

18-4 For the process $M^{n+} + ne^- \rightleftarrows M(s)$

$$E = E^0 - \frac{0.0592}{n}\log\frac{1}{[M^{n+}]} = E^0 - \frac{0.0592}{n}pM$$

The potential E should be linearly related to pM. A *Nernstian* response occurs where the slope of the plot of E vs. pM is $-0.0592/n$.

18-5 The pH-dependent potential that develops across a glass membrane arises from the difference in positions of dissociation equilibria that arise on each of the two surfaces. These equilibria are described by the equation

$$\underset{\text{membrane}}{H^+Gl^-} \rightleftarrows \underset{\text{soln}}{H^+} + \underset{\text{membrane}}{Gl^-}$$

The surface exposed to the solution having the higher hydrogen ion activity then becomes positive with respect to the other surface. This charge difference, or potential, serves as the analytical parameter when the pH of the solution on one side of the membrane is held constant.

18-6 In order for a glass membrane to be pH sensitive, it is necessary that the two surfaces are hydrated so that the equilibrium shown in the previous answer can be established.

18-7 Uncertainties that may be encountered in pH measurements include: (1) the acid error in highly acidic solutions, (2) the alkaline error in strongly basic solutions, (3) the error that arises when the ionic strength of the calibration standards differ from that of the analyte solution, (4) uncertainties in the pH of the standard buffers, (5) nonreproducible junction potentials when samples of low ionic strength are measured, and (6) dehydration of the working surface.

18-8 Owing to variables that cannot be controlled, it is necessary to calibrate the response of the membrane against one or more standards. It must then be assumed that the junction potential associated with the external reference electrode does not change when the standard is replaced by the test solution. The uncertainty associated with this assumption translates into uncertainties in the second decimal place of the measured p-value.

Chapter 18

18-9 The alkaline error arises when a glass electrode is employed to measure the pH of solutions having pH values in the 10 to 12 range or greater. In the presence of alkali ions, the glass surface becomes responsive to both hydrogen and alkali ions. Low pH values arise as a consequence.

18-10 A gas-sensing probe functions by permitting the gas to penetrate a hydrophobic membrane and altering the composition of liquid on the inner side of the membrane. The changes are registered by an indicator/reference electrode pair in contact with the inner solution. Thus, there is no direct contact between the electrodes and the test solution as there is with membrane electrodes.

18-11 (a) The *asymmetry potential* in a membrane arises from differences in the composition or structure of the inner and outer surfaces of the membrane. These differences may arise from contamination of one of the surfaces, wear and abrasion, and strains set up during manufacture.

(b) The *boundary potential* for a membrane electrode is a potential that develops when the membrane separates two solutions that have different concentrations of a cation or an anion that the membrane binds selectively. For an aqueous solution, the following equilibria develop when the membrane is positioned between two solutions of A^+:

$$\underset{\text{membrane}_1}{A^+M^-} \rightleftarrows \underset{\text{soln}_1}{A^+} + \underset{\text{membrane}_1}{M^-}$$

$$\underset{\text{membrane}_2}{A^+M^-} \rightleftarrows \underset{\text{soln}_2}{A^+} + \underset{\text{membrane}_2}{M^-}$$

where the subscripts refer to the two sides of the membrane. A potential develops across this membrane if one of these equilibria proceeds further to the right than the other, and this potential is the boundary potential. For example if the concentration of A^+ is greater in solution 1 than in solution 2, the negative charge on side 1 of the membrane will be less than that of side 2 because the equilibrium on side 1 will lie further to the left. Thus, a greater fraction of the negative charge on side 1 will be neutralized by A^+.

Chapter 18

(c) The *junction potential* in a glass/calomel system develops at the interface between the saturated KCl solution in the salt bridge and the solution of the sample. It is caused by the charge separation created by the differences in the rates at which ions migrate across this interface.

(d) The membrane in a solid-state electrode for F^- is crystalline LaF_3, which when immersed in aqueous solution LaF_3 dissociates according to the equation

$$LaF_3(s) \rightleftharpoons La^{3+} + 3F^-$$

Thus, a boundary potential develops across this membrane when it separates two solutions of different F^- ion concentration. The source of this potential is described in part (b) of this answer.

CHAPTER 19

19-1 The direct potentiometric measurement of pH provides a measure of the equilibrium activity of hydronium ions present in a solution of the sample. A potentiometric titration provides information on the amount of reactive protons, both ionized and nonionized, that are present in a sample.

19-2 (a)
$$E_{Ag} = 0.799 - 0.0592 \log \frac{1}{[Ag^+]} \quad K_{sp} = [Ag^+][IO_3^-] = 3.1 \times 10^{-8}$$

$$E_{Ag} = 0.799 - 0.0592 \log \frac{[IO_3^-]}{K_{sp}}$$

When $[IO_3^-] = 1.00$, E_{Ag} is equal to $E^0_{AgIO_3}$ for the reduction of $AgIO_3$, that is,

$$E^0_{AgIO_3} = 0.799 - 0.0592 \log \frac{1.00}{3.1 \times 10^{-8}} = \underline{0.354 \text{ V}}$$

(b) SCE $\|$ IO_3^- (x M), $AgIO_3$(sat'd) $|$ Ag

(c) $E_{cell} = E_{AgIO_3} - E_{SCE} = 0.354 - 0.0592 \log [IO_3^-] - 0.244$

$= 0.110 + 0.0592 \text{ pIO}_3$

$$\text{pIO}_3 = \frac{E_{cell} - 0.110}{0.0592}$$

(d)
$$\text{pIO}_3 = \frac{0.294 - 0.110}{0.0592} = \underline{3.11}$$

19-3 (a) Proceeding as in the previous solution, we write

$$E = -0.126 - \frac{0.0592}{2} \log \frac{1}{K_{sp}} - \frac{0.0592}{2} \log [I^-]^2$$

$$= -0.126 - \frac{0.0592}{2} \log \frac{1}{7.9 \times 10^{-9}} - \frac{0.0592}{2} \log [I^-]^2$$

When $[I^-] = 1.00$, E becomes $E^0_{PbI_2}$ and

264

$$E^0 = -0.126 - 0.240 - \frac{0.0592}{2} \log (1.00)^2 = -0.366$$

$$PbI_2(s) + 2e^- \rightleftarrows Pb(s) + 2I^- \quad E^0_{PbI_2} = \underline{\underline{-0.366 \text{ V}}}$$

(b) SCE ∥ I⁻ (x M), PbI$_2$(s) | Pb

(c)
$$E_{cell} = -0.366 - \frac{0.0592}{2} \log [I^-]^2 - 0.244$$

$$= -0.610 - 0.0592 \log [I^-] = -0.610 + 0.0592 \text{ pI}$$

$$\underline{\underline{pI = \frac{E_{cell} + 0.610}{0.0592}}}$$

(d)
$$pI = \frac{-0.348 + 0.610}{0.0592} = \underline{\underline{4.43}}$$

19-4 (a) SCE ∥ SCN⁻ (x M), AgSCN(sat'd) | Ag

(b) SCE ∥ I⁻ (x M), AgI(sat'd) | Ag

(c) SCE ∥ SO$_3^{2-}$ (x M), Ag$_2$SO$_3$(sat'd) | Ag

(d) SCE ∥ PO$_4^{3-}$ (x M), Ag$_3$PO$_4$(sat'd) | Ag

19-5 (a) $AgSCN(s) \rightleftarrows Ag^+ + SCN^- \quad K_{sp} = 1.1 \times 10^{-12}$

To obtain E^0_{AgSCN}, we proceed as in Solution 19-3a. Thus,

$$E^0_{AgSCN} = 0.799 - 0.0592 \log \frac{1}{1.1 \times 10^{-12}} = 0.091 \text{ V}$$

Then, we proceed as in Solution 19-3c and write

$$E_{cell} = 0.091 - 0.0592 \log [SCN^-] - 0.244 = -0.153 + 0.0592 \text{ pSCN}$$

$$\underline{\underline{pSCN = \frac{E_{cell} + 0.153}{0.0592}}}$$

Chapter 19

(b) $AgI(s) + e^- \rightleftarrows Ag(s) + I^-$ $\quad E^0_{AgI} = -0.151$ V (Appendix 5)

$$E_{cell} = -0.151 - 0.0592 \log [I^-] - 0.244 = -0.395 + 0.0592 \, pI$$

$$pI = \frac{E_{cell} + 0.395}{0.0592}$$

(c) $Ag_2SO_3(s) \rightleftarrows 2Ag^+ + SO_3^{2-}$ $\quad K_{sp} = 1.5 \times 10^{-14}$

$$E^0_{Ag_2SO_3} = 0.799 - \frac{0.0592}{2} \log \frac{1}{1.5 \times 10^{-14}} = 0.390 \text{ V}$$

$$E_{cell} = 0.390 - \frac{0.0592}{2} \log [SO_3^{2-}] - 0.244 = 0.146 + \frac{0.0592}{2} pSO_3$$

$$pSO_3 = \frac{2(E_{cell} - 0.146)}{0.0592}$$

(d) $Ag_3PO_4(s) \rightleftarrows 3Ag^+ + PO_4^{3-}$ $\quad K_{sp} = 1.3 \times 10^{-20}$

$$E^0_{Ag_3PO_4} = 0.799 - \frac{0.0592}{3} \log \frac{1}{1.3 \times 10^{-20}} = 0.407 \text{ V}$$

$$E_{cell} = 0.407 - \frac{0.0592}{3} \log [PO_4^{3-}] - 0.244 = 0.163 + \frac{0.0592}{3} pPO_4$$

$$pPO_4 = \frac{3(E_{cell} - 0.163)}{0.0592}$$

19-6 (a) $\quad pSCN = \dfrac{0.122 + 0.153}{0.0592} = \underline{\underline{4.65}}$

(b) $\quad pI = \dfrac{-0.211 + 0.395}{0.0592} = \underline{\underline{3.11}}$

(c) $\quad pSO_3 = \dfrac{2(0.300 - 0.146)}{0.0592} = \underline{\underline{5.20}}$

(d) $\text{pPO}_4 = \dfrac{3(0.244 - 0.163)}{0.0592} = \underline{\underline{4.10}}$

19-7 $Ag_2CrO_4(s) + 2e^- \rightleftarrows 2Ag(s) + CrO_4^{2-}$ $E^0 = 0.446$ V

$0.402 = 0.446 - 0.0296 \log [CrO_4^{2-}] - 0.244 = 0.202 + 0.0296 \, \text{pCrO}_4$

$\text{pCrO}_4 = (0.402 - 0.202)/0.0296 = \underline{\underline{6.76}}$

19-8 (a) $E_{cell} = 0.854 - 0.0296 \log (1/7.40 \times 10^{-3}) - 0.244$

$= 0.854 - 0.063 - 0.244 = \underline{\underline{0.547 \text{ V}}}$

(b) $E_{cell} = 0.788 - 0.0296 \log (1/7.40 \times 10^{-3}) - 0.244 = \underline{\underline{0.481 \text{ V}}}$

(c) $E_{cell} = 0.615 - 0.0296 \log 0.0250 - 0.244 = \underline{\underline{0.418 \text{ V}}}$

(d) $[Hg(OAc)_2] = 2.00 \times 10^{-3} - [Hg^{2+}] \approx 2.00 \times 10^{-3}$

$[OAc^-] = 0.100 - 2 \times 2.00 \times 10^{-3} + 2[Hg^{2+}] \approx 0.0960$

$\dfrac{[Hg(OAc)_2]}{[Hg^{2+}][OAc^-]^2} = 2.7 \times 10^8 = \dfrac{2.00 \times 10^{-3}}{[Hg^{2+}](0.0960)^2}$

$[Hg^{2+}] = 2.00 \times 10^{-3}/[2.7 \times 10^8 \times (0.0960)^2] = 8.04 \times 10^{-10}$

$E_{cell} = 0.854 - 0.0296 \log (1/8.04 \times 10^{-10}) - 0.244 = \underline{\underline{0.341 \text{ V}}}$

19-9 (a) $E_{cell} = 0.210 - 0.0296 \log ([Y^{4-}]/[HgY^{2-}]) - 2.44$

$[Y^{4-}] = c_{EDTA} \times \alpha_4$ (Equation 14-2, page 283)

At $[H^+] = 1.00 \times 10^{-4}$, $\alpha_4 = 3.6 \times 10^{-9}$ (Table 14-2)

$[Y^{4-}] = 0.0200 \times 3.6 \times 10^{-9} = 7.2 \times 10^{-11}$

$E_{cell} = 0.210 - 0.0296 \log (7.2 \times 10^{-11}/2.00 \times 10^{-4}) - 0.244 = \underline{\underline{0.157 \text{ V}}}$

Chapter 19

(b) At pH = 8.00, α_4 = 5.4×10^{-3}

$$[Y^{4-}] = 0.0200 \times 5.4 \times 10^{-3} = 1.08 \times 10^{-4}$$

$$E_{cell} = 0.210 - 0.0296 \log(1.08 \times 10^{-4}/2.00 \times 10^{-4}) - 0.244 = \underline{-0.026 \text{ V}}$$

19-10 See solution to Problem 19-9.

(a) At pH = 6.0, α_4 = 2.2×10^{-5}

$$[Y^{4-}] = 2.2 \times 10^{-5} \times 0.0100 = 2.2 \times 10^{-7}$$

$$E_{cell} = 0.210 - 0.0296 \log(2.2 \times 10^{-7}/2.00 \times 10^{-4}) - 0.244 = \underline{0.054 \text{ V}}$$

(b) At pH = 10, α_4 = 0.35 and $[Y^{4-}] = 0.35 \times 0.01 = 0.0035$

$$E_{cell} = 0.210 - 0.0296 \log(0.0035/2.00 \times 10^{-4}) - 0.244 = \underline{-0.071 \text{ V}}$$

19-11 Substituting into Equation 19-5 gives

$$\text{pH} = -\frac{E_{cell} - K}{0.0592/1} \quad \text{and} \quad 4.006 = -\frac{0.2094 - K}{0.0592}$$

$$K = 4.006 \times 0.0592 + 0.2094 = 0.44656$$

(a) pH = $-(-0.3011 - 0.44656)/0.0592 = \underline{12.629}$

a_{H^+} = antilog(-12.629) = $\underline{2.35 \times 10^{-13}}$

(b) pH = $-(0.1163 - 0.44656)/0.0592 = \underline{5.579}$

a_{H^+} = antilog(-5.579) = $\underline{2.64 \times 10^{-6}}$

(c) For part (a),

If $E = -0.3011 + 0.002 = -0.2991$ V

pH = $-(-0.2991 - 0.44656)/0.0592 = \underline{12.596}$

a_{H^+} = $\underline{2.54 \times 10^{-13}}$

Chapter 19

If $E = -0.3011 - 0.002 = -0.3031$ V

$$\text{pH} = -(-0.3031 - 0.44656)/0.0592 = \underline{\underline{12.663}}$$

$$a_{H^+} = \underline{\underline{2.17 \times 10^{-13}}}$$

Proceeding in the same way for part (b) we obtain

$$\text{pH} = \underline{\underline{5.545}} \text{ and } \underline{\underline{5.612}}$$

$$a_{H^+} = \underline{\underline{2.85 \times 10^{-6}}} \text{ and } \underline{\underline{2.44 \times 10^{-6}}}$$

19-12 Applying Equation 19-5, gives

$$\text{pMg} = -(E_{\text{cell}} - K)/0.0296$$

$$= -\log 9.62 \times 10^{-3} = 2.017 = -(0.367 - K)/0.0296$$

$$K = 2.017 \times 0.0296 + 0.367 = 0.4267$$

(a) $\text{pMg} = -(+0.244 - 0.4267)/0.0296 = \underline{\underline{6.172}}$

(b) If +0.242 V and +0.246 V are substituted for +0.244 V, we obtain $\underline{\underline{6.240}}$ and $\underline{\underline{6.105}}$ and

$$[\text{Mg}^{2+}] = \text{antilog}(-6.240) \text{ to } (-6.105)$$

$$= 5.75 \times 10^{-7} \text{ to } 7.85 \times 10^{-7}$$

19-13 $E_{\text{cell}} = -0.721 = E^0_{\text{Cd}^{2+}} - 0.0296 \log(1/[\text{Cd}^{2+}]) - 0.244$

$$= -0.403 - 0.0296 \log\left([\text{A}^-]^2/K_{\text{sp}}\right) - 0.244$$

$$= -0.647 - 0.0296 \log(0.0250)^2 + 0.0296 \log K_{\text{sp}}$$

$$0.0296 \log K_{\text{sp}} = -0.721 + 0.647 + 0.0296 \log(0.0250)^2 = -0.1688$$

$$\log K_{\text{sp}} = -5.704 \text{ and } K_{\text{sp}} = \text{antilog}(-5.704) = \underline{\underline{2.0 \times 10^{-6}}}$$

Chapter 19

19-14
$$E_{cell} = 0.000 - \frac{0.0592}{2}\log\frac{1.00}{[H^+]^2} - 0.244 = -0.797$$

$$\frac{0.0592}{2}\log[H^+]^2 = -0.797 + 0.244 = -0.553$$

$$[H^+] = \text{antilog}(-0.553/0.0592) = 4.56 \times 10^{-10}$$

$$K_a = 4.56 \times 10^{-10}[A^-]/[HA] = 4.56 \times 10^{-10} \times 0.170/0.250 = \underline{\underline{3.1 \times 10^{-10}}}$$

19-15
$$\text{no. mmol HA} = 27.22 \text{ mL NaOH} \times 0.1025\frac{\text{mmol NaOH}}{\text{mL NaOH}} \times \frac{1 \text{ mmol HA}}{\text{mmol NaOH}} = 2.7901$$

$$\frac{0.3798 \text{ g HA}}{2.7901 \text{ mmol HA}} \times 10^3 \frac{\text{mmol}}{\text{mol}} = \underline{\underline{136 \frac{\text{g HA}}{\text{mol}}}} = \mathcal{M}_{HA}$$

19-16 Examine the plot of the data and determine the pH when the acid was half neutralized-- that is, at 27.22/2 = 13.61 mL. At this point, $[HA] \approx A^-$ and

$$K_a = \frac{[H_3O^+][A^-]}{[HA]}$$

Thus, $pK_a \approx pH$

19-17 $2Ce^{4+} + HNO_2 + H_2O \rightarrow 2Ce^{3+} + NO_3^- + 3H^+$

initial no. mmol HNO_2 = 40.00 × 0.0500 = 2.000

5.00 mL

$$\text{no. mmol NO}_3^- = (5.00 \times 0.0800) \text{ mmol Ce}^{4+} \times 1 \text{ mmol NO}_3^-/2 \text{ mmol Ce}^{4+}$$

$$= 0.2000$$

$$c_{NO_3^-} \approx [NO_3^-] = 0.2000 \text{ mmol NO}_3^-/80.00 = 2.500 \times 10^{-3}$$

$$c_{HNO_2} \approx [HNO_2] = (2.000 - 0.2000)/80.00 = 2.25 \times 10^{-2}$$

$$[H^+] = 0.1000$$

$$E_{NO_3^-} = 0.94 - \frac{0.0592}{2} \log \frac{[HNO_2]}{[NO_3^-][H^+]}$$

$$= 0.94 - \frac{0.0592}{2} \log \frac{2.25 \times 10^{-2}}{2.5 \times 10^{-3}(0.1000)^3}$$

$$E_{cell} = 0.94 - 0.117 - 0.244 = \underline{\underline{0.58 \text{ V}}}$$

The remaining preequivalence point values, which are obtained in the same way, are found in the table that follows.

50.00 mL Proceeding as in the section labeled "Equivalence-Point Potential" in Section 16C-2, we write

$$2E = 2 \times 0.94 - 0.0592 \log \left([HNO_2]/[NO_3^-][H^+]^3\right)$$

$$E = 1.44 - 0.0592 \log ([Ce^{3+}]/[Ce^{4+}])$$

$$3E = 3.32 - 0.0592 \log \frac{[HNO_2][Ce^{3+}]}{[NO_3^-][Ce^{4+}](0.1000)^3}$$

At equivalence, $[Ce^{4+}] = 2[HNO_2]$ and $[Ce^{3+}] = 2[NO_3^-]$

$$E = 3.32/3 - (0.0592/3) \log 10^3 = 1.047 \text{ V}$$

$$E_{cell} = 1.047 - 0.244 = \underline{\underline{0.80 \text{ V}}}$$

51.00 mL

$$[Ce^{3+}] = \frac{2.000 \text{ mmol } HNO_2 \times 2 \text{ mmol } Ce^{3+}/\text{mmol } HNO_2}{(75.00 + 50.00) \text{ mL}} = 3.200 \times 10^{-2}$$

$$[Ce^{4+}] = (51.00 - 50.00) \times 0.08000/125.00 = 6.400 \times 10^{-4}$$

$$E = 1.44 - 0.0592 \log (3.200 \times 10^{-2}/6.400 \times 10^{-4}) = 1.339$$

$$E_{cell} = 1.339 - 0.244 = \underline{\underline{1.10 \text{ V}}}$$

The remaining post-equivalence point data are found in the table that follows.

Chapter 19

mL Reagent	E vs. SCE, V	mL Reagent	E vs. SCE, V
5.00	0.58	49.00	0.66
10.00	0.59	50.00	0.80
15.00	0.60	51.00	1.10
25.00	0.61	55.00	1.14
40.00	0.63	60.00	1.15

19-18 E_{Ag^+} = $0.799 - 0.0592 \log (1/[Ag^+])$ = $0.799 - 0.0592 \log ([SeCN^-]/K_{sp})$

E_{AgSeCN} = $0.799 + 0.0592 \log 4.20 \times 10^{-16} - 0.0592 \log [SeCN^-]$

= $-0.1113 - 0.0592 \log [SeCN^-]$

5.00 mL

$[SeCN^-]$ = $(50.00 \times 0.0800 - 5.00 \times 0.1000)/55$ = 6.364×10^{-2}

E_{cell} = $-0.1113 - 0.0592 \log 6.364 \times 10^{-2} - 0.244$ = $\underline{\underline{-0.284 \text{ V}}}$

Proceeding in the same way, we obtain the remaining preequivalence point data shown in the table that follows.

40.00 mL

$[Ag^+]$ = $\sqrt{4.20 \times 10^{-16}}$ = 2.05×10^{-8} = $[SeCN^-]$

E_{cell} = $0.799 - 0.0592 \log (1/2.05 \times 10^{-8}) - 0.244$ = $\underline{\underline{0.100 \text{ V}}}$

41.00 mL

$c_{AgNO_3} \approx [Ag^+]$ = (1.00×0.100) mmol $AgNO_3/91.00$ mL = 1.099×10^{-3}

E_{cell} = $0.799 - 0.0592 \log (1/1.099 \times 10^{-3}) - 0.244$ = $\underline{\underline{0.380 \text{ V}}}$

The remaining data are found in the table that follows.

Chapter 19

mL Reagent	E vs. SCE, V	mL Reagent	E vs. SCE, V
5.00	-0.284	39.00	-0.181
15.00	-0.272	40.00	0.100
25.00	-0.255	41.00	0.380
30.00	-0.243	45.00	0.420
35.00	-0.223	50.00	0.437

19-19 An operational definition of pH is a definition describing in detail how a pH is determined experimentally. To obtain an operational pH_μ the potential E_s of an electrode system immersed in buffers prepared by a carefully prescribed method and having a pH of pH_s, is measured. The potential E_U for an unknown is then measured with the same electrode system. The operational definition of pH_μ of the unknown is then

$$pH_\mu = pH_s - \frac{(E_U - E_s)}{0.0592}$$

19-20 (a) Employing Equation 19-1, we write

$$E_r = \frac{R_s}{R_M + R_s} \times 100\% = \frac{450 \text{ M}\Omega}{50 \text{ M}\Omega + 450 \text{ M}\Omega} \times 100\% = \underline{\underline{90\%}}$$

Similarly

(b) $E_r = \frac{450}{500 + 450} \times 100\% = \underline{\underline{47\%}}$

(c) $E_r = \frac{450}{1000 + 450} \times 100\% = \underline{\underline{31\%}}$

(d) $E_r = \frac{450}{5000 + 450} \times 100\% = \underline{\underline{8.3\%}}$

CHAPTER 20

20-1 (a) *Concentration polarization* is a condition in which the current in an electrochemical cell is limited by the rate at which reactants are brought to or removed from the surface of one or both electrodes. *Kinetic polarization* is a condition in which the current in an electrochemical cell is limited by the rate at which electrons are transferred between the electrode surfaces and reactants in solution. For either type of polarization, the current is no longer proportional to the cell potential.

(b) An amperostat is an instrument that provides a constant current; a potentiostat is an instrument that provides an electrical output having a constant potential.

(c) Both the *coulomb* and the *Faraday* are units of quantity of charge, or electricity. The former is the quantity transported by one ampere of current in one second; the latter is equal to 96,495 coulombs or one mole of electrons.

(d) A *working electrode* is the electrode at which the desired electrochemical reaction is carried out. A *counter electrode* is used with the working electrode to complete the electrical circuit in a cell. It is connected to the working electrode by an external metallic conductor.

(e) The *electrolysis circuit* consists of a working electrode and a counter electrode. The *control circuit* regulates the applied potential such that the potential between the working electrode and a reference electrode in the control circuit is constant and at a desired level.

20-2 (a) *Current density* is the current at an electrode divided by the surface area of that electrode. Ordinarily, it has units of amperes per square centimeter.

(b) *Ohmic potential*, or *IR drop*, of a cell is the product of the current in the cell in amperes and the electrical resistance of the cell in ohms.

(c) A *coulometric titration* is an electroanalytical method in which a constant current of known magnitude generates a reagent that reacts with the analyte. The time required to generate enough reagent to complete the reaction is measured.

(d) In a *controlled-potential electrolysis* the potential applied to a cell is continuously adjusted to maintain a constant potential between the working electrode and a reference electrode.

(e) *Current efficiency* is a measure of agreement between the number of faradays of current and the number of moles or reactant oxidized or reduced at a working electrode.

(f) *Overvoltage* is the increased potential required to offset the decrease in current brought about by polarization.

20-3 Mass transport in an electrochemical cell results from one or more of the following: (1) *diffusion*, which arises from concentration differences between the electrode surface and the bulk of the solution; (2) *migration*, which results from electrostatic attraction or repulsion between the species and an electrode; and (3) *convection*, which results from stirring, vibration, or temperature difference.

20-4 A current in an electrochemical cell always causes the cell potential to become less positive or more negative.

20-5 Both kinetic and concentration polarization cause the potential of a cell to be more negative than the thermodynamic potential. Concentration polarization arises from the slow rate at which reactants or products are transported to or away from the electrode surfaces. Kinetic polarization arises from the slow rate of the electrochemical reactions at the electrode surfaces.

20-6 Variables that decrease concentration polarization include elevated temperatures, vigorous stirring, high reactant concentrations, absence of other electrolytes, and large electrode surface areas.

20-7 Kinetic polarization is often encountered when the product of a reaction is a gas, particularly when the electrode is a soft metal such as mercury, zinc, or copper. It is likely to occur at low temperatures and high current densities.

20-8 Potentiometric measurements are made under conditions of zero (or minuscule) current; electrogravimetric and coulometric methods depend upon the presence of a current.

20-9 Temperature, current density, complexation of the analyte, and codeposition of a gas influence the physical properties of an electrogravimetric deposit.

20-10 A cathode depolarizer is a substance that is reduced more readily than a potentially interfering species. The codeposition of hydrogen, for example, is prevented through the introduction of nitrate ion as a cathode depolarizer.

Chapter 20

20-11 (a) An *amperostat* is an instrument that provides a constant current to an electrolysis cell.

(b) A *potentiostat* controls the applied potential to maintain a constant potential between the working electrode and a reference electrode.

20-12 In *amperostatic coulometry*, the cell is operated so that the current in the cell is held constant. In *potentiostatic coulometry*, the potential of the working electrode is maintained constant.

20-13 The species produced at the counter electrode is a potential interference by reacting with the product at the working electrode. Isolation of the one from the other is ordinarily necessary.

20-14 An *auxiliary reagent* is generally required in a coulometric titration to permit the analyte to be oxidized or reduced with 100% current efficiency. As a titration proceeds, the potential of the working electrode will inevitably rise as concentration polarization of the analyte begins. Unless an auxiliary reagent is present to terminate this rise by producing a species that reacts with the analyte, some other species will be oxidized or reduced thus lowering the current efficiency and producing erroneous results.

20-15 (a)
$$0.020 \text{ C} \times \frac{1 \text{ F}}{96,485 \text{ C}} \times \frac{1 \text{ mol e}^-}{\text{F}} \times \frac{1 \text{ mol cation}}{\text{mol e}^-} \times 6.02 \times 10^{23} \frac{\text{cation}}{\text{mol cation}} =$$

$$\underline{\underline{1.2 \times 10^{17} \text{ cations}}}$$

(b)
$$0.020 \text{ C} \times \frac{1 \text{ F}}{96,485 \text{ C}} \times \frac{1 \text{ mol e}^-}{\text{F}} \times \frac{1 \text{ mol cation}}{2 \text{ mol e}^-} \times 6.02 \times 10^{23} \frac{\text{cation}}{\text{mol cation}} =$$

$$\underline{\underline{6.2 \times 10^{16} \text{ cations}}}$$

(c)
$$0.020 \text{ C} \times \frac{1 \text{ F}}{96,485 \text{ C}} \times \frac{1 \text{ mol e}^-}{\text{F}} \times \frac{1 \text{ mol cation}}{3 \text{ mol e}^-} \times 6.02 \times 10^{23} \frac{\text{cation}}{\text{mol cation}} =$$

$$\underline{\underline{4.2 \times 10^{16} \text{ cations}}}$$

20-16 (a)
$$E_{\text{cathode}} = 0.337 - \frac{0.0592}{2} \log \frac{1}{0.150} = 0.313 \text{ V}$$

$$E_{anode} = 1.229 - \frac{0.0592}{4} \log \frac{1}{1.00 \times (1.00 \times 10^{-3})^4} = 1.051 \text{ V}$$

$$E_{applied} = E_{cathode} - E_{anode} = 0.313 - 1.051 = \underline{\underline{-0.738 \text{ V}}}$$

(b) $$E_{cathode} = -0.136 - \frac{0.0592}{2} \log \frac{1}{0.120} = -0.163 \text{ V}$$

$$E_{anode} = 1.229 - \frac{0.0592}{4} \log \frac{1}{770/760)(1.00 \times 10^{-4})^4} = 0.992 \text{ V}$$

$$E_{applied} = -0.163 - 0.992 = \underline{\underline{-1.155 \text{ V}}}$$

(c) $[H^+] = $ antilog $(-3.40) = 3.98 \times 10^{-4}$

$$E_{cathode} = 0.000 - \frac{0.0592}{2} \log \frac{765/760}{(3.98 \times 10^{-4})^2} = -0.201 \text{ V}$$

$$E_{anode} = 0.073 - 0.0592 \log 0.0864 = 0.136 \text{ V}$$

$$E_{applied} = -0.201 - 0.136 = \underline{\underline{-0.337 \text{ V}}}$$

(d) cathode $Cu^{2+} + 2e^- \rightarrow Cu(s)$

anode $2Tl^+ + 6OH^- \rightarrow Tl_2O_3(s) + 3H_2O + 4e^-$

$$E_{cathode} = 0.337 - \frac{0.0592}{2} \log \frac{1}{0.100} = 0.307 \text{ V}$$

$$E_{anode} = 0.020 - \frac{0.0592}{4} \log [Tl^+]^2 [OH^-]^6$$

$$= 0.020 - \frac{0.0592}{4} \log (4 \times 10^{-3})^2 (1.00 \times 10^{-6})^6 = 0.624 \text{ V}$$

$$E_{applied} = 0.307 - 0.624 = \underline{\underline{-0.317 \text{ V}}}$$

20-17 $E_{cathode} = -0.763 - 0.0296 \log (1/3.75 \times 10^{-3}) = -0.835$

Chapter 20

$$E_{anode} = -0.277 - 0.0296 \log(1/6.40 \times 10^{-2}) = -0.312$$

$$E_{cell} = -0.835 - (-0.312) - 0.078 \times 5 = \underline{\underline{-0.913 \text{ V}}}$$

20-18 $E_{cathode} = -0.403 - 0.0296 \log(1/7.50 \times 10^{-2}) = -0.436$

$$E_{anode} = -0.136 - 0.0296 \log(1/8.22 \times 10^{-4}) = -0.227$$

$$E_{cell} = -0.436 - (-0.227) - 0.072 \times 3.95 = \underline{\underline{-0.493 \text{ V}}}$$

20-19 (a) $E_{cathode} = 0.337 - \dfrac{0.0592}{2} \log \dfrac{1}{0.200} = 0.316 \text{ V}$

$$E_{anode} = 1.229 - \dfrac{0.0592}{4} \log \dfrac{1}{(1.00 \times 10^{-4})^4 \times 740/760} = 0.992 \text{ V}$$

$$E_{cell} = 0.316 - 0.992 = \underline{\underline{-0.676 \text{ V}}}$$

(b) $IR = -0.10 \times 3.60 = \underline{\underline{-0.36 \text{ V}}}$

(c) Recall that the overpotential in an electrolytic cell requires the application of a larger or more negative potential. That is, 0.50 V must be subtracted from the cell potential.

$$E_{applied} = -0.676 - 0.36 - 0.50 = \underline{\underline{-1.54 \text{ V}}}$$

(d) $E_{cathode} = 0.337 - \dfrac{0.0592}{2} \log \dfrac{1}{8.00 \times 10^{-6}} = 0.186 \text{ V}$

$$E_{applied} = 0.186 - 0.992 - 0.36 - 0.50 = \underline{\underline{-1.67 \text{ V}}}$$

20-20 (a) $E_{cathode} = -0.250 - \dfrac{0.0592}{2} \log \dfrac{1}{0.200} = -0.271 \text{ V}$

$$E_{anode} = 1.229 - \dfrac{0.0592}{4} \log \dfrac{1}{(1.00)(1.00 \times 10^{-2})^4} = 1.111 \text{ V}$$

$$E_{cell} = -0.271 - 1.111 = \underline{\underline{-1.38 \text{ V}}}$$

(b) $IR = (-1.10)(3.15) = \underline{\underline{-3.46 \text{ V}}}$

(c) current density at anode = 1.10 A/80 cm² = <u>0.0138 A/cm²</u>

current density at cathode = 1.10 A/120 cm² = <u>0.0092 A/cm²</u>

(d) As in 20-19(c), the overpotential is subtracted from the cell potential or

$$E_{applied} = -1.38 - 3.46 - 0.52 = \underline{-5.36 \text{ V}}$$

(e) $$E_{cathode} = -0.250 - \frac{0.0592}{2} \log \frac{1}{2.00 \times 10^{-4}} = -0.359 \text{ V}$$

$$E_{applied} = -0.359 - 1.111 - 3.46 - 0.52 = \underline{-5.45 \text{ V}}$$

20-21 (a) $$E_{cathode} = -0.31 - 0.0592 \log \frac{(0.320)^2}{0.150} = -0.30 \text{ V}$$

$$E_{anode} = 1.229 - \frac{0.0592}{4} \log \frac{1}{1.00(1.00 \times 10^{-10})^4} = 0.637 \text{ V}$$

$$E_{cell} = -0.30 - 0.637 = \underline{-0.94 \text{ V}}$$

(b) $IR = -0.12 \times 2.90 = \underline{-0.35 \text{ V}}$

(c) $E_{applied} = -0.94 - 0.35 - 0.80 = \underline{-2.09 \text{ V}}$ [see Solution 20 – 19(c)]

(d) Reduction of 0.150 mol/L $Ag(CN)_2^-$ to $Ag(s)$ produces

$2 \times 0.150 = 0.300$ mol CN^-/L and $[CN^-] \approx 0.300 + 0.320 = 0.620$.

$$E_{cathode} = -0.31 - 0.0592 \log \frac{(0.620)^2}{1.00 \times 10^{-5}} = -0.581 \text{ V}$$

$$E_{applied} = -0.581 - 0.637 - 0.35 - 0.80 = \underline{-2.37 \text{ V}}$$

20-22 Cd begins to form when

$$E = -0.403 - 0.0296 \log (1/0.0750) = -0.436 \text{ V}$$

(a) $[Co^{2+}]$ concentration when Cd first forms is given by

$$-0.436 = -0.277 - 0.0296 \log (1/[Co^{2+}])$$

Chapter 20

$$\log [Co^{2+}] = (-0.436 + 0.277)/0.0296 = -5.372$$

$$[Co^{2+}] = \text{antilog}(-5.372) = \underline{\underline{4.2 \times 10^{-6}}}$$

(b) $E_{cathode} = -0.277 - 0.0296 \log(1/1.00 \times 10^{-5}) = \underline{\underline{-0.425 \text{ V}}}$

20-23 pH = 2.50 $[H^+] = 3.16 \times 10^{-3}$

BiO^+ more readily reduced than Co^{2+}

(a) Deposition of Co begins when

$$E_{cathode} = -0.277 - 0.0296 \log(1/0.040) = -0.318 \text{ V}$$

$$-0.318 = 0.320 - \frac{0.0592}{3} \log \frac{1}{[BiO^+](3.16 \times 10^{-3})^2}$$

$$(0.0592/3) \log [BiO^+] = -0.318 - 0.320 - 0.0197 \log (3.16 \times 10^{-3})^2 = -0.5393$$

$$\log [BiO^+] = -0.5393 \times 3/0.0592 = -27.33$$

$$[BiO^+] = \text{antilog}(-27.33) = 4.6 \times 10^{-28} = \underline{\underline{5 \times 10^{-28} \text{ M}}}$$

(b) $$E = 0.320 - \frac{0.0592}{3} \log \frac{1}{1.00 \times 10^{-6}(3.16 \times 10^{-3})^2} = \underline{\underline{0.103 \text{ V}}}$$

20-24 (a) Bi deposits at a lower potential, that is

$$[H^+] = \text{antilog}(-1.50) = 3.16 \times 10^{-2}$$

$$E_{cathode} = 0.320 - \frac{0.0592}{3} \log \frac{1}{0.200(3.16 \times 10^{-2})^2} = \underline{\underline{0.247 \text{ V}}}$$

(b) Sn deposits when

$$E_{cathode} = -0.136 - 0.0296 \log(1/0.200) = -0.157 \text{ V}$$

$$-0.157 = 0.320 - \frac{0.0592}{3} \log \frac{1}{[BiO^+](3.16 \times 10^{-2})^2}$$

$$= 0.320 + 0.0197 \log (3.16 \times 10^{-2})^2 + 0.0197 \log [BiO^+]$$

$$\log [\text{BiO}^+] = (-0.157 - 0.320 + 0.059)/0.0197 = -21.22$$

$$[\text{BiO}^+] = \text{antilog}(-21.22) = \underline{\underline{6.0 \times 10^{-22}}}$$

(c) When $[\text{BiO}^+] = 10^{-6}$

$$E_{\text{cathode}} = 0.320 - \frac{0.0592}{3} \log \frac{1}{1.00 \times 10^{-6}(3.16 \times 10^{-2})^2} = 0.142 \text{ V}$$

Sn begins to form when $E_{\text{cathode}} = -0.157$ [see part (b)]

$$\text{range vs. SCE} = 0.142 - 0.244 \quad \text{to} \quad -0.157 - 0.244 \quad \text{or}$$

$$= \underline{\underline{-0.102 \quad \text{to} \quad -0.401 \text{ V}}}$$

20-25 (a) Br⁻ separation begins when

$$E_{\text{anode}} = 0.073 - 0.0592 \log 0.250 = 0.109 \text{ V}$$

When $[\text{I}^-] = 10^{-5}$

$$E_{\text{anode}} = -0.151 - 0.0592 \log 10^{-5} = 0.145 \text{ V}$$

The equilibrium [Br⁻] corresponding to this potential is

$$0.145 = 0.073 - 0.0592 \log [\text{Br}^-]$$

$$[\text{Br}^-] = \text{antilog}[(0.073 - 0.145)/0.0592] = 0.061$$

Thus a <u>separation is impossible</u>.

(b) Proceeding in the same way for Cl⁻

$$0.145 = +0.222 - 0.0592 \log [\text{Cl}^-]$$

$$[\text{Cl}^-] = \text{antilog}[(0.222 - 0.145)/0.0592] = 20$$

Thus if $[\text{Cl}^-] < 20$ M, formation of AgCl will not occur and <u>separation is feasible</u>.

Chapter 20

(c) To separate I⁻ quantitatively

$$E \text{ (vs. SCE)} = 0.244 - 0.145 = \underline{\underline{0.099 \text{ V}}}$$

To prevent Cl⁻ from precipitating

$$E \text{ (vs. SCE)} = 0.244 - (0.222 - 0.0592 \log 0.250) = \underline{\underline{-0.014 \text{ V}}}$$

Thus, a cell consisting of a saturated calomel electrode and the analyte half-cell would be galvanic, and a <u>separation would be feasible</u> by allowing the cell to discharge to 0.00 V.

20-26 Deposition of A is complete when

$$E_A = E_A^0 - \frac{0.0592}{n} \log \frac{1}{1.00 \times 10^{-5}} = E_A^0 - \frac{0.296}{n_A} \text{ V}$$

Deposition of B begins when

$$E_B = E_B^0 - \frac{0.0592}{n} \log \frac{1}{1.00 \times 10^{-1}} = E_B^0 - \frac{0.0592}{n_B} \text{ V}$$

Boundary condition is that $E_A = E_B$. Thus,

$$E_A^0 - 0.296/n_A = E_B^0 - 0.0592/n_B$$

or $\quad E_A^0 - E_B^0 = 0.296/n_A - 0.0592/n_B$

(a) $E_A^0 - E_B^0 = 0.296 - 0.0592 = \underline{\underline{0.237 \text{ V}}}$

(b) $E_A^0 - E_B^0 = 0.296/2 - 0.0592 = \underline{\underline{0.0888 \text{ V}}}$

(c) $E_A^0 - E_B^0 = 0.296/3 - 0.0592 = \underline{\underline{0.0395 \text{ V}}}$

(d) $E_A^0 - E_B^0 = 0.296 - 0.0592/2 = \underline{\underline{0.266 \text{ V}}}$

(e) $E_A^0 - E_B^0 = 0.296/2 - 0.0592/2 = \underline{\underline{0.118 \text{ V}}}$

(f) $E_A^0 - E_B^0 = 0.296/3 - 0.0592/2 = \underline{\underline{0.0691 \text{ V}}}$

Chapter 20

(g) $E_A^0 - E_B^0 = 0.296/1 - 0.0592/3 = \underline{\underline{0.276 \text{ V}}}$

(h) $E_A^0 - E_B^0 = 0.296/2 - 0.0592/3 = \underline{\underline{0.128 \text{ V}}}$

(i) $E_A^0 - E_B^0 = 0.296/3 - 0.0592/3 = \underline{\underline{0.0789 \text{ V}}}$

20-27 (a)
$$0.500 \text{ g Co} \times \frac{1 \text{ mol Co}}{58.93 \text{ g Co}} \times \frac{2 \text{ mol e}^-}{\text{mol Co}} \times \frac{1 \text{ F}}{\text{mol e}^-} \times 96485 \frac{\text{C}}{\text{F}} = 1.637 \times 10^3 \text{ C}$$

$$1.637 \times 10^3 \text{ C} \times \frac{1 \text{ A} \cdot \text{s}}{\text{C}} \times \frac{1}{0.961 \text{ A}} \times \frac{1 \text{ min}}{60 \text{ s}} = \underline{\underline{28.4 \text{ min}}}$$

(b) $3\text{Co}^{2+} + 4\text{H}_2\text{O} \rightleftarrows \text{Co}_3\text{O}_4(s) + 8\text{H}^+ + 2\text{e}^-$ $(3/2) \text{ mol Co}^{2+} \equiv 1 \text{ mol e}^-$

$$0.500 \text{ g Co} \times \frac{1 \text{ mol Co}}{58.93 \text{ g Co}} \times \frac{1 \text{ mol e}^-}{(3/2) \text{ mol Co}} \times \frac{1 \text{ F}}{\text{mol e}^-} \times 96485 \frac{\text{C}}{\text{F}} = 545.8 \text{ C}$$

$$545.8 \text{ C} \times \frac{1 \text{ A} \cdot \text{s}}{\text{C}} \times \frac{1}{0.961 \text{ A}} \times \frac{1 \text{ min}}{60 \text{ s}} = \underline{\underline{9.47 \text{ min}}}$$

20-28 (a)
$$0.500 \text{ g Tl} \times \frac{1 \text{ mol Tl}}{204.38 \text{ g Tl}} \times \frac{3 \text{ mol e}^-}{1 \text{ mol Tl}} \times \frac{1 \text{ F}}{\text{mol e}^-} \times 96485 \frac{\text{C}}{\text{F}} = 7.081 \times 10^2 \text{ C}$$

$$7.081 \times 10^2 \text{ C} \times \frac{1 \text{ A} \cdot \text{s}}{\text{C}} \times \frac{1}{1.20 \text{ A}} \times \frac{1 \text{ min}}{60 \text{ s}} = \underline{\underline{9.84 \text{ min}}}$$

(b) $2\text{Tl}^+ + 3\text{H}_2\text{O} \rightarrow \text{Tl}_2\text{O}_3 + 6\text{H}^+ + 4\text{e}^-$ $2 \text{ mol Tl}^+ \equiv 4 \text{ mol e}^-$

$$0.500 \text{ g Tl}^+ \times \frac{1 \text{ mol Tl}^+}{204.38 \text{ g Tl}^+} \times \frac{4 \text{ mol e}^-}{2 \text{ mol Tl}^+} \times \frac{1 \text{ F}}{\text{mol e}^-} \times 96485 \frac{\text{C}}{\text{F}} = 472.1 \text{ C}$$

$$472.1 \text{ C} \times \frac{1 \text{ A} \cdot \text{s}}{\text{C}} \times \frac{1}{1.20 \text{ A}} \times \frac{1 \text{ min}}{60 \text{ s}} = \underline{\underline{6.56 \text{ min}}}$$

Chapter 20

(c) Proceeding as in part (a)

$$0.500 \text{ g} \times \frac{1 \text{ mol Tl}}{204.38 \text{ g}} \times \frac{1 \text{ mol e}^-}{1 \text{ mol Tl}} \times \frac{1 \text{ F}}{\text{mol e}^-} \times \frac{96485 \text{ C}}{\text{F}} \times \frac{1 \text{ A} \cdot \text{s}}{\text{C}} \times \frac{1}{1.20 \text{ A}} \times \frac{1 \text{ min}}{60 \text{ s}} =$$

$$\underline{\underline{3.28 \text{ min}}}$$

20-29
$$(5 \times 60 + 24) \text{ s} \times 0.401 \text{ A} \times \frac{1 \text{ C}}{1 \text{ A} \cdot \text{s}} \times \frac{1 \text{ F}}{96485 \text{ C}} \times \frac{1 \text{ eq HA}}{\text{F}} = 1.3466 \times 10^{-3} \text{ eq HA}$$

$$0.1516 \text{ g HA} / 1.3466 \times 10^{-3} \text{ eq HA} = \underline{\underline{112.6 \text{ g/eq}}}$$

20-30 1 mol NaCN ≡ 1 mol H$^+$ ≡ 1 mol e$^-$ 3 min 22 s = 202 s

$$\frac{43.4 \times 10^{-3} \text{ A} \times 202 \text{ s} \times \frac{1 \text{ C}}{\text{A} \cdot \text{s}} \times \frac{1 \text{ F}}{96485 \text{ C}} \times \frac{1 \text{ mol NaCN}}{\text{F}} \times \frac{49.01 \text{ g NaCN}}{\text{mol NaCN}}}{10 \text{ mL sample} \times 10^{-3} \text{ L/mL}} = \underline{\underline{0.445 \text{ g/L}}}$$

20-31 1 mol CaCO$_3$ ≡ 1 mol HgNH$_3$Y^{2-} ≡ 2 mol e$^-$

$$31.6 \times 10^{-3} \text{ A} \times 2.02 \text{ min} \times 60 \frac{\text{s}}{\text{min}} \times \frac{1 \text{ C}}{\text{A} \cdot \text{s}} \times \frac{1 \text{ F}}{96485 \text{ C}} \times \frac{1 \text{ mol e}^-}{1 \text{ F}} \times \frac{1 \text{ mol CaCO}_3}{2 \text{ mol e}^-} =$$

$$1.9847 \times 10^{-5} \text{ mol CaCO}_3$$

$$\frac{1.9847 \times 10^{-5} \text{ mol CaCO}_3 \times 100.09 \text{ g CaCO}_3/\text{mol}}{25.00 \text{ mL H}_2\text{O} \times 1.000 \text{ g H}_2\text{O}/\text{mL H}_2\text{O}} \times 10^6 \text{ ppm} = \underline{\underline{79.5 \text{ ppm CaCO}_3}}$$

20-32 Proceeding as in Problem 20-31,

$$1 \text{ mol H}_2\text{S} \equiv 1 \text{ mol I}_2 \equiv 2 \text{ mol e}^-$$

$$\frac{36.32 \times 10^{-3} \times 10.12 \times 60}{96485} \text{ F} \times \frac{1 \text{ mol e}^-}{\text{F}} \times \frac{1 \text{ mol H}_2\text{S}}{2 \text{ mol e}^-} = 1.1428 \times 10^{-4} \text{ mol H}_2\text{S}$$

$$\frac{1.1428 \times 10^{-4} \text{ mol H}_2\text{S} \times 34.08 \text{ g H}_2\text{S}/\text{mol}}{100 \text{ mL H}_2\text{O} \times 1.000 \text{ g H}_2\text{O}/\text{mL H}_2\text{O}} \times 10^6 \text{ ppm} = \underline{\underline{38.9 \text{ ppm H}_2\text{S}}}$$

20-33 1 mol $C_6H_5NO_2$ ≡ 4 mol e^-

$$26.74 \text{ C} \times \frac{1 \text{ F}}{96485 \text{ C}} \times \frac{1 \text{ mol } e^-}{1 \text{ F}} \times \frac{1 \text{ mol } C_6H_5NO_2}{4 \text{ mol } e^-} = 6.929 \times 10^{-5} \text{ mol } C_6H_5NO_2$$

$$\frac{6.929 \times 10^{-5} \text{ mol } C_6H_5NO_2 \times 123.11 \text{ g } C_6H_5NO_2/\text{mol}}{210 \text{ mg sample} \times 10^{-3} \text{ g/mg}} \times 100\% = \underline{\underline{4.06\% \ C_6H_5NO_2}}$$

20-34 1 mol C_6H_5OH ≡ 3 mol Br_2 ≡ 6 mol e^- 7 min 33 s = 453 s

$$0.0313 \frac{\text{C}}{\text{s}} \times 453 \text{ s} \times \frac{1 \text{ F}}{96485 \text{ C}} \times \frac{1 \text{ mol } e^-}{\text{F}} \times \frac{1 \text{ mol } C_6H_5OH}{6 \text{ mol } e^-} = 2.449 \times 10^{-5} \text{ mol } C_6H_5OH$$

$$\frac{2.449 \times 10^{-5} \text{ mol } C_6H_5OH \times 94.11 \text{ g } C_6H_5OH/\text{mol}}{100 \text{ mL sample} \times 1.00 \text{ g sample/mL}} \times 10^6 \text{ ppm} = \underline{\underline{23.0 \text{ ppm } C_6H_5OH}}$$

20-35 1 mol CCl_4 ≡ 1 mol e^- 1 mol $CHCl_3$ ≡ 3 mol e^-

$$11.63 \text{ C} \times \frac{1 \text{ F}}{96485 \text{ C}} \times \frac{1 \text{ mol } e^-}{\text{F}} \times \frac{1 \text{ mol } CCl_4}{\text{mol } e^-} = 1.205 \times 10^{-4} \text{ mol } CCl_4$$

$$\frac{68.6}{96485} \text{F} \times \frac{1 \text{ mol } e^-}{\text{F}} \times \frac{1 \text{ mol } CHCl_3}{3 \text{ mol } e^-} = 2.370 \times 10^{-4} \text{ mol } CHCl_3$$

original no. mol $CHCl_3$ = $2.370 \times 10^{-4} - 1.205 \times 10^{-4}$ = 1.165×10^{-4}

$$\frac{1.205 \times 10^{-4} \text{ mol } CCl_4 \times 153.82 \text{ g } CCl_4/\text{mol } CCl_4}{0.750 \text{ g sample}} \times 100\% = \underline{\underline{2.471\% \ CCl_4}}$$

$$\frac{1.165 \times 10^{-4} \text{ mol } CHCl_3 \times 119.38 \text{ g } CHCl_3/\text{mol}}{0.750 \text{ g sample}} \times 100\% = \underline{\underline{1.854\% \ CHCl_3}}$$

20-36 Reactions:

$$2CHCl_3 + 6H^+ + 6e^- + 6Hg(l) \rightarrow 2CH_4(g) + 3Hg_2Cl_2(s)$$

$$CH_2Cl_2 + 2H^+ + 2e^- + 2Hg(l) \rightarrow CH_4(g) + Hg_2Cl_2(s)$$

Chapter 20

Let x = mass CHCl$_3$, then $0.1309 \text{ g} - x$ = mass CH$_2$Cl$_2$

$$\frac{x \text{ g CHCl}_3}{119.38 \text{ g/mol CHCl}_3} \times \frac{3 \text{ mol e}^-}{\text{mol CHCl}_3} + \frac{(0.1309 - x) \text{ g CH}_2\text{Cl}_2}{84.93 \text{ g/mol CH}_2\text{Cl}_2} \times \frac{2 \text{ mol e}^-}{\text{mol CH}_2\text{Cl}_2} =$$

$$\text{total mol e}^- = \frac{306.7 \text{ C}}{96485 \text{ C/mol e}^-} = 3.1787 \times 10^{-3} \text{ mol e}^-$$

$$2.5130 \times 10^{-2} x + 2.3549 \times 10^{-2} (0.1309 - x) = 3.1787 \times 10^{-3}$$

mass CHCl$_3$ = x = 6.087×10^{-2} g

$$\% \text{ CHCl}_3 = \left(\frac{6.087 \times 10^{-2} \text{ g}}{0.1309 \text{ g}}\right) \times 100\% = \underline{\underline{46.5\%}}$$

$\% \text{ CH}_2\text{Cl}_2$ = $100.0\% - 46.50\%$ = $\underline{\underline{53.50\%}}$

20-37 1 mol C$_6$H$_5$NH$_2$ ≡ 3 mol Br$_2$ ≡ 6 mol e$^-$

$$(3.76 - 0.27) \text{ min} \times \frac{60 \text{ s}}{\text{min}} \times 1.51 \times 10^{-3} \frac{\text{C}}{\text{s}} \times \frac{\text{mol e}^-}{96485 \text{ C}} \times \frac{1 \text{ mol C}_6\text{H}_5\text{NH}_2}{6 \text{ mol e}^-} =$$

$$5.462 \times 10^{-7} \text{ mol C}_6\text{H}_5\text{NH}_2$$

$$5.462 \times 10^{-7} \text{ mol C}_6\text{H}_5\text{NH}_2 \times 93.128 \frac{\text{g C}_6\text{H}_5\text{NH}_2}{\text{mol C}_6\text{H}_5\text{NH}_2} \times 10^6 \text{ µg/g} = \underline{\underline{50.9 \text{ µg C}_6\text{H}_5\text{NH}_2}}$$

20-38 1 mol Sn^{4+} + 2 mol e$^-$ → 1 mol Sn^{2+} ≡ 1 mol C$_6$H$_4$O$_2$

$(8.34 - 0.691) \text{ min} \times 60 \text{ (s/min)} \times 1.062 \times 10^{-3} \text{ C/s}$ = 0.4874 C

$$0.4874 \text{ C} \times \frac{1 \text{ mol e}^-}{96485 \text{ C}} \times \frac{1 \text{ mol C}_6\text{H}_4\text{O}_2}{2 \text{ mol e}^-} \times \frac{108.10 \text{ g C}_6\text{H}_4\text{O}_2}{\text{mol C}_6\text{H}_4\text{O}_2} = \underline{\underline{2.73 \times 10^{-4} \text{ g C}_6\text{H}_4\text{O}_2}}$$

CHAPTER 21

21-1 (a) *Voltammetry* is an analytical technique that is based upon measuring the current that develops in a microelectrode as the applied potential is varied. *Polarography* is a particular type of voltammetry in which the microelectrode is a dropping mercury electrode.

(b) In *linear scan polarography*, current in a cell containing a dropping mercury electrode is monitored continuously as applied potential is increased at a constant rate. In *pulse polarography*, an excitation signal is used that consists of a series of voltage pulses that increase in size linearly as a function of time. The pulses are applied during the last few milliseconds of the life of a drop. The current increases during this period.

(c) As shown in Figure 21-19 (page 486) and 21-22 (page 488) *differential pulse polarography* and *square-wave polarography* differ in the type of pulse sequence used.

(d) In the *dropping mercury electrode*, mercury is forced through a fine capillary tubing leading to a continuous stream of identical drops that typically have a lifetime of two to six seconds. A *hanging drop* consists of a single mercury drop that is not dislodged until the end of the experiment.

(e) A *residual current* in polarography is a nonfaradaic charging current that arises from the flow of electrons required to charge individual drops of mercury as they form and fall. A *limiting current* is a constant faradaic current that is limited in magnitude by the rate at which a reactant is brought to the surface of a microelectrode.

(f) In voltammetry, a *limiting current* is a constant current that is independent of applied potential. Its magnitude is limited by the rate at which a reactant is brought to the surface of the electrode by diffusion, migration, and convection. A *diffusion current* is a limiting current when analyte transport by migration and convection has been eliminated.

(g) *Turbulent flow* is a type of liquid flow that has no regular pattern. *Laminar flow* is a type of liquid flow in which layers of liquid slide by one another in a direction that is parallel to a solid surface.

(h) The difference between the *half-wave potential* and the *standard potential* for a reversible reaction is shown by Equation 21-10. That is,

Chapter 21

$$E_{1/2} = E_A^0 - \frac{0.0592}{n} \log \frac{k_A}{k_P} - E_{ref}$$

where k_A and k_P are constants that are proportional to the diffusion coefficients of the analyte and product and E_{ref} is the potential of the reference electrode.

21-2 (a) A *voltammogram* is a plot of current as a function of voltage applied to a microelectrode.

(b) *Hydrodynamic voltammetry* is a type of voltammetry in which the analyte solution is vigorously stirred or caused to flow by a microelectrode thus causing the reactant to be brought to the electrode surface by both convection and diffusion.

(c) The *Nernst diffusion layer* is the thin layer of stagnant solution that is immediately adjacent to the surface of an electrode in hydrodynamic voltammetry.

(d) A *mercury film electrode* is formed by depositing a thin film of mercury on a small solid metal or graphite surface.

(e) The *half-wave potential* is the potential on a voltammetric wave at which the current is one half of the limiting current.

21-3 Most organic electrode process involve hydrogen ions. A typical reaction is

$$R + nH^+ + ne^- \rightarrow RH_n$$

where R and RH_n are the oxidized and reduced forms of the organic molecule. Unless buffered solutions are used, marked pH changes will occur at the electrode surface as the reaction proceeds. Because the reduction potential (and sometimes the product) is affected by pH, drawn out and poorly defined waves are observed in the absence of a buffer.

21-4 The advantages of the dropping mercury electrode compared with other types of microelectrodes include: (1) its high hydrogen overvoltage, (2) its continuously produced fresh metal surface, and (3) its reproducible average currents that are immediately realized at any given applied potential. The disadvantages include: (1) its poor anodic potential range, (2) its relatively large residual current, (3) its tendency to yield current maxima, and (4) its inconvenience.

21-5 A plot of E_{applied} versus $\log \frac{i}{i_d - i}$ will yield a straight line having a slope of $-0.0592/n$.

Thus n is readily computed from the slope.

21-6 For the reduction of quinone (Q) to hydroquinone (H$_2$Q), Equation 21-2 takes the form

$$E_{\text{appl}} = E_Q^0 - \frac{0.0592}{n} \log \frac{c_{\text{H}_2\text{Q}}^0}{c_Q^0 \left(c_{\text{H}^+}^0\right)^2} - E_{\text{ref}}$$

where $c_{\text{H}^+}^0$ is the concentration of H$^+$ at the electrode surface. Substituting Equations 21-6 and 21-8 into this equation gives

$$E_{\text{appl}} = E_Q^0 - \frac{0.0592}{n} \log \frac{k_Q}{k_{\text{H}_2\text{Q}} \left(c_{\text{H}^+}^0\right)^2} - \frac{0.0592}{n} \log \frac{i}{i_l - i} - E_{\text{ref}}$$

When $i = i_l/2$, E_{appl} is the half-wave potential. With the added assumption that $k_Q \approx k_{\text{H}_2\text{Q}}$, the foregoing equation becomes

$$E_{1/2} = E_Q^0 - E_{\text{ref}} - \frac{0.0592}{n} \log \frac{1}{\left(c_{\text{H}^+}^0\right)^n} \quad (1)$$

$$= 0.599 - 0.244 - 0.0592 \text{ pH}$$

$$= 0.0355 - 0.0592 \text{ pH}$$

(a) At pH 7.00,

$$E_{1/2} = 0.355 - 0.0592 \times 7.00 = \underline{\underline{-0.059 \text{ V}}}$$

(b) Similarly at pH 5.00, $E_{1/2} = \underline{\underline{+0.059 \text{ V}}}$

21-7 In linear scan polarography the residual currents result from trace impurities but more important from a charging current in which each mercury drop is charged and carries this charge down to the bottom of the cell where the drop breaks off from the capillary. In current sampled polarography, current measurements are made at only the end of the drop lifetime so that large fluctuations in the current are eliminated. This technique does not eliminate the charging current but the decrease in fluctuations make it possible to determine the residual current more accurately.

Chapter 21

21-8 (a)
$$31.3 \times 10^{-6} \text{ A} \times 5 \text{ min} \times \frac{60 \text{ s}}{\text{min}} \times \frac{1 \text{ C}}{\text{A} \cdot \text{s}} \times \frac{1 \text{ mol e}^-}{96485 \text{ C}} = 9.732 \times 10^{-8} \text{ mol e}^-$$

$$9.732 \times 10^{-8} \text{ mol e}^- \times \frac{1 \text{ mol Cd}^{3+}}{2 \text{ mol e}^-} = 4.866 \times 10^{-8} \text{ mol Cd deposits}$$

$$20.0 \text{ mL} \times 10^{-3} \frac{\text{L}}{\text{mL}} \times 3.65 \times 10^{-3} \frac{\text{mol Cd}^{2+}}{\text{L}} = 7.30 \times 10^{-5} \text{ mol Cd}^{3+} \text{ originally}$$

$$\% \text{ decrease in } [\text{Cd}^{2+}] = \frac{4.866 \times 10^{-8}}{7.30 \times 10^{-5}} \times 100\% = \underline{\underline{0.067\%}}$$

(b) At 10 min, % decrease = $0.067 \times 10/5$ = $\underline{\underline{0.13\%}}$

(c) At 30 min, % decrease = $0.067 \times 30/5$ = $\underline{\underline{0.40\%}}$

21-9
$$c_{std} = 2.00 \times 10^{-3} \frac{\text{mmol}}{\text{mL}} \times \frac{112.4 \text{ mg Cd}}{\text{mmol}} = 0.2248 \frac{\text{mg Cd}}{\text{mL}}$$

When $V_{std} = 0$, $(i_d)_x = k V_x c_x / 50.0$ (V_x = volume sample)

When standard present, $(i_d)_{std} = k(V_x c_x + 224.8 V_s)/50.0$

(a) Solving these two equations simultaneously and substituting the data for (a) gives

$$c_x = \frac{0.2248 \, V_s \, (i_d)_x}{V_x [(i_d)_{std} - (i_d)_x]} = \frac{0.2248 \times 5.00 \times 79.7}{15.0 \, (95.7 - 79.7)} = \underline{\underline{0.369 \frac{\text{mg Cd}}{\text{mL}}}}$$

Proceeding in the same way, we obtain

(b) $\underline{0.346}$ (c) $\underline{0.144}$ (d) $\underline{0.314}$

21-10 At about +0.1 V, the following anodic reaction begins:

$$2\text{Hg} + 2\text{Br}^- \rightarrow \text{Hg}_2\text{Br}_2(s) + 2e^-$$

A limiting current for this reaction then occurs in the region of approximately +0.17 to +0.18 V, whereupon oxidation of the mercury leads to very large currents. That is,

$$2\text{Hg} \rightarrow \text{Hg}_2^{2+} + 2e^-$$

The wave at 0.12 V is useful for determination of Br⁻. The diffusion current should be directly proportional to [Br⁻].

21-11 If we assume that the diffusion coefficient for Ox and R are about the same, equation (1) in Solution 21-6 becomes

$$E_{1/2} = E^0_{Ox} - E_{ref} - \frac{0.0592}{4} \log \frac{1}{\left(c^0_{H^+}\right)^4} = K - 0.0592 \text{ pH}$$

When pH 2.5, $E_{1/2} = -0.349$. Thus

$$K = -0.349 + 0.0592 \times 2.5 = -0.201$$

(a) At pH = 1.00, $E_{1/2} = -0.201 - 0.0592 \times 1.00 = \underline{-0.260 \text{ V}}$

(b) At pH = 3.5, $E_{1/2} = -0.201 - 0.0592 \times 3.5 = \underline{-0.408 \text{ V}}$

(c) At pH = 7.0, $E_{1/2} = -0.201 - 0.0592 \times 7.0 = \underline{-0.615 \text{ V}}$

21-12 Stripping methods are generally more sensitive than other voltammetric procedures because the analyte can be removed from a relatively large volume of solution and concentrated on a tiny electrode for a long period (often many minutes). After concentration, the potential is reversed and all of the analyte that has been stored on the electrode is rapidly oxidized or reduced producing a large current.

21-13 The purpose of the electrodeposition step in stripping analysis is to concentrate the analyte in the thin film of mercury on the surface of the working electrode.

21-14 The advantage of voltammetry at ultramicroelectrodes include: (1) their low *IR* loss that permits their use in low-dielectric-constant solvents, such as hydrocarbons, (2) their small size that permits electrochemical studies on exceedingly small samples, such as the inside of living organs, and (3) their great speed of equilibration even in unstirred solution.

CHAPTER 22

22-1 (a) Transitions among various electronic energy levels.

(b) Transitions between the ground state of a molecule and the various vibrational and rotational energy states of the ground electronic state.

(c) Transitions from the lowest vibrational levels of the various excited electronic states to the various vibrational states of the ground state.

(d) Transitions from atoms in excited electronic states to lower energy electronic states.

(e) Transitions from molecules in excited electronic states to lower energy electronic states.

(f) Emission line spectra are produced by transitions of electrons in free atoms that have been excited to higher electronic states by absorption of thermal, electrical, or electromagnetic energy. Emission accompanies relaxation of the excited atom to a lower electronic energy level.

22-2 (a) The *ground state* is the lowest energy state of an atom, ion, or molecule of a species at room temperature.

(b) *Excited electronic states* are energy states of an atom or molecule that have greater energy than the ground state as a consequence of an electron being promoted to an energy level having a higher quantum number.

(c) A *photon* is a bundle or a particle of radiant energy whose magnitude is given by $h\nu$, where h is Planck's constant and ν is the frequency of the radiation.

(d) A *band spectrum* is a spectrum made up of lines that are so closely spaced that resolution into individual lines is so difficult that the spectrum often appears to be continuous.

(e) A *continuous spectrum* is one that is produced by heated bodies. Even at highest resolution, a continuous spectrum cannot be dispersed into lines.

(f) *Line spectra* are associated with gaseous atoms or elementary ions and consist of a relatively few narrow (approximately 0.005 nm) absorption or emission peaks that have wavelengths that are characteristic of the atom or ions.

(g) *Resonance fluorescence* is a type of fluorescence in which the emitted radiation is identical in frequency to the excitation frequency.

(h) *Phosphorescence* is a type of emission that is produced when a molecule is excited by absorption of electromagnetic radiation and then returns to the ground state by emitting electromagnetic radiation of a wavelength that differs from that of the excitation beam. Phosphorescence differs from fluorescence in the respect that the emission occurs after a delay of 10^{-5} s or more.

(i) $\%T = (P/P_0) \times 100\%$ where T is the transmittance, P_0 is the power of a beam of radiation that is incident upon an absorbing medium, and P is the power of the beam after having passed through a layer of the medium.

(j) The *absorbance A* of a solution is given by

$$A = \log(P_0/P) = \log(1/T)$$

where the terms used in defining A are found in the answer to part (i).

(k) The *molar absorptivity* ε is defined by the equation

$$\varepsilon = A/bc$$

where A is the absorbance of a medium having a path length of b cm and an analyte concentration of c molar.

(l) *Absorptivity a* is defined by the equation

$$a = A/bc$$

where A is the absorbance of a medium having a cell length of b, where b may be expressed in cm or any other specified unit of length, and a concentration of c where c may be expressed in any specified concentration unit (such as g/L).

(m) *Stray radiation* is unwanted radiation that appears at the exit slit of a monochromator as a result of scattering by dust particles and reflection off of interior surfaces of the monochromator. Its wavelength usually differs from the wavelength to which the monochromator is set.

(n) The *wavenumber* of radiation is the reciprocal of the wavelength in centimeters.

Chapter 22

(o) *Relaxation* is a process whereby an excited species loses energy and returns to a lower energy state.

(p) The *Stokes shift* is the difference in wavelength between the radiation used to excite fluorescence and the wavelength of the fluorescence itself.

22-3 $\nu = c/\lambda = 3.00 \times 10^{10} \text{ cm s}^{-1}/\lambda \text{ (cm)} = (3.00 \times 10^{10}/\lambda) \text{ s}^{-1}$

$= (3.00 \times 10^{10}/\lambda) \text{ Hz}$

(a) $\nu = 3.00 \times 10^{10} \text{ cm s}^{-1}/(2.65 \text{ Å} \times 10^{-8} \text{ cm/Å}) = \underline{1.13 \times 10^{18} \text{ Hz}}$

(b) $\nu = 3.00 \times 10^{10} \text{ cm s}^{-1}/(211.0 \text{ nm} \times 10^{-7} \text{ cm/nm}) = \underline{1.421 \times 10^{15} \text{ Hz}}$

(c) $\nu = 3.00 \times 10^{10} \text{ cm s}^{-1}/(694.3 \text{ nm} \times 10^{-7} \text{ cm/nm}) = \underline{4.318 \times 10^{14} \text{ Hz}}$

(d) $\nu = 3.00 \times 10^{10} \text{ cm s}^{-1}/(10.6 \text{ μm} \times 10^{-4} \text{ cm/μm}) = \underline{2.83 \times 10^{13} \text{ Hz}}$

(e) $\nu = 3.00 \times 10^{10} \text{ cm s}^{-1}/(19.6 \text{ μm} \times 10^{-4} \text{ cm/μm}) = \underline{1.53 \times 10^{13} \text{ Hz}}$

(f) $\nu = 3.00 \times 10^{10} \text{ cm s}^{-1}/1.86 \text{ cm} = \underline{1.61 \times 10^{10} \text{ Hz}}$

22-4 (a) $\lambda = \dfrac{c}{\nu} = \dfrac{3.00 \times 10^{10} \text{ cm s}^{-1}}{118.6 \text{ MHz} \times 10^{6} \text{ Hz/MHz}} \times \dfrac{1 \text{ Hz}}{\text{s}^{-1}} = \underline{252.8 \text{ cm}}$

(b) $\lambda = 3.00 \times 10^{10} \text{ cm s}^{-1}/114.10 \times 10^{3} \text{ s}^{-1} = \underline{2.628 \times 10^{5} \text{ cm}}$

(c) $\lambda = \dfrac{c}{\nu} = \dfrac{3 \times 10^{10} \text{ cm s}^{-1}}{105 \text{ MHz} \times 10^{6} \text{ Hz/MHz}} \times \dfrac{1 \text{ Hz}}{\text{s}^{-1}} = \underline{286 \text{ cm}}$

(d) $\lambda = 1/1210 \text{ cm}^{-1} = \underline{8.26 \times 10^{-4} \text{ cm}}$

22-5 (a) $\bar{\nu} = 1/(3 \text{ μm} \times 10^{-4} \text{ cm/μm}) = \underline{3.33 \times 10^{3} \text{ cm}^{-1}}$ to

$1/(15 \times 10^{-4}) = \underline{6.67 \times 10^{2} \text{ cm}^{-1}}$

(b) $\nu = 3.00 \times 10^{10} \text{ (cm/s)} \times 3.33 \times 10^{3} \text{ cm}^{-1} = \underline{1.00 \times 10^{14} \text{ Hz}}$ to

$3 \times 10^{10}/(6.667 \times 10^{2}) = \underline{2.00 \times 10^{13} \text{ Hz}}$

294

Chapter 22

22-6 $\bar{\nu}$ = $1/(185 \text{ nm} \times 10^{-7} \text{ cm/nm})$ = $\underline{\underline{5.41 \times 10^4 \text{ cm}^{-1}}}$ to

$1/3000 \times 10^{-7}$ = $\underline{\underline{3.33 \times 10^3 \text{ cm}^{-1}}}$

ν = $3.00 \times 10^{10} \text{ cm s}^{-1} \times 5.41 \times 10^4 \text{ cm}^{-1}$ = $\underline{\underline{1.62 \times 10^{15} \text{ Hz}}}$ to

$3.00 \times 10^{10} \times 3.33 \times 10^3$ = $\underline{\underline{1.00 \times 10^{14} \text{ Hz}}}$

22-7 (a) T = $100 \times \text{antilog}(-0.064)$ = $\underline{\underline{86.3\%}}$

Proceeding in the same way, we obtain

(b) $\underline{\underline{17.2\%}}$ (c) $\underline{\underline{48.1\%}}$ (d) $\underline{\underline{61.8\%}}$ (e) $\underline{\underline{36.6\%}}$ (f) $\underline{\underline{38.6\%}}$

22-8 (a) A = $-\log(19.4/100)$ = $\underline{\underline{0.712}}$

Proceeding in the same way, we obtain

(b) $\underline{\underline{0.064}}$ (c) $\underline{\underline{0.565}}$ (d) $\underline{\underline{1.35}}$ (e) $\underline{\underline{1.00}}$ (f) $\underline{\underline{0.098}}$

22-9 (a) T = $100 \times \text{antilog}(-0.064 \times 2)$ = $\underline{\underline{74.5\%}}$

Similarly,

(b) $\underline{\underline{2.95\%}}$ (c) $\underline{\underline{23.1\%}}$ (d) $\underline{\underline{38.2\%}}$ (e) $\underline{\underline{13.4\%}}$ (f) $\underline{\underline{14.9\%}}$

22-10 (a) A = $-\log \dfrac{0.194}{2}$ = $\underline{\underline{1.013}}$

Similarly,

(b) $\underline{\underline{0.365}}$ (c) $\underline{\underline{0.866}}$ (d) $\underline{\underline{1.65}}$ (e) $\underline{\underline{1.30}}$ (f) $\underline{\underline{0.399}}$

22-11 (a) $\%T$ = $\text{antilog}(-0.416) \times 100\%$ = $\underline{\underline{38.4\%}}$

ε = A/bc = $0.416/(1.40 \times 1.25 \times 10^{-4})$ = $\underline{\underline{2.38 \times 10^3}}$

c = $1.25 \times 10^{-4} \dfrac{\text{mol}}{\text{L}} \times \dfrac{250 \text{ g}}{\text{mol}} \times \dfrac{1 \text{ L}}{1000 \text{ g}} \times 10^6 \text{ ppm}$ = $\underline{\underline{31.2 \text{ ppm}}}$

a = $0.416/31.3$ = $\underline{\underline{1.33 \times 10^{-2}}}$

Chapter 22

(b) $A = -\log 0.455 = \underline{\underline{0.342}}$

$\varepsilon = 0.342/(2.10 \times 8.15 \times 10^{-3}) = \underline{\underline{20.0}}$

$c = 8.15 \times 10^{-3} \times 250 \times 10^{6}/10^{3} = \underline{\underline{2.04 \times 10^{3} \text{ ppm}}}$

$a = 0.342/2.04 \times 10^{3} = \underline{\underline{1.68 \times 10^{-4}}}$

(c) $\%T = (\text{antilog} - 1.424) \times 100\% = \underline{\underline{3.77\%}}$

$c = A/ab = 1.424/(0.137 \times 0.996) = 10.44 = \underline{\underline{10.4 \text{ ppm}}}$

$c = \dfrac{10.44 \text{ g}}{10^{6} \text{ g soln}} \times \dfrac{1000 \text{ g soln}}{\text{L soln}} \times \dfrac{1 \text{ mol analyte}}{250 \text{ g analyte}} = 4.176 \times 10^{-4} \text{ mol/L} = \underline{\underline{4.18 \times 10^{-5} \text{ M}}}$

$\varepsilon = 1.424/(0.996 \times 4.176 \times 10^{-5}) = \underline{\underline{3.42 \times 10^{4}}}$

(d) $A = -\log 0.196 = \underline{\underline{0.708}}$

$b = 0.708/c\varepsilon = 0.708/(2.50 \times 10^{-4} \times 5.42 \times 10^{3}) = \underline{\underline{0.523 \text{ cm}}}$

$c = 2.50 \times 10^{-4} \times 250 \times 10^{6}/10^{3} = \underline{\underline{62.5 \text{ ppm}}}$ [see part (a)]

$a = 0.708/(0.523 \times 62.5) = \underline{\underline{2.17 \times 10^{-2}}}$

(e) $c = \dfrac{3.33 \text{ g analyte}}{10^{6} \text{ g soln}} \times \dfrac{1000 \text{ g soln}}{\text{L soln}} \times \dfrac{1 \text{ mol analyte}}{250 \text{ g}} = 1.332 \times 10^{-5} \dfrac{\text{mol}}{\text{L}}$

$= \underline{\underline{1.33 \times 10^{-5} \text{ M}}}$

$A = \varepsilon bc = 3.46 \times 10^{3} \times 2.50 \times 1.332 \times 10^{-5} = 0.1152 = \underline{\underline{0.115}}$

$\%T = (\text{antilog} -0.1152) \times 100\% = \underline{\underline{76.7\%}}$

$a = 0.1152/(2.50 \times 3.33) = \underline{\underline{0.0138}}$

(f) $A = 1.214 \times 10^{3} \times 1.25 \times 7.77 \times 10^{-4} = 1.179 = \underline{\underline{1.18}}$

$\%T = (\text{antilog} -1.179) \times 100\% = \underline{\underline{6.62\%}}$

$c = 7.77 \times 10^{-4} \times 250 \times 10^{6} / 10^{3} = 194.2 = \underline{\underline{194 \text{ ppm}}}$ [see part (a)]

$a = 1.179/(1.25 \times 194.2) = \underline{\underline{4.86 \times 10^{-3}}}$

(g) $A = -\log 0.483 = 0.3161 = \underline{\underline{0.316}}$

$c = 6.72 \times 10^{3}/(10^{6} \times 250) = 2.688 \times 10^{-5} = \underline{\underline{2.69 \times 10^{-5} \text{ M}}}$ [see part (a)]

$\varepsilon = 0.3161/(0.250 \times 2.68 \times 10^{-5}) = \underline{\underline{4.70 \times 10^{4}}}$

$a = 0.316/(0.250 \times 6.72) = \underline{\underline{0.188}}$

(h) $\%T = (\text{antilog} -0.842) \times 100\% = \underline{\underline{14.4\%}}$

$c = 0.842/(7.73 \times 10^{3} \times 2.00) = \underline{\underline{5.44 \times 10^{-5} \text{ M}}}$

$c = 5.44 \times 10^{-5} \times 250 \times 10^{6}/10^{3} = \underline{\underline{13.6 \text{ ppm}}}$ [see part (a)]

$a = 0.842/(2.00 \times 13.6) = \underline{\underline{0.0310}}$

(i) $A = -\log 0.763 = 0.11748 = \underline{\underline{0.117}}$

$c = A/ab = 0.11748/(0.0631 \times 1.10) = 1.692 = \underline{\underline{1.69 \text{ ppm}}}$

$c = 1.692 \times 10^{3}/(10^{6} \times 250) = 6.768 \times 10^{-6} = \underline{\underline{6.77 \times 10^{-6} \text{ M}}}$ [see part (a)]

$\varepsilon = 0.11748/(1.10 \times 6.768 \times 10^{-6}) = \underline{\underline{1.58 \times 10^{4}}}$

(j) $A = -\log 0.0654 = \underline{\underline{1.184}}$

$b = A\varepsilon c = 1.184/(9.82 \times 10^{2} \times 8.64 \times 10^{-3}) = \underline{\underline{0.140 \text{ cm}}}$

$c = 8.64 \times 10^{-3} \times 250 \times 10^{6}/10^{3} = \underline{\underline{2160 \text{ ppm}}}$

$a = 1.184/(0.140 \times 2160) = \underline{\underline{3.92 \times 10^{-3}}}$

Chapter 22

22-12
$$c = \frac{4.48 \text{ mg KMnO}_4}{L} \times \frac{10^{-3} \text{ g KMnO}_4}{\text{mg KMnO}_4} \times \frac{1 \text{ mol KMnO}_4}{158.03 \text{ g KMnO}_4}$$

$$= 2.835 \times 10^{-5} \frac{\text{mol KMnO}_4}{L}$$

$A = -\log 0.309 = 0.5100$

$$\varepsilon = \frac{0.5100}{1.00 \times 2.835 \times 10^{-5}} = \underline{\underline{1.80 \times 10^4 \text{ L cm}^{-1} \text{ mol}^{-1}}}$$

22-13
$$3.75 \frac{\text{mg}}{100 \text{ mL}} \times \frac{1 \text{ mmol A}}{220 \text{ mg A}} = 1.705 \times 10^{-4} \frac{\text{mmol A}}{\text{mL}}$$

$A = -\log 0.396 = 0.4023$

$\varepsilon = 0.4023 / (1.50 \times 10^{-4}) = \underline{\underline{1.57 \times 10^3}}$

22-14 $\varepsilon = 9.32 \times 10^3 \text{ L cm}^{-1} \text{ mol}^{-1}$

(a) $A = 9.32 \times 10^3 \times 1.00 \times 6.24 \times 10^{-5} = \underline{\underline{0.582}}$

(b) $T = \text{antilog}(-0.582) \times 100\% = \underline{\underline{26.2\%}}$

(c) $c = A/\varepsilon b = 0.582/(9.32 \times 10^3 \times 5.00) = \underline{\underline{1.25 \times 10^{-5} \text{ M}}}$

22-15 (a) $A_1 = \varepsilon b c = 7.00 \times 10^3 \times 1.00 \times 2.50 \times 10^{-5} = \underline{\underline{0.175}}$

(b) $A_2 = 2 \times A_1 = 2 \times 0.175 = \underline{\underline{0.350}}$

(c) $T_1 = \text{antilog}(-0.175) = \underline{\underline{0.668}}$ or $\underline{\underline{66.8\%}}$

$T_2 = \text{antilog}(-0.350) = \underline{\underline{0.447}}$ or $\underline{\underline{44.7\%}}$

(d) $A = -\log(0.668/2) = \underline{\underline{0.476}}$

Chapter 22

22-16 $\varepsilon = 7.00 \times 10^3 \text{ L cm}^{-1} \text{ mol}^{-1}$

$c_{Fe} = 3.8 \dfrac{\text{mg Fe}}{\text{L}} \times 10^{-3} \dfrac{\text{g Fe}}{\text{mg Fe}} \times \dfrac{1 \text{ mol Fe}}{55.85 \text{ g Fe}} \times \dfrac{2.50 \text{ mL}}{50.0 \text{ mL}} = 3.40 \times 10^{-6} \text{ M}$

$A = 7.00 \times 10^3 \times 2.50 \times 3.40 \times 10^{-6} = \underline{\underline{0.0595}}$

22-17 (a) $T = (\text{antilog} -0.464) \times 100\% = \underline{\underline{34.4\%}}$

(b) In 2.50-cm cell, $A = 0.464 \times 2.50 / 1.00 = 1.160$

$T = (\text{antilog} -1.160) \times 100\% = \underline{\underline{6.92\%}}$

(c) $c_{\text{complex}} = c_{Zn^{2+}} = 1.60 \times 10^{-4}$

$\varepsilon = \dfrac{0.464}{1.00 \times 1.60 \times 10^{-4}} = \underline{\underline{2.90 \times 10^3}}$

22-18 $[H_3O^+][In^-]/[HIn] = 8.00 \times 10^{-5}$

$[H_3O^+] = [In^-]$ $[HIn] = c_{HIn} - [In^-] \approx c_{HIn}$

(a) At $c_{HIn} = 3.00 \times 10^{-4}$

$\dfrac{[In^-]^2}{(3.00 \times 10^{-4}) - [In^-]} = 8.00 \times 10^{-5}$

$[In^-]^2 + 8.00 \times 10^{-5} [In^-] - 2.40 \times 10^{-8} = 0$

$[In^-] = 1.20 \times 10^{-4}$

$[HIn] = 3.00 \times 10^{-4} - 1.20 \times 10^{-4} = 1.80 \times 10^{-4}$

$A_{430} = 1.20 \times 10^{-4} \times 0.775 \times 10^3 + 1.80 \times 10^{-4} \times 8.04 \times 10^3 = \underline{\underline{1.54}}$

$A_{600} = 1.20 \times 10^{-4} \times 6.96 \times 10^3 + 1.80 \times 10^{-4} \times 1.23 \times 10^3 = \underline{\underline{1.06}}$

The remaining calculations are performed in the same way and yield the following results.

Chapter 22

c_{ind}, M	[In$^-$]	[HIn]	A_{430}	A_{600}
3.00×10^{-4}	1.20×10^{-4}	1.80×10^{-4}	1.54	1.06
2.00×10^{-4}	9.27×10^{-5}	1.07×10^{-4}	0.935	0.777
1.00×10^{-4}	5.80×10^{-5}	4.20×10^{-5}	0.383	0.455
0.500×10^{-4}	3.48×10^{-5}	1.52×10^{-5}	0.149	0.261
0.250×10^{-4}	2.00×10^{-5}	5.00×10^{-6}	0.056	0.145

(b)

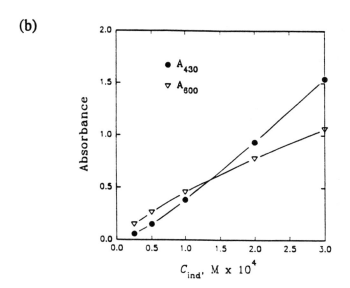

22-19 $[Cr_2O_7^{2-}]/[CrO_4^{2-}]^2[H^+]^2 = 4.2 \times 10^{14}$

$[H^+]$ = antilog (-5.60) = 2.51×10^{-6}

$[Cr_2O_7^{2-}] = c_{K_2Cr_2O_7} - [CrO_4^{2-}]/2$

$$\frac{c_{K_2Cr_2O_7} - [CrO_4^{2-}]/2}{[CrO_4^{2-}]^2 \times (2.51 \times 10^{-6})^2} = 4.2 \times 10^{14}$$

$c_{K_2Cr_2O_7} - 0.500 \, [CrO_4^{2-}] = 2.65 \times 10^3 \, [CrO_4^{2-}]^2$

$[CrO_4^{2-}]^2 + 1.887 \times 10^{-4} \, [CrO_4^{2-}] - 3.774 \times 10^{-4} \, c_{K_2Cr_2O_7} = 0$

When $c_{K_2Cr_2O_7} = 4.00 \times 10^{-4}$

$$[CrO_4^{2-}]^2 + 1.887 \times 10^{-4} [CrO_4^{2-}] - 1.510 \times 10^{-7} = 0$$

$$[CrO_4^{2-}] = 3.055 \times 10^{-4}$$

$$[Cr_2O_7^{2-}] = 4.00 \times 10^{-4} - 3.055 \times 10^{-4}/2 = 2.473 \times 10^{-4}$$

$A_{345} = 1.84 \times 10^3 \times 3.055 \times 10^{-4} + 10.7 \times 10^2 \times 2.473 \times 10^{-4} = 0.826$

$A_{370} = 4.81 \times 10^3 \times 3.055 \times 10^{-4} + 7.28 \times 10^2 \times 2.473 \times 10^{-4} = 1.649$

$A_{400} = 1.88 \times 10^3 \times 3.055 \times 10^{-4} + 1.89 \times 10^2 \times 2.473 \times 10^{-4} = 0.621$

Proceeding in the same way, we obtain

$c_{K_2Cr_2O_7}$	$[CrO_4^{2-}]$	$[Cr_2O_7^{2-}]$	A_{345}	A_{370}	A_{400}
4.00×10^{-4}	3.055×10^{-4}	2.473×10^{-4}	0.826	1.649	0.621
3.00×10^{-4}	2.551×10^{-4}	1.725×10^{-4}	0.654	1.352	0.512
2.00×10^{-4}	1.961×10^{-4}	1.019×10^{-4}	0.470	1.018	0.388
1.00×10^{-4}	1.216×10^{-4}	3.920×10^{-5}	0.266	0.613	0.236

(b)
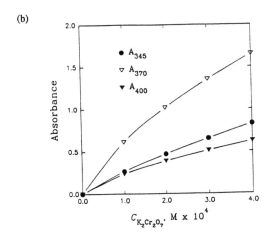

CHAPTER 23

23-1 The *effective bandwidth* of a filter is the width in units of wavelength of the band of transmitted radiation measured at one half the height of the band.

23-2 $n\lambda = d(\sin i + \sin r)$ (Equation 23 – 4)

$$d = \frac{n\lambda}{\sin i + \sin r} = \frac{(1)(500 \text{ nm})}{\sin 60 + \sin 10}$$

$$= \frac{500 \text{ nm} \times 10^{-6} \text{ mm/nm}}{0.866 + 0.174} = 4.81 \times 10^{-4} \text{ mm/line}$$

$$\frac{1}{4.81 \times 10^{-4} \text{ mm/line}} = \underline{\underline{2079 \text{ lines/mm}}}$$

23-3 Rearranging Equation 23-4 leads to

$$\lambda_n = \frac{d}{n}(\sin i + \sin r)$$

$$d = \frac{1}{72 \text{ lines/mm}} = 0.01389 \text{ mm}$$

(a) $\lambda_n = \dfrac{0.01389 \text{ mm} \times 10^3 \text{ μm/mm}}{n}(\sin 50 + \sin 0) = 10.64 \text{ μm/n}$

For $n = 1$, $\lambda_1 = \underline{\underline{10.64 \text{ μm}}}$ For $n = 2$, $\lambda_2 = \underline{\underline{5.32 \text{ μm}}}$

Proceeding in the same way

(b) $\lambda_1 = 0.01389 \times 10^3 (\sin 50 + \sin 15) = \underline{\underline{14.24 \text{ μm}}}$

$\lambda_2 = \underline{\underline{7.12 \text{ μm}}}$

23-4 A *spectroscope* consists of a monochromator that has been modified so that the focal plane contains a movable eyepiece that permits visual detection of the various emission lines of the elements. A *spectrograph* is a monochromator equipped with a photographic film or plate holder located along its focal plane; spectra are recorded photographically. A *spectrophotometer* is a monochromator equipped with a photoelectric detector that is located behind an exit slit located on the focal plane of the device.

23-5 Quantitative analyses can tolerate rather wide slits since measurements are usually carried out at maxima where the slope of the spectrum $dA/d\lambda$ is relatively constant. On the other hand, qualitative analyses require narrow slits so that any fine structure in the spectrum will be resolved.

23-6 $\lambda_{max} T = 2.90 \times 10^3 \qquad \lambda_{max} = 2.90 \times 10^3 / T$

(a) $\lambda_{max} = 2.90 \times 10^3 / 4000 = 0.725\ \mu m$ or $\underline{\underline{725\ nm}}$

(b) $\lambda_{max} = 2.90 \times 10^3 / 3000 = 0.967\ \mu m$ or $\underline{\underline{967\ nm}}$

(c) $\lambda_{max} = 2.90 \times 10^3 / 2000 = \underline{\underline{1.45\ \mu m}}$

(d) $\lambda_{max} = 2.90 \times 10^3 / 1000 = \underline{\underline{2.90\ \mu m}}$

23-7 $E_t = \alpha T^4 = \underline{\underline{5.69 \times 10^{-8}\ W/m^2\ K^{-4}\ T^4}}$

(a) $E_t = \alpha (4000\ K)^4 = \underline{\underline{1.46 \times 10^7\ W/m^2}}$

(b) $E_t = \alpha (3000\ K)^4 = \underline{\underline{4.61 \times 10^6\ W/m^2}}$

(c) $E_t = \alpha (2000\ K)^4 = \underline{\underline{9.10 \times 10^5\ W/m^2}}$

Chapter 23

(d) $E_t = \alpha(1000 \text{ K})^4 = \underline{\underline{5.69 \times 10^4 \text{ W}/\text{m}^2}}$

23-8 (a) $\lambda_{max} = \dfrac{2.90 \times 10^3}{2870} = \underline{\underline{1.01 \text{ μm}}}$

$\lambda'_{max} = \dfrac{2.90 \times 10^3}{3000} = \underline{\underline{0.967 \text{ μm}}}$

(b) $E_t = 5.69 \times 10^{-8} \times (2870)^4 = \underline{\underline{3.86 \times 10^6 \text{ W}/\text{m}^2}}$

$E'_t = 5.69 \times 10^{-8} \times (3000)^4 = \underline{\underline{4.61 \times 10^6 \text{ W}/\text{m}^2}}$

23-9 (a) *Hydrogen and deuterium lamps* differ only in the gases they contain. The latter produces radiation of somewhat higher intensity.

(b) *Filters* provide low resolution wavelength selection often suitable for quantitative work but not for qualitative analysis. *Monochromators* produce high resolution (narrow bandwidths) for both qualitative and quantitative work.

(c) A *phototube* is a vacuum tube equipped with a photoemissive cathode. It has a high electrical resistance and requires a potential of 90 V or more to produce a photocurrent. The currents are generally small enough to require considerable amplification before they can be measured. A *photovoltaic cell* consists of a photosensitive semiconductor sandwiched between two electrodes. A current is generated between the electrodes when radiation is absorbed by the semiconducting layer. The current is generally large enough to be measured directly with a microammeter. The advantages of a phototube are greater sensitivity and wavelength range as well as better reproducibility. The advantages of the photocell are its simplicity, low cost, and general ruggedness. In addition, it does not require an external power supply or elaborate electric circuitry. Its use is limited to visible radiation, however, and it addition it suffers from fatigue whereby its electrical output decreases gradually with time.

(d) *Phototubes* consist of a single photoemissive surface (cathode) and an anode in an evacuated envelope. They exhibit low dark current but have no inherent amplification (gain). *Photomultipliers* have built-in gain but suffer from somewhat larger dark current.

(e) A *photometer* is an instrument for absorption measurements that consists of a source, a filter, and a photoelectric detector. A *colorimeter* differs from a photometer in the respect that the human eye serves as the detector. The photometer offers the advantage of greater precision and the ability to discriminate between colors provided they are not too much alike. The main advantages of a colorimeter are simplicity, low cost, and the fact that no power supply is needed.

(f) *Spectrophotometers* have monochromators for multiple wavelength operation and for procuring spectra while *photometers* utilize filters for fixed wavelength operation.

(g) A *single-beam spectrophotometer* employs a fixed beam of radiation that irradiates first the solvent and then the analyte solution. In a *double-beam instrument*, the solvent and solution are irradiated simultaneously or nearly so. The advantages of the double-beam instruments are freedom from problems arising from fluctuations in the source intensity from drift in electronic circuits; in addition, it is more easily adapted to automatic spectral recording. The single-beam instrument offers the advantages of simplicity and lower cost.

Chapter 23

(h) Diode array spectrometers detect the entire spectral range essentially simultaneously and can produce a spectrum in less than a second. *Spectrometers* require several minutes to perform the same task. Accordingly the diode array instruments can be used to monitor processes that occur on very fast time scales. Conventional spectrometers require several minutes to perform the same task.

23-10 (a) $\%T = P/P_0 \times 100\% = I/I_0 \times 100\%$

$= 24.9\,\mu A / 73.6\,\mu A \times 100\% = \underline{\underline{33.8\%}}$

(b) $A = -\log T = -\log 24.9/73.6 = \underline{\underline{0.471}}$

(c) $A = \varepsilon bc \qquad 1/3\,A = \varepsilon bc/3$

$A_{1/3} = 1/3\,(0.471) = 0.157$

$T = 10^{-0.157} = \underline{\underline{0.697}}$

(d) $A_2 = 2A = 2(0.471) = 0.942$

$T = 10^{-0.942} = \underline{\underline{0.114}}$ or 11.4%

23-11 (a) $T = 179\,\text{mV}/685\,\text{mV} = \underline{\underline{0.261}}$

$A = -\log T = -\log 0.261 = \underline{\underline{0.583}}$

(b) $A = 0.582/2 = \underline{\underline{0.291}}$

$T = 10^{-0.291} = \underline{\underline{0.511}}$

(c) $A = 2 \times 0.583 = 1.166$

$T = 10^{-1.166} = \underline{\underline{0.0682}}$

23-12 In a *deuterium lamp*, the input energy from the power source produces an excited deuterium molecule that dissociates into two atoms in the ground state and a photon of radiation. As the excited deuterium molecule relaxes, its quantized energy is distributed between the energy of the photon and the kinetic energies of the two deuterium atoms. The latter can vary from nearly zero to the original energy of the excited molecule. Therefore, the energy of the radiation, which is the difference between the quantized energy of the excited molecule and the kinetic energies of the atoms, can also vary continuously over the same range. Consequently, the emission spectrum is continuous.

23-13 A *photon detector* produces a current or voltage as a result of the emission of electrons from a photosensitive surface upon collision with photons. On the other hand, a *heat detector* consists of a darkened surface to absorb infrared energy to produce a temperature increase and a thermal transducer to produce an electrical signal whose magnitude is related to the temperature and thus the intensity of the infrared energy.

23-14 The power of an infrared beam is measured with a heat detector, which consists of a tiny blackened surface that is warmed as a consequence of radiation absorption. This surface is attached to a transducer, which converts the heat signal to an electrical one. Heat transducers are of three types: (1) a *thermopile*, which consists of several dissimilar metal junctions that develop a potential that depends upon the difference in temperature of the junctions; (2) a *bolometer*, which is fashioned from a conductor whose electrical resistance depends upon temperature; and (3) a *pneumatic detector* whose internal pressure is temperature-dependent.

23-15 Photons from the infrared region of the spectrum do not have sufficient energy to cause photoemission from the cathode of a photomultiplier.

Chapter 23

23-16 *Tungsten/halogen lamps* contain a small amount of iodine in the evacuated quartz envelope that contains the tungsten filament. The iodine prolongs the life of the lamp and permits it to operate at a higher temperature. The iodine combines with gaseous tungsten that sublimes from the filament and causes the metal to be redeposited, thus adding to the life of the lamp.

23-17 Basically an *absorption photometer* and a *fluorescence photometer* consist of the same components. The basic difference is in the location of the detector. The detector in a fluorometer is positioned at an angle of 90 deg to the direction of the beam from the source so that emission is detected rather than transmission. In addition, a filter is often positioned in from of the detector to remove radiation from the excitation beam that may result from scattering or other nonfluorescence processes. In a transmission photometer, the detector is positioned in a line with the source, the filter, and the detector.

23-18 The basic difference between an *absorption spectrometer* and an *emission spectrometer* is that the former requires a separate radiation source and a sample compartment that holds containers for the sample and its solvent. With an emission spectrometer, the sample container is a hot flame, a heated surface, or an electric arc or spark that also serves as the radiation source.

23-19 The performance characteristics of an interference filter include the wavelength of its transmittance peak, the percent transmission at the peak, and the effective bandwidth.

23-20 (a) The *dark current* is the small current that develops in a radiation transducer in the absence of radiation.

(b) A *transducer* is a device that converts a physical or chemical quantity into an electrical signal.

Chapter 23

(c) *Scattered radiation* in a monochromator is unwanted radiation that reaches the exit slit as a result of reflection and scattering. Its wavelength is usually different from that of the radiation reaching the slit directly from the dispersing element.

(d) An *n-type semiconductor* is a material consisting of silicon or germanium that has been doped with an element from Group 5A of the periodic table to impart extra electrons into the crystal lattice of the semiconducting element. These electrons become the mobile charge carriers when the material passes a current.

(e) The *majority carrier* in a semiconductor is the mobile charge carrier in either *n*-type or *p*-type semiconductors. For *n*-type, the majority carrier is the electron while in *p*-type, the majority carrier is a positively charged hole.

23-21 (a) $\lambda_{max} = 2t\eta/n \qquad t = \dfrac{\lambda_{max}/n}{2\eta}$

$t = \dfrac{4.54\ \mu m\ (1)}{2\ (1.34)} = \underline{1.69\ \mu m}$

(b) $\lambda_n = (2t\eta)/n$

For **n** = 1, 2, 3, ···

$\lambda_1 = 2t\eta = \underline{4.54\ \mu m}$

$\lambda_2 = 1/2\ \lambda_1 = \underline{2.27\ \mu m}$

$\lambda_3 = 1/3\ \lambda_1 = \underline{1.51\ \mu m}$ (etc.)

CHAPTER 24

24-1 (a) *Spectrophotometers* use a grating or a prism to provide narrow bands of radiation while *photometers* utilize filters for this purpose. The advantages of spectrophotometers are greater versatility and the ability to procure entire spectra. The advantages of photometers are simplicity, ruggedness, and low cost.

(b) A *single-beam spectrophotometer* employs a fixed beam of radiation that irradiates first the solvent and then the analyte solution. In a *double-beam instrument*, the solvent and solution are irradiated simultaneously or nearly so. The advantages of double-beam instruments are freedom from problems arising from fluctuations in the source intensity and from drift in electronic circuits; in addition, they are more easily adapted to automatic spectral recording. The single-beam instrument offers the advantages of simplicity and lower cost.

(c) *Diode-array spectrophotometers* detect the entire spectral range essentially simultaneously and can produce a spectrum in less than a second. *Conventional spectrophotometers* require several minutes to perform the same task. Accordingly diode array instruments can be used to monitor processes that occur on very fast time scales. Their resolution is usually lower than that of a conventional spectrophotometer.

24-2 As a minimum requirement, the radiation emitted by the source of a single-beam instrument must be stable for however long it takes to make the 0% T adjustment, the 100% T adjustment, and the measurement of T for the sample.

24-3 (a) The 0% transmittance is made with no light reaching the detector and compensates for any dark currents.

(b) The 100% transmittance adjustment is made with a blank in the light path and compensates for any absorption or reflection losses.

24-4 Electrolyte concentration, pH, temperature.

24-5 The standard-addition method permits absorbance measurements, free of matrix effects in the sample. Adherence to Beer's law throughout the concentration range is a minimum requirement.

24-6 $A = \varepsilon bc = 9.32 \times 10^3 \times 1.00 \times c$

$c \geq A/9.32 \times 10^3 = 0.15/9.32 \times 10^3 = \underline{\underline{1.6 \times 10^{-5} \text{ M}}}$

$c \leq 0.80/9.32 \times 10^3 = \underline{\underline{8.6 \times 10^{-5} \text{ M}}}$

24-7 $A = -\log 0.80 = 0.097$ and $-\log 0.050 = 1.30$

$A = \varepsilon bc = 6.17 \times 10^3 \times 1.00 \times c$

$c \geq 0.097/6.17 \times 10^3 = \underline{\underline{1.6 \times 10^{-5} \text{ M}}}$

$c \leq 1.30/6.17 \times 10^3 = \underline{\underline{2.1 \times 10^{-4} \text{ M}}}$

24-8 $A = -\log T$ $A_{10\%} = -\log 0.100 = 1.00$ $A_{90\%} = -\log 0.900 = 0.0458$

$\log \varepsilon = 2.75$ and $\varepsilon = 562$

$c_{10\%} = \dfrac{A}{b\varepsilon} = \dfrac{1.00}{(1.5)(562)} = \underline{\underline{1.2 \times 10^{-3} \text{ M}}}$

$c_{90\%} = \dfrac{0.0458}{(1.5)(562)} = \underline{\underline{5.4 \times 10^{-5} \text{ M}}}$

24-9 $\log \varepsilon = 3.812$ $\varepsilon = 6.49 \times 10^3$

$c_{0.100} = \dfrac{0.100}{(1.25)(6.49 \times 10^3)} = \underline{\underline{1.23 \times 10^{-5} \text{ M}}}$

$c_{2.000} = \dfrac{2.000}{(1.25)(6.49 \times 10^3)} = \underline{\underline{2.47 \times 10^{-4} \text{ M}}}$

24-10 (a) $T = 179 \text{ mV}/685 \text{ mV} = 0.261$ or $\underline{\underline{26.1\%}}$

$A = -\log 0.261 = \underline{\underline{0.583}}$

(b) $A = 0.583/2 = \underline{\underline{0.291}}$

$T = \text{antilog}(-0.583/2) = 0.511$ or $\underline{\underline{51.1\%}}$

Chapter 24

(c) $A = 2 \times 0.583 = 1.166$

$T = \text{antilog}(-1.166) = \underline{\underline{0.0682}}$ or $\underline{\underline{6.82\%}}$

24-11 (a) $T = 24.9\,\mu A / 73.6\,\mu A = 0.338$ or $\underline{\underline{33.8\%}}$

(b) $A = -\log 0.338 = \underline{\underline{0.471}}$

(c) $A = 0.471/2 = 0.234$ and $T = \text{antilog}(-0.471/2) = \underline{\underline{0.581}}$

(d) $A = 2 \times 0.471 = 0.942$ and $T = \text{antilog}(-0.942) = \underline{\underline{0.114}}$

24-12

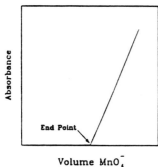

A green filter should be used because the red permanganate solution absorbs green light.

24-13

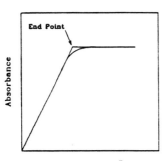

A green filter is used because the red $Fe(SCN)^{2+}$ absorbs green light.

24-14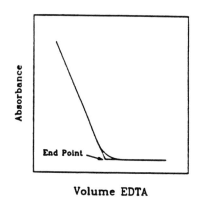

24-15 The data must be corrected for dilution so

$$A_{corr} = A_{500} \times \frac{10.00 \text{ mL} + V}{10.00 \text{ mL}}$$

For 1.00 mL

$$A_{corr} = 0.147 \times \frac{10.00 \text{ mL} + 1.00 \text{ mL}}{10.00 \text{ mL}} = \underline{\underline{0.162}}$$

A_{corr} is calculated for each volume in the same way and the following table of results are obtained.

Vol, mL	A_{500}	A_{corr}	Vol, mL	A_{500}	A_{corr}
0	0	0	5.00	0.347	0.521
1.00	0.147	0.162	6.00	0.325	0.520
2.00	0.271	0.325	7.00	0.306	0.520
3.00	0.375	0.488	8.00	0.289	0.520
4.00	0.371	0.519			

These data are plotted below. The point of intersection of the linear portion of the plot can be determined graphically or evaluated by performing least-squares on the linear portions and solving the two linear simultaneous equations. Least-squares analysis gives the following results.

Chapter 24

Points 1 to 4	Points 5 to 9
b_1 = slope = 1.626×10^{-1}	$b_2 = 1.300 \times 10^{-4}$
a_1 = intercept = 3.000×10^{-4}	$a_2 = 5.1928 \times 10^{-1}$
$y = a_1 + b_1 x$	$y = a_2 + b_2 x$

$$x = \frac{a_2 - a_1}{b_1 - b_2} = \underline{\underline{3.19 \text{ mL}}}$$

$$\frac{3.19 \text{ mL} \times 2.44 \times 10^{-4} \frac{\text{mmol Nitroso R}}{\text{mL}} \times \frac{1 \text{ mmol Pd(II)}}{2 \text{ mmol Nitroso R}}}{10.00 \text{ ml solution}} = \underline{\underline{3.89 \times 10^{-5} \text{ M}}}$$

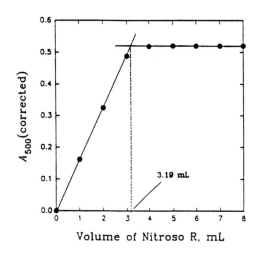

24-16 For the unknown alone, we can write Beer's law in the form

$A_x = \varepsilon b c_x V_x / V_t$ where V_t is the total volume of solution.

For the solution after standard addition

$$A_s = \varepsilon b (c_x V_x + c_s V_s)/V_t$$

Dividing the first equation by the second gives

$$\frac{A_x}{A_s} = \frac{c_x V_x}{c_x V_x + c_s V_s}$$

$$A_x c_x V_x + A_x c_s V_s = A_s c_x V_x$$

This equation rearranges to

$$c_x (A_s V_x - A_x V_x) = A_x c_s V_s$$

$$c_x = \frac{A_x c_s V_s}{V_x (A_s - A_x)}$$

Substituting numerical data gives

$$c_x = \frac{0.398 \times 3.00 \times 5.00}{25.00 \,(0.510 - 0.398)} = 2.132 \text{ ppm Co}$$

$$\frac{2.132 \text{ g Co}}{10^6 \text{ g soln}} \times 500 \text{ g soln} = 1.066 \times 10^{-3} \text{ g Co}$$

$$\text{percent Co} = \frac{1.066 \times 10^{-3} \text{ g Co}}{4.97 \text{ g sample}} \times 100\% = \underline{\underline{0.0214\%}}$$

24-17 Applying the equation we developed in Solution 24-16, we write

$$c_x = \frac{0.231 \times 2.75 \times 5.00}{(0.549 - 0.231)\, 50.0} = \underline{\underline{0.200 \text{ ppm Fe}}}$$

24-18 $A_{365} = 3529 \times 1.00 \times c_{Co} + 3228 \times 1.00 \times c_{Ni}$

$A_{700} = 428.9 \times 1.00 \times c_{Co} + 0.00 \times 1.00 \times c_{Ni}$

$$c_{Co} = \frac{A_{700}}{428.9} \qquad c_{Ni} = \frac{A_{365} - 3529\, c_{Co}}{3228}$$

(a)
$$c_{Co} = \frac{0.0235}{428.9} = \underline{\underline{5.48 \times 10^{-5} \text{ M}}}$$

$$c_{Ni} = \frac{0.617 - 3529 \times 5.48 \times 10^{-5}}{3228} = \underline{\underline{1.31 \times 10^{-4} \text{ M}}}$$

Chapter 24

Proceeding in the same way, we obtain

(b) $c_{Co} = \underline{\underline{1.66 \times 10^{-4} \text{ M}}}$ $c_{Ni} = \underline{\underline{5.19 \times 10^{-5} \text{ M}}}$

(c) $c_{Co} = \underline{\underline{2.20 \times 10^{-4} \text{ M}}}$ $c_{Ni} = \underline{\underline{4.41 \times 10^{-5} \text{ M}}}$

(d) $c_{Co} = \underline{\underline{3.43 \times 10^{-5} \text{ M}}}$ $c_{Ni} = \underline{\underline{1.46 \times 10^{-4} \text{ M}}}$

24-19 $A_{510} = 0.446 = 36400 \times 1.00 \times c_{Co} + 5520 \times 1.00 \times c_{Ni}$

$A_{656} = 0.326 = 1240 \times 1.00 \times c_{Co} + 17500 \times 1.00 \times c_{Ni}$

Simultaneous solution of the two equations gives

$c_{Co} = 9.53 \times 10^{-6} \text{ M}$ $c_{Ni} = 1.80 \times 10^{-5} \text{ M}$

$$c_{Co} = \frac{50.0 \text{ mL} \times 9.53 \times 10^{-6} \frac{\text{mmol}}{\text{mL}} \times \frac{50.0 \text{ mL}}{25.0 \text{ mL}} \times \frac{0.05893 \text{ g Co}}{\text{mmol}}}{0.425 \text{ g}} \times 10^6 \text{ ppm} = \underline{\underline{132 \text{ ppm}}}$$

$$c_{Ni} = \frac{50.0 \text{ mL} \times 1.80 \times 10^{-5} \frac{\text{mmol}}{\text{mL}} \times \frac{50.0 \text{ mL}}{25.0 \text{ mL}} \times \frac{0.05869 \text{ g Ni}}{\text{mmol}}}{0.425 \text{ g}} \times 10^6 \text{ ppm} = \underline{\underline{249 \text{ ppm}}}$$

24-20 The data are plotted in the figure that follows. The isosbestic point is estimated to be at 526 nm.

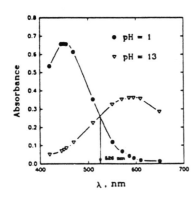

24-21 In Section 10F (page 217), we find

$$\alpha_0 = \frac{[H_3O^+]}{[H_3O^+] + K_{HIn}} \qquad \text{(Equation 10-16)}$$

Chapter 24

$$\alpha_1 = 1 - \alpha_0$$

At 450 nm and $b = 1.00$

$$A_{450} = \varepsilon_{HIn} \times 1.00 \times [HIn] + \varepsilon_{In^-} \times 1.00 \times [In^-]$$

$$= \varepsilon_{HIn} \alpha_0 c_{In} + \varepsilon_{In^-} \alpha_1 c_{In}$$

$$= (\varepsilon_{HIn} \alpha_0 + \varepsilon_{In^-} \alpha_1) c_{In}$$

where c_{In} is the analytical concentration of the indicator ($c_{In} = [HIn] + [In^-]$).

We may assume at pH 1.00 all of the indicator is present as HIn; at pH 13.0 it is all present as In^-. Therefore from the data in Problem 24-20, we may write

$$\varepsilon_{HIn} = \frac{A_{450}}{b \, c_{In}} = \frac{0.658}{1.00 \times 8.00 \times 10^{-5}} = 8.225 \times 10^3$$

$$\varepsilon_{In^-} = \frac{A_{450}}{b \, c_{In}} = \frac{0.076}{1.00 \times 8.00 \times 10^{-5}} = 9.5 \times 10^2$$

(a) pH = 4.92 $[H_3O^+] = 1.20 \times 10^{-5}$

$$\alpha_0 = \frac{1.20 \times 10^{-5}}{1.20 \times 10^{-5} + 4.80 \times 10^{-6}} = 0.714$$

$$\alpha_1 = 1.000 - 0.714 = 0.286$$

$$A_{450} = (8.225 \times 10^3 \times 0.714 + 9.5 \times 10^2 \times 0.286) \, 8.00 \times 10^{-5} = \underline{\underline{0.492}}$$

Chapter 24

Proceeding in the same way, we obtain

	pH	$[H_3O^+]$	α_0	α_1	A_{450}
(a)	4.92	1.20×10^{-5}	0.714	0.286	0.492
(b)	5.46	3.47×10^{-6}	0.419	0.581	0.320
(c)	5.93	1.18×10^{-6}	0.197	0.803	0.190
(d)	6.16	6.92×10^{-7}	0.126	0.874	0.149

24-22 The approach is identical to that of Solution 24-21. At 595 nm and

$$\text{at pH} = 1.00, \quad \varepsilon_{HIn} = \frac{A_{595}}{bc_{In}} = \frac{0.032}{(1.00)(8.00 \times 10^{-5} \text{ M})} = 4.0 \times 10^2$$

$$\text{at pH} = 13.00, \quad \varepsilon_{In^-} = \frac{0.361}{(1.00)(8.00 \times 10^{-5} \text{ M})} = 4.51 \times 10^3$$

(a) At pH = 5.30 and with 1.00-cm cells, $[H_3O^+] = 5.01 \times 10^{-6}$ and

$$\alpha_0 = \frac{[H_3O^+]}{[H^+] + K_{HIn}} = \frac{5.01 \times 10^{-6}}{5.01 \times 10^{-6} + 4.80 \times 10^{-6}} = 0.511$$

$$\alpha_1 = 1 - \alpha_0 = 0.489$$

$$A_{595} = (\varepsilon_{HIn}\alpha_0 + \varepsilon_{In^-}\alpha_1) c_{In}$$

$$= [(4.0 \times 10^2)(0.511) + (4.51 \times 10^3)(0.489)](1.25 \times 10^{-4}) = \underline{0.301}$$

Similarly for parts (b) and (c)

	pH	$[H_3O^+]$	α_0	α_1	Absorbance
(a)	5.30	5.01×10^{-6}	0.511	0.489	0.301
(b)	5.70	2.00×10^{-6}	0.294	0.706	0.413
(c)	6.10	7.94×10^{-7}	0.142	0.858	0.491

24-23 In these solutions the concentrations of the two absorbers HIn and In⁻ must be determined by the analysis of mixtures, so

$$A_{450} = \varepsilon'_{HIn} b [HIn] + \varepsilon'_{In^-} b [In^-]$$

$$A_{595} = \varepsilon''_{HIn} b [HIn] + \varepsilon''_{In^-} b [In^-]$$

From the data given in Problem 24-20

$$\varepsilon'_{HIn} = \frac{0.658}{(8.00 \times 10^{-5})(1.00)} = 8.22 \times 10^3$$

$$\varepsilon''_{HIn} = \frac{0.032}{(8.00 \times 10^{-5})(1.00)} = 4.0 \times 10^2$$

$$\varepsilon'_{In^-} = \frac{0.076}{(8.00 \times 10^{-5})(1.00)} = 9.5 \times 10^2$$

$$\varepsilon''_{In^-} = \frac{0.361}{(8.00 \times 10^{-5})(1.00)} = 4.51 \times 10^3$$

Thus, $A_{450} = 0.344 = (8.22 \times 10^3)[HIn] + (9.5 \times 10^2)[In^-]$

$A_{595} = 0.310 = (4.0 \times 10^2)[HIn] + (4.51 \times 10^3)[In^-]$

Solving these equations gives

$[HIn] = 3.42 \times 10^{-5}$ M   and $[In^-] = 6.57 \times 10^{-5}$ M

$$K_{HIn} = \frac{[H_3O^+][In^-]}{[HIn]}$$

$$[H_3O^+] = K_{HIn} \frac{[HIn]}{[In^-]} = \frac{(4.80 \times 10^{-6})(3.42 \times 10^{-5})}{6.57 \times 10^{-5}} = 2.50 \times 10^{-6}$$

$$pH = -\log [H_3O^+] = -\log (2.50 \times 10^{-6}) = \underline{5.60}$$

The results for all solutions are shown below.

Chapter 24

Solution	[HIn]	[In⁻]	pH
A	3.42×10^{-5}	6.57×10^{-5}	5.60
B	5.69×10^{-5}	4.19×10^{-5}	5.19
C	7.70×10^{-5}	2.33×10^{-5}	4.80
D	1.72×10^{-5}	8.27×10^{-5}	6.00

24-24 (a) In this case,

$$[\text{HIn}]/[\text{In}^-] = 3.00 \quad \text{and} \quad ([\text{HIn}] + [\text{In}^-]) = 8.00 \times 10^{-5}\,\text{M}$$

so $[\text{HIn}] = 6.00 \times 10^{-5}\,\text{M}$ and $[\text{In}^-] = 2.00 \times 10^{-5}\,\text{M}$

For each wavelength λ

$$A_\lambda = \varepsilon_{\text{HIn}} b\,[\text{HIn}] + \varepsilon_{\text{In}^-} b\,[\text{In}^-]$$

At $\lambda = 595$ nm

$$\varepsilon_{\text{HIn}} = 4.0 \times 10^3 \quad \text{and} \quad \varepsilon_{\text{In}^-} = 4.51 \times 10^3 \quad \text{(see Solution 24-22)}$$

$$A_{595} = (4.0 \times 10^2)(1.00)(6.00 \times 10^{-5}) + (4.51 \times 10^3)(1.00)(2.00 \times 10^{-5})$$

$$= \underline{\underline{0.114}}$$

Results for other wavelengths are tabulated and plotted below.

(b) Here, $[\text{HIn}] = [\text{In}^-] = 4.00 \times 10^{-5}\,\text{M}$

Absorbance data are computed in the same way as in part (a).

(c) Here, $[HIn] = 2.00 \times 10^{-5}$ M and $[In^-] = 6.00 \times 10^{-5}$ M

λ, nm	ε_{HIn}	ε_{In^-}	A (3:1)	A (1:1)	A (1:3)
420	6688	625	0.414	0.293	0.171
445	8213	850	0.510	0.363	0.215
450	8225	950	0.513	0.367	0.222
455	8200	1063	0.513	0.371	0.228
470	7675	1450	0.490	0.365	0.241
510	4413	2788	0.321	0.288	0.256
550	1488	4050	0.170	0.222	0.273
570	850	4400	0.139	0.210	0.281
585	550	4500	0.123	0.202	0.281
595	400	4513	0.114	0.197	0.279
610	238	4438	0.103	0.187	0.271
650	175	3550	0.082	0.149	0.217

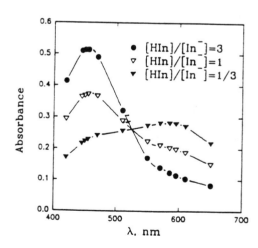

Chapter 24

24-25 (a) $A_\lambda = A_P + A_Q$

For example, $A_{600} = 0.264 + 0.100 = 0.364$

In the same way absorbances at each of the given wavelengths are calculated, tabulated, and plotted below.

λ	A	λ	A	λ	A	λ	A
400	0.628	480	0.670	560	0.381	640	0.389
420	0.679	500	0.625	580	0.340	660	0.403
440	0.695	520	0.546	600	0.364	680	0.405
460	0.692	540	0.459	620	0.381	700	0.409

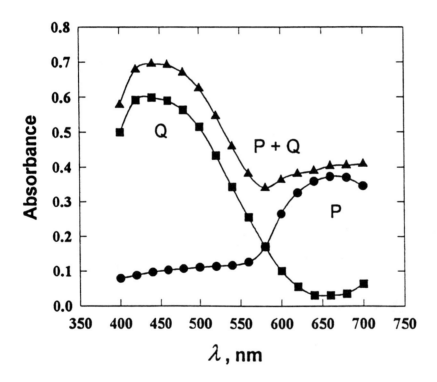

(b) At 440 nm, $A_P = \varepsilon_P b c_P$

$$\varepsilon_P = \frac{A_P}{bc_P} = \frac{0.096}{(1.00)(8.55 \times 10^{-5})} = 1.12 \times 10^3$$

$$\varepsilon_Q = \frac{A_Q}{bc_Q} = \frac{0.599}{(1.00)(2.37 \times 10^{-4} \text{ M})} = 2.53 \times 10^3$$

$$A = \varepsilon_P bc_P + \varepsilon_Q bc_Q$$
$$= (1.12 \times 10^3)(1.00)(4.00 \times 10^{-5}) + (2.53 \times 10^3)(1.00)(3.60 \times 10^{-4})$$
$$= \underline{\underline{0.956}}$$

(c) At 620 nm,

$$\varepsilon_P = \frac{A_P}{bc_P} = \frac{0.326}{(1.00)(8.55 \times 10^{-5})} = 3.81 \times 10^3$$

$$\varepsilon_Q = \frac{A_Q}{bc_Q} = \frac{0.055}{(1.00)(2.37 \times 10^{-4})} = 2.32 \times 10^2$$

$$A = (3.81 \times 10^3)(1.00)(1.61 \times 10^{-4}) + (2.32 \times 10^2)(1.00)(7.35 \times 10^{-4})$$
$$= \underline{\underline{0.784}}$$

24-26 In Solution 24-25(b), we found that at 440 nm

$$\varepsilon'_P = 1.12 \times 10^3 \quad \text{and} \quad \varepsilon'_Q = 2.53 \times 10^3$$

At 620 nm

$$\varepsilon''_P = 3.81 \times 10^3 \quad \text{and} \quad \varepsilon''_Q = 2.32 \times 10^2$$

In 1.00 cells

(a) $0.357 = 1.12 \times 10^3 \, c_P + 2.53 \times 10^3 \, c_Q$

$0.803 = 3.81 \times 10^3 \, c_P + 2.32 \times 10^2 \, c_Q$

Solving the two simultaneous equations gives

$$c_P = \underline{\underline{2.08 \times 10^{-4} \text{ M}}} \quad c_Q = \underline{\underline{4.91 \times 10^{-5} \text{ M}}}$$

Chapter 24

Proceeding in the same way, we find

	c_P, M	c_Q, M
(a)	2.08×10^{-4}	4.91×10^{-5}
(b)	1.00×10^{-4}	2.84×10^{-4}
(c)	8.37×10^{-5}	6.10×10^{-5}
(d)	6.87×10^{-5}	3.29×10^{-4}
(e)	2.11×10^{-4}	9.65×10^{-5}
(f)	8.02×10^{-5}	4.12×10^{-5}

24-27 (a)

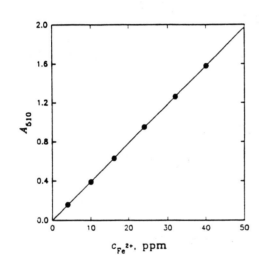

Chapter 24

(b) Proceeding as in Example 4-9 (page 000)

$c_{Fe} = x_i$	$A = y_i$	x_i^2	y_i^2	$x_i y_i$
4.00	0.160	16.00	0.0256	0.64
10.00	0.390	100.00	0.1521	3.90
16.00	0.630	256.00	0.3969	10.08
24.00	0.950	576.00	0.9025	22.80
32.00	1.260	1024.00	1.5876	40.32
40.00	1.580	1600.00	2.4964	63.20
126.00	4.970	3572.00	5.5611	140.94

$S_{xx} = 3572.00 - (126)^2/6 = 926.00$

$S_{yy} = 5.5611 - (4.970)^2/6 = 1.4442833$

$S_{xy} = 140.94 - 126.00 \times 4.970/6 = 36.57$

$m = 36.57/926.0 = 0.03949244$

$b = (4.97/6) - 0.03949244 \times 126/6 = -0.001008$

$\underline{A_{510} = 0.03949\, c_{Fe} - 0.001008}$

(c) $s_r = \sqrt{\dfrac{1.444283 - (0.03949244)^2 \times 926}{6-2}} = 0.00333 = 0.0033$

$s_m = \sqrt{(0.00333)^2/926} = \underline{\underline{1.1 \times 10^{-4}}}$

$s_b = 0.00333\sqrt{\dfrac{1}{6 - (126)^2/3572}} = \underline{\underline{2.7 \times 10^{-3}}}$

24-28 (a) $c_{Fe(II)} = \dfrac{A_{510} + 1.01 \times 10^{-3}}{3.95 \times 10^{-2}} = \dfrac{0.143 + 1.01 \times 10^{-3}}{3.95 \times 10^{-2}} = 3.65$ ppm Fe(II)

$s_c = \dfrac{s_r}{m}\sqrt{\dfrac{1}{M} + \dfrac{1}{N} + \dfrac{(\bar{y}_c - \bar{y})^2}{m^2 S_{xx}}}$

325

For $N = 6$, $M = 1$

$$s_c = \frac{3.33 \times 10^{-3}}{3.949 \times 10^{-2}} \sqrt{\frac{1}{1} + \frac{1}{6} + \frac{[0.143 - (4.97/6)]^2}{(3.9492 \times 10^{-2})^2 \times 926}} = 0.1035 = 0.10$$

$$\frac{0.103}{3.65} \times 100\% = \underline{\underline{2.8\%}}$$

$$c_{Fe(II)} = \underline{\underline{3.65 \text{ ppm} (\pm 2.8\%)}}$$

For $N = 6$, $M = 3$

$$s_c = 0.0769$$

$$\frac{0.0769}{3.65} \times 100\% = \underline{\underline{2.1\%}}$$

$$c_{Fe(II)} = \underline{\underline{3.65 \text{ ppm} (\pm 2.1\%)}}$$

Proceeding in the same way, we obtain

	$c_{Fe(II)}$, ppm	s_c, rel %	
		1 Result	3 Results
(a)	3.65	2.8	2.1
(b)	17.1	0.54	0.36
(c)	1.75	6.1	4.6
(d)	25.6	0.36	0.24
(e)	38.3	0.27	0.20
(f)	13.8	0.68	0.46

CHAPTER 25

25-1 (a) *Resonance fluorescence* is observed when excited atoms emit radiation of the same wavelength as that used to excite them.

(b) *Vibrational relaxation* occurs when excited species collide with molecules such as the solvent and in doing so loses energy without emission of electromagnetic radiation. The energy of the excited species is decreased by an amount equal to the quantity of vibration energy transferred.

(c) *Internal conversion* is the nonradiative relaxation of a molecule from the lowest vibrational level of an excited electronic state to a high vibrational level of a lower electronic state.

(d) The *quantum yield* of fluorescence is the ratio of the number of fluorescing molecules to the total number of excited molecules.

(e) The *Stokes shift* is the difference in wavelength between the radiation used to excite fluorescence and the wavelength of the emitted radiation.

(f) *Self-quenching* occurs when the fluorescent radiation from an excited analyte molecule is absorbed by an unexcited analyte molecule. This process results in a decrease in fluorescence intensity.

25-2 For spectrofluorometry, the analytical signal F is given by $F = 2.3 K' \varepsilon b c P_0$. The magnitude of F, and thus sensitivity, can be enhanced by increasing the source intensity P_0 or the transducer sensitivity.

For spectrophotometry, the analytical signal A is given by $A = \log P_0/P$. Increasing P_0 or the detector's response to P_0 is accompanied by a corresponding increase in P. Thus, the ratio does not change nor does the analytical signal. Consequently, no improvement in sensitivity accompanies such changes.

25-3 (a) Fluorescein because of its greater structural rigidity due to the bridging $-O-$ groups.

(b) o,o'-Dihydroxyazobenzene because the $-N=N-$ group provides rigidity that is absent in the $-\underset{|}{N}-\underset{|}{N}-$ group.

Chapter 25

25-4 Compounds that fluoresce have structures that slow the rate of nonradiative relaxation to the point where there is time for fluorescence to occur. Compounds that do not fluoresce have structures that permit rapid relaxation by nonradiative processes.

25-5 Organic compounds containing aromatic rings often exhibit fluorescence. Rigid molecules or multiple ring systems tend to favor large quantum yields of fluorescence while flexible molecules generally have lower quantum yields.

25-6 Excitation of fluorescence usually involves transfer of an electron to a high vibrational state of an upper electronic state. Relaxation to a lower vibrational state of this electronic state goes on much more rapidly than fluorescence relaxation. When fluorescence relaxation occurs it is to a high vibrational state of the ground state or to a high vibrational state of an electronic state that is above the ground state. Such transitions involve less energy than the excitation energy. Therefore, the emitted radiation is longer in wavelength than the excitation wavelength.

25-7 See Figure 25-8. A fluorometer usually consists of a light source, a filter or a monochromator for selecting the excitation wavelength, a sample container, an emission filter or monochromator, and a detector. There may also be a reference detector for monitoring and correcting for fluctuations in the light source intensity. Emission is usually detected at right angles to the incident radiation to maximize the fluorescence signal.

25-8 Most fluorescence instruments are double beamed in order to compensate for fluctuations in the analytical signal arising from variations in the power supply to the source.

25-9 Filter fluorometers are usually more sensitive than spectrofluorometers because filters have a higher radiation throughput than do monochromators. They also allow positioning of the detector closer to the sample than in spectrofluorometers, thus increasing the magnitude of the signal.

25-10 (a) Plotting the data gives a straight line that passes through the origin. The slope is estimated to be 22.3.

(b) Proceeding as in Example 4-9, page 63, we may write

$x_i = c_{NADH}$, μmol/L	$y_i = I_r$	x_i^2	y_i^2	$x_i y_i$
0.200	4.52	0.04	20.4304	0.904
0.400	9.01	0.16	81.1801	3.604
0.600	13.71	0.36	187.9641	8.226
0.800	17.91	0.64	320.7681	14.328

$\sum x_i = 2.000$ $\sum y_i = 45.15$ $\sum x_i^2 = 1.20$ $\sum y_i^2 = 610.3427$ $\sum x_i y_i = 27.062$

$\bar{x}_i = 2.000/4 = 0.500$ and $\bar{y}_i = 45.15/4 = 11.29$

$S_{xx} = 1.20 - (2.000)^2/4 = 0.200$

$S_{yy} = 610.3427 - (45.15)^2/4 = 100.712075$

$S_{xy} = 27.062 - (2.000 \times 45.15)/4 = 4.4870$

$m = 4.487/0.200 = 22.435$

$b = \dfrac{45.15}{4} - 22.435 \times \dfrac{2.00}{4} = 0.0700$

Thus the equation is

$y = 22.4x + 0.0700$ or $\underline{\underline{I_r = 22.4 c_{NADH} + 0.0700}}$

(c) First we calculate s_r from Equation 4-15:

$s_r = \sqrt{\dfrac{100.712075 - (22.435)^2 \times 0.200}{4-2}} = 0.152$

$s_m = \sqrt{(0.152)^2/0.200} = \underline{\underline{0.340}}$

$s_b = 0.152 \sqrt{\dfrac{1}{4 - (2.00)^2/1.20}} = \underline{\underline{0.186}}$

Chapter 25

(d) c_{NADH} = $(I_r - 0.0700)/22.4$

= $(12.16 - 0.0700)/22.4$ = <u>0.540 µmol NADH/L</u>

(e) For the single measurements, M in Equation 4-18 is 1 and

$$s_c = \frac{0.152}{22.435}\sqrt{\frac{1}{1} + \frac{1}{4} + \frac{(12.16 - 11.29)^2}{(22.435)^2 \, 0.200}} = 0.0076$$

$$(s_c)_r = \frac{0.0076}{0.540} \times 100\% = \underline{\underline{1.4\%}}$$

(f) For the mean of three measurements

$$s_c = \frac{0.152}{22.435}\sqrt{\frac{1}{3} + \frac{1}{4} + \frac{(12.16 - 11.29)^2}{(22.435)^2 \, 0.200}} = 0.0052$$

$$(s_c)_r = \frac{0.0052}{0.540} \times 100\% = 0.96\%$$

25-11 (a) A plot of the data shows a linear relationship between the fluorometer reading F and volume V_s of standard.

(b) Proceeding as in Example 4-9 (page 63, text), we obtain

$$S_{xx} = \Sigma x_i^2 - (\Sigma x_i)^2/N = 224.0 - (24.0)^2/4 = 80.0$$

(Equation 4 – 10)

$$S_{yy} = \Sigma y_i^2 - (\Sigma y_i)^2/N = 883.872 - (53.60)^2/4 = 115.632$$

(Equation 4 – 11)

$$S_{xy} = \Sigma x_i y_i - (\Sigma x_i \Sigma y_i)^2/N = 417.76 - 24.0 \times 53.6/4 = 96.16$$

(Equation 4 – 12)

$$m = S_{xy}/S_{xx} = 96.16/80.0 = 1.202$$

$$b = \bar{y} - m\bar{x} = 53.6/4 - 1.202 \times 24.0/4 = 6.188$$

$$y = 6.188 + 1.202x = \underline{\underline{F = 1.202 \, V_{Zn^{2+}} + 6.188}}$$

(c) Substituting into Equations 4-15, 4-16, and 4-17 yields

$$s_r = \sqrt{\frac{S_{yy} - m^2 S_{xx}}{N-2}} = \sqrt{\frac{115.632 - (1.202)^2 \times 80.0}{2}} = 0.1544 = \underline{\underline{0.15}}$$

$$s_m = \sqrt{s_r^2 / S_{xx}} = \sqrt{(0.1544)^2 / 80} = 0.017$$

$$s_b = \sqrt{\frac{1}{N - (\Sigma x_i)^2 / \Sigma x_i^2}} = \underline{\underline{0.13}}$$

(d) Substituting into Equation 24-2 (page 574) gives

$$c_{Zn^{2+}} = \frac{bc_s}{mV_x} = \frac{6.188}{1.202} \times \frac{1.10}{5.00} = \underline{\underline{1.13 \text{ ppm } Zn^{2+}}}$$

(e)
$$s_{c_x} = c_x \sqrt{\left(\frac{s_m}{m}\right)^2 + \left(\frac{s_b}{b}\right)^2} = 1.13 \sqrt{\left(\frac{0.017}{1.202}\right)^2 + \left(\frac{0.15}{6.19}\right)^2}$$

$$= \underline{\underline{0.032 \text{ ppm } Zn^{2+}}}$$

25-12 (a) The fluoride ion must react with and decompose the fluorescent complex to give nonfluorescing products. Thus, if we symbolize the complex as AlG_x^{3+}, we may write

$$\underset{\text{fluorescent}}{AlG_x^{3+}} + 6F^- \rightleftarrows \underset{\text{nonfluorescent}}{AlF_6^{3-}} + \underset{\text{nonfluorescent}}{xG}$$

(b) A plot of the meter reading R versus the volume of standard V_s is a straight line with a negative slope.

(c) Proceeding as in Example 4-9 (page 63), we obtain

$$\Sigma x_i = 6.00 \text{ and } \Sigma x_i^2 = 14.00$$

$$\Sigma y_i = 193.6 \text{ and } \Sigma y_i^2 = 10244.46$$

$$\Sigma x_i y_i = 224.3$$

Chapter 25

Substituting these values into Equations 4-10, 4-11, and 4-12 gives

$$S_{xx} = \Sigma x_i^2 - (\Sigma x_i)^2/n = 14.00 - (6.0)^2/4 = 5.00$$

$$S_{yy} = \Sigma y_i^2 - (\Sigma y_i)^2/n = 10244.46 - (193.6)^2/4 = 874.22$$

$$S_{xy} = \Sigma x_i y_i - \Sigma x_i \Sigma y_i/n = 224.3 - 6.0 \times 193.6/4 = -66.1$$

$$m = S_{xy}/S_{xx} = -66.1/5.00 = -13.22$$

$$b = \bar{y} - m\bar{x} = 193.6/4 - (-13.22 \times 6.0/4) = 68.23$$

$$\underline{\underline{R = 68.23 - 13.22\, V_x}}$$

(d)
$$s_y = \sqrt{\frac{S_{yy} - m^2 S_{xx}}{N-2}} = \sqrt{\frac{874.22 - (-13.22)^2 \times 5.00}{4-2}} = 0.435 = \underline{\underline{0.44}}$$

$$s_m = \sqrt{s_y^2/S_{xx}} = \sqrt{(0.435)^2/5.00} = \underline{\underline{0.19}}$$

$$s_b = s_y \sqrt{\frac{1}{N - (\Sigma x_i)^2/\Sigma x_i^2}} = 0.435 \sqrt{\frac{1}{4 - (6.0)^2/14.00}} = \underline{\underline{0.36}}$$

(e)
$$c_x = -\frac{b}{m}\frac{c_s}{V_x} = -\frac{68.23}{-13.22} \times \frac{10.0}{5.00} = \underline{\underline{10.3 \text{ ppb F}^-}} \quad \text{(Equation 24-2)}$$

(f)
$$s_{c_x} = c_x \sqrt{\left(\frac{s_m}{m}\right)^2 + \left(\frac{s_b}{b}\right)^2} \quad \text{(Equation 24-3)}$$

$$= 10.3 \sqrt{\left(\frac{0.19}{-13.22}\right)^2 + \left(\frac{0.36}{68.23}\right)^2} = \underline{\underline{0.16 \text{ ppb F}^-}}$$

CHAPTER 26

26-1 In *atomic emission spectroscopy* the radiation source is the sample itself. The energy for excitation of analyte atoms is supplied by a flame, an oven, a plasma, or an electric arc or spark. The signal is the measured intensity of the source at the wavelength of interest. In *atomic absorption spectroscopy* the radiation source is usually a line source such as a hollow cathode lamp, and the output signal is the absorbance calculated from the incident power of the source and the resulting power after the light has passed through the atomized sample in the source.

26-2 (a) *Atomization* is a process in which a sample, usually in solution is volatilized and decomposed to produce an atomic vapor.

(b) *Pressure broadening* is an increase in the widths of atomic emission or absorption lines brought about by collision of atoms that causes slight changes in their ground state energies.

(c) *Doppler broadening* is an increase in the width of atomic absorption or emission lines caused by the Doppler effect in which atoms moving toward a detector absorb or emit wavelengths that are slightly shorter than those absorbed or emitted by atoms moving at right angles to the detector. The effect is reversed for atoms moving away from the detector.

(d) A *nebulizer* is a device that converts a liquid into a mist or an aerosol by the flow of gas around the end of a capillary tube the other end of which is immersed in the liquid.

(e) A *plasma* is a conducting gas that contains a large concentration of ions and/or electrons.

(f) A *hollow cathode lamp* consists of a tungsten wire anode and a cylindrical cathode sealed in a glass tube that contains argon at a pressure of 1 to 5 torr. The cathode is constructed from or supports the element whose emission spectrum is desired.

(g) *Sputtering* is a process in which atoms of an element are dislodged from the surface of a cathode by bombardment by a stream of inert gas ions that have been accelerated toward the cathode by a high electrical potential.

Chapter 26

- (h) An *ionization suppressor* is a compound containing an easily ionized cation, such as potassium; it is introduced into a solution of an analyte to provide a high concentration of electrons during atomization. In this way, interference due to ionization of the analyte is minimized.

- (i) A *spectral interference* in atomic spectroscopy occurs when an absorption or emission peak of an element in the sample matrix overlaps that of the analyte.

- (j) A *chemical interference* in atomic spectroscopy is encountered when a species interacts with the analyte in such a way as to alter the spectral emission or absorption characteristics of the analyte.

- (k) A *radiation buffer* is a substance that is added in large excess to both standards and samples in atomic spectroscopy to prevent the presence of that substance in the sample matrix from having an appreciable effect on the results.

- (l) A *releasing agent* in atomic spectroscopy is a substance that contains a cation that reacts preferentially with species that tend to form nonvolatile compounds with the analyte and thus interfere in an analysis.

- (m) A *protective agent* in atomic spectroscopy is a substance, such as EDTA or 8-hydroxyquinoline, that forms stable but volatile complexes with analyte cation and thus prevents interference by anions that form nonvolatile species with the analyte.

26-3 In atomic emission spectroscopy, the analytical signal is produced by *excited* atoms or ions, whereas in atomic absorption and fluorescence the signal results from absorption by *unexcited* species. Typically, the number of unexcited species exceeds that of the excited ones by several orders of magnitude. The ratio of unexcited to excited atoms in a hot medium varies exponentially with temperature. Thus a small change in temperature brings about a large change in the number of excited atoms. The number of unexcited atoms changes very little, however, because they are present in an enormous excess. Therefore, emission spectroscopy is more sensitive to temperature changes than is absorption or fluorescence spectroscopy.

26-4 In atomic absorption spectroscopy the source radiation must be modulated to create an ac signal at the detector. The detector is then made to reject the dc signal from the flame and measure the modulated signal from the source. In this way, the radiation from the flame is prevented from interfering with the analyte signal.

26-5 (a) Sulfate ion forms complexes with Fe(III) that are not readily atomized. Thus, the concentration of iron atoms in a flame is lower in the presence of sulfate ions.

(b) Sulfate interference could be overcome (1) by adding a releasing agent that forms complexes with sulfate that are more stable than the iron complexes, (2) by adding a protective agent, such as EDTA, that forms a highly stable but volatile complex with the Fe(III), and (3) by employing a high temperature flame (oxygen/acetylene or nitrous oxide/acetylene).

26-6 The temperature in a hollow cathode lamp is significantly less than that in a flame. As a consequence, Doppler broadening is less pronounced in the former, and narrower lines results.

26-7 The lack of linearity is a consequence of ionization of the uranium, which decreases the relative concentration of uranium atoms. Alkali metal atoms ionize readily and thus give a high concentration of electrons, which represses the uranium ionization.

26-8 When an internal standard is used in emission methods, the ratio of the intensity of the analyte line to the intensity of an internal standard line serves as the analytical parameter. This procedure tends to compensate for random errors arising from fluctuations in flame temperatures.

26-9 By linear interpolation

$$0.400 + (0.600 - 0.400)\frac{(0.502 - 0.396)}{(0.599 - 0.396)} = \underline{\underline{0.504 \text{ ppm Pb}}}$$

26-10 (a)

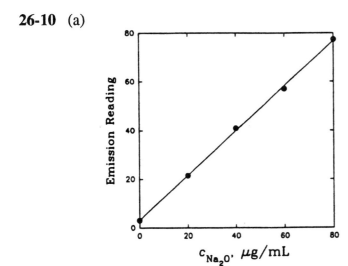

Chapter 26

(b) Proceeding as in Example 4-9, page 63,

$$\Sigma x_i = 0 + 20.0 + 40.0 + 60.0 + 80.0 = 200.00$$

$$\Sigma x_i^2 = 0 + 400.0 + 1600.0 + 3600.0 + 6400.0 = 12000.00$$

$$\Sigma y_i = 3.1 + 21.5 + 40.9 + 57.1 + 77.3 = 199.9$$

$$\Sigma y_i^2 = 9.61 + 462.25 + 1672.81 + 3260.41 + 5975.29 = 11380.37$$

$$\Sigma x_i y_i = 0 \times 3.1 + 20.0 \times 21.5 + 40.0 \times 40.9 + 60.0 \times 57.1 + 80.0 \times 77.3$$
$$= 11676.00$$

$$S_{xx} = \Sigma x_i^2 - (\Sigma x_i)^2/N = 12000 - (200.0)^2/5 = 4000.0$$

$$S_{yy} = \Sigma y_i^2 - (\Sigma y_i)^2/N = 11380.37 - (199.9)^2/5 = 3388.368$$

$$S_{xy} = \Sigma x_i y_i - \Sigma x_i \Sigma y_i/N = 11676.00 - (200.0 \times 199.9)/5 = 3680.00$$

$$m = \frac{3680.00}{4000.00} = 0.9200$$

$$b = \Sigma y_i/N - 0.9200 \Sigma x_i/N = 199.90/5 - 0.9200 \times 200.0/5 = 3.180$$

$$y = mx + b = \underline{0.920x + 3.18}$$

where y = photometric reading and $x = \mu g\ Na_2O/mL$

(c)
$$s_r = \sqrt{\frac{S_{yy} - m^2 S_{xx}}{N - 2}} = \sqrt{\frac{3388.368 - (0.920)^2 \times 4000}{5 - 2}} = 0.961 = \underline{0.96}$$

$$s_m = \sqrt{s_r^2/S_{xx}} = \sqrt{(0.961)^2/4000} = 0.0152 = \underline{0.015}$$

$$s_b = \sqrt{\frac{1}{N - (\Sigma x_i)^2/\Sigma x_i^2}} = \sqrt{\frac{1}{5 - (200)^2/12000}} = 0.77$$

(d) $\bar{x}_c = (\bar{y}_c - b)/m = (\bar{y}_c - 3.18)/0.920$

For blank, $\bar{y}_c = (5.1 + 4.8 + 4.9)/3 = 4.933$

$$\bar{x}_c = (4.933 - 3.18)/0.920 = 1.91 \ \mu g \ Na_2O/mL$$

$$s_c = \frac{s_r}{m}\sqrt{\frac{1}{M}+\frac{1}{N}+\frac{(\bar{y}_c-\bar{y})^2}{m^2 S_{xx}}}$$

$$= \frac{0.961}{0.92}\sqrt{\frac{1}{3}+\frac{1}{5}+\frac{(4.93-199.9/5)^2}{(0.920)^2\ 4000}} = 0.99 \ \mu g \ Na_2O/mL$$

For sample A,

$$\bar{y}_c = (28.6 + 28.2 + 28.9)/3 = 28.567$$

$$\bar{x}_c = (28.567 - 3.18)/0.920 = 27.59 \ \mu g \ Na_2O/mL$$

$$s_c = \frac{0.61}{0.92}\sqrt{\frac{1}{3}+\frac{1}{5}+\frac{(28.57-199.9/5)^2}{(0.920)^2\ 4000}}$$

$$= 0.79 \ \mu g \ Na_2O/mL$$

$$\mu g/mL \ Na_2O \text{ in sample} = 27.59 - 1.91 = 25.68$$

The absolute standard deviation for the difference s_d is given by

$$s_d = \sqrt{(0.99)^2 + (0.79)^2} = \underline{\underline{1.27 \ \mu g \ Na_2O/mL}}$$

$$\text{percent } Na_2O = \frac{(25.68 \ \mu g \ Na_2O/mL) \times 100 \ mL}{1.00 \ g \ sample \times 10^6 \ \mu g/g} \times 100\% = \underline{\underline{0.257\%}}$$

The relative standard deviation of this result is determined by the relative standard deviation of 25.68 μg Na_2O/mL. Then

$$\frac{s_{c_{Na_2O}}}{c_{Na_2O}} = \frac{1.27}{25.68} \times 1000 \ ppt = \underline{\underline{49 \ ppt}}$$

and the absolute standard deviation is given by

$$s_{c_{Na_2O}} = (49/1000) \times 0.257 = \underline{\underline{0.013\% \ Na_2O}}$$

or percent $Na_2O = \underline{\underline{0.26 \pm 0.01}}$

Chapter 26

The data for samples B and C were treated in the same way giving.

	Blank	A	B	C
Ave meter reading, $\bar{x}_i$	4.933	28.567	40.700	72.600
Ave concen Na$_2$O, µg/mL, $\bar{y}_c$	1.91	27.59	40.78	75.46
Std dev, µg Na$_2$O/mL	0.99	0.79	0.76	1.05
Concn Na$_2$O corrected for blank	0.00	25.68	38.87	73.55
percent Na$_2$O		0.257	0.389	0.736
Rel std dev, %		4.9	3.2	2.0
Abs std dev, % Na$_2$O		0.013	0.012	0.015

26-11 (a)

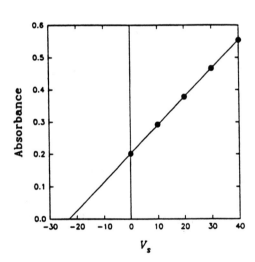

(b) Rearranging Beer's law gives

$$c_{Cr} = \frac{A}{ab} = \frac{A}{k}$$

The concentration of chromium for each solution is given by

$$c_{Cr} = \frac{c_x V_x + c_s V_s}{V_t} = \frac{A}{k}$$

338

where c_s and c_x are the concentrations of chromium in the unknown and the standard and V_s and V_x are the volumes of standard and unknown. $V_t = V_s + V_x$. Rearranging the above equation gives

$$A = kc_s\frac{V_s}{V_t} + kc_x\frac{V_x}{V_t}$$

(c) The equation derived in part (b) can be written in the form

$$A = mV_s + b$$

where

$$m = \text{slope} = \frac{kc_s}{V_t}$$

$$b = \text{intercept} = \frac{kc_x V_x}{V_t}$$

(d) Dividing the intercept by the slope gives

$$\frac{b}{m} = \frac{kc_x V_x / V_t}{kc_s / V_t} \quad \text{and} \quad c_x = \frac{b}{m} \times \frac{c_s}{V_x}$$

(e) Proceeding as in Example 4-9 (page 63), we find

$$S_{xx} = \Sigma x_i^2 - (\Sigma x_i)^2/N = 3000.00 - (100)^2/5 = 1000.00$$

$$S_{yy} = \Sigma y_i^2 - (\Sigma y_i)^2/N = 0.793554 - (1.892)^2/5 = 0.0776212$$

$$S_{xy} = \Sigma x_i y_i - \Sigma x_i \Sigma y_i / N = 46.65 - 100 \times 1.892/5 = 8.810$$

$$m = S_{xy}/S_{xx} = 8.810/1000 = \underline{\underline{8.81 \times 10^{-3}}}$$

$$b = \bar{y} - m\bar{x} = 1.892/5 - (8.81 \times 10^{-3}) \times 100/5 = \underline{\underline{0.202}}$$

$$y = mx + b$$

$$\underline{\underline{A = 8.81 \times 10^{-3} V_s + 0.202}}$$

Chapter 26

(f) $$s_r = \sqrt{\frac{S_{yy} - m^2 S_{xx}}{N-2}} = \sqrt{\frac{0.0776212 - (8.81 \times 10^{-3})^2 \times 1000}{5-2}}$$

$$= 1.3 \times 10^{-3}$$

$$s_m = \sqrt{\frac{(1.30 \times 10^{-3})^2}{1000}} = \underline{\underline{4.11 \times 10^{-5}}}$$

$$s_b = 1.3 \times 10^{-3} \sqrt{\frac{1}{5 - 10000/3000}} = \underline{\underline{1.01 \times 10^{-3}}}$$

(g) Substituting into the equation derived in part (d)

$$c_x = \frac{0.202 \times 12.2}{8.81 \times 10^{-3} \times 10.0} = \underline{\underline{28.0 \text{ ppm Cr}}}$$

CHAPTER 27

27-1 (a) The *order of a reaction* is the numerical sum of the exponents of the concentration terms in the rate law for the reaction.

(b) A reaction is *pseudo-first-order* when all reactants but one are in large excess so that their concentrations are essentially constant during the course of the reaction. The rate of the reaction then depends upon the concentration of the *isolated* reactant A. Under these conditions, rate = $k[A]$ where k contains the concentration(s) of the reagents that are in excess.

(c) *Enzymes* are high-molecular-weight organic molecules that catalyze reactions of biochemical importance.

(d) A *substrate* is the reagent that is acted upon by an enzyme.

(e) The *Michaelis constant* is an equilibrium-like constant defined by the equation $K_m = (k_{-1} + k_2)/k_1$, where k_1 and k_{-1} are the rate constants for the forward and reverse reactions in the formation of an enzyme/substrate complex, which is the intermediate in an enzyme-catalyzed reaction. The term k_2 is the rate constant for the decomposition of the complex to give products.

(f) *Differential methods* are those in which the concentration of the analyte is determined from a differential form of the rate law. For example, the expression $-d[A]/dt = k[A]_0 e^{-kt}$ allows the computation of $[A]_0$ if the rate is measured at time t and k is known from other experiments.

(g) *Integral methods* used integrated forms of rate equations to compute concentrations from kinetic data.

(h) An *indicator reaction* (or *follow-up reaction*) is one used to monitor the appearance of a reactant or the disappearance of a product. For example, the reaction of 1,10-phenanthhroline with Fe^{2+} is often used to monitor the progress of reactions that produce Fe^{2+}. The indicator reaction must not affect the rate of the reaction of interest.

27-2 The "separation" of the components of mixture is essentially carried out by finding the differences in their rates of reaction with a common reagent.

27-3 Pseudo-first-order conditions are used in kinetic methods because under these conditions the reaction rate is directly proportional to the concentration of the analyte.

Chapter 27

27-4 (a) Measurements are made relatively early in the reaction before side reactions can occur.

(b) Measurements do not depend upon the determination of absolute concentration but rather depend upon differences in concentration.

(c) Selectivity is often enhanced in reaction-rate methods, particularly in enzyme-based methods.

27-5
$$[A]_t = [A]_0 e^{-kt} \qquad \ln\frac{[A]_t}{[A]_0} = -kt$$

For $t = t_{1/2}$, $[A]_t = [A]_0/2$ $\quad \ln\dfrac{[A]_0/2}{[A]_0} = \ln(1/2) = -kt_{1/2}$

$$\ln 2 = kt_{1/2}$$

$$t_{1/2} = \ln 2/k = \underline{\underline{0.693/k}}$$

27-6 (a) $\tau = 1/k = 1/0.497 \text{ s}^{-1} = \underline{\underline{2.01 \text{ s}}}$

(b) $\tau = \dfrac{1}{6.62 \text{ h}^{-1}} = \dfrac{1}{6.62 \text{ h}^{-1}} \times \dfrac{1 \text{ h}}{60 \text{ min}} \times \dfrac{1 \text{ min}}{60 \text{ s}} = \underline{\underline{544 \text{ s}}}$

(c) $\ln\dfrac{[A]_0}{[A]_t} = kt \qquad \tau = \dfrac{1}{k} = \dfrac{t}{\ln [A]_0/[A]_t}$

$$\tau = \frac{3650 \text{ s}}{\ln(3.16 \text{ M}/0.496 \text{ M})} = \underline{\underline{1.97 \times 10^3 \text{ s}}}$$

(d) Assume 1:1 stoichiometry so that

$$[P]_\infty = [A]_0$$

Substituting into Equation 27-9, gives

$$[P]_t = [P]_\infty (1 - e^{-kt})$$

$$[P]_t/[P]_\infty = 1 - e^{-kt}$$

$$1 - [P]_t/[P]_\infty = e^{-kt}$$

$$\ln\left(1 - \frac{[P]_t}{[P]_\infty}\right) = -kt \qquad \tau = \frac{1}{k} = \frac{-t}{\ln(1 - [P]_t/[P]_\infty)}$$

$$\tau = -\frac{9.62 \text{ s}}{\ln(1 - 0.0423 \text{ M}/0.176 \text{ M})} = \underline{\underline{35.0 \text{ s}}}$$

(e) $t_{1/2} = 0.693/k = 0.693\,\tau$ (see answer to Problem 27–5)

$$\tau = \frac{1}{k} = \frac{t_{1/2}}{0.693} = \frac{32.4 \text{ years} \times 365 \text{ days/year} \times 24 \text{ h/day} \times 3600 \text{ s/h}}{0.693}$$

$$= \underline{\underline{1.47 \times 10^9 \text{ s}}}$$

(f) $\tau = \dfrac{1}{k} = \dfrac{t_{1/2}}{0.693} = \dfrac{0.478 \text{ s}}{0.693} = \underline{\underline{0.690 \text{ s}}}$

27-7 (a) $\ln\dfrac{[A]_t}{[A]_0} = -kt \qquad k = -\dfrac{1}{t}\ln\dfrac{[A]_t}{[A]_0}$

$$k = -\frac{1}{0.0100 \text{ s}} \ln(0.666) = -\frac{1}{0.0100 \text{ s}}(-0.4065) = \underline{\underline{40.6 \text{ s}^{-1}}}$$

(b) $k = -\dfrac{1}{0.100 \text{ s}}(-0.4065) = \underline{\underline{4.06 \text{ s}^{-1}}}$

(c) $k = -\dfrac{1}{1.00 \text{ s}}(-0.40650) = \underline{\underline{0.406 \text{ s}^{-1}}}$

(d) $k = -\dfrac{1}{5280 \text{ s}}(-0.4065) = \underline{\underline{7.70 \times 10^{-5} \text{ s}^{-1}}}$

(e) $k = -\dfrac{1}{26.8 \times 10^{-6} \text{ s}}(-0.4065) = \underline{\underline{1.52 \times 10^4 \text{ s}^{-1}}}$

(f) $k = -\dfrac{1}{8.86 \times 10^{-9} \text{ s}}(0.4065) = \underline{\underline{4.59 \times 10^7 \text{ s}^{-1}}}$

27-8 (a) $\ln[A]/[A]_0 = -kt = -t/\tau$ where $\tau = 1/k$

Chapter 27

$$n = \text{no. lifetimes} = t/\tau = -\ln[A]/[A]_0 = -\ln 0.90 = \underline{\underline{0.105}}$$

(at 10.0% completion $[A]/[A]_0 = 0.90$)

(b) $n = -\ln 0.50 = \underline{\underline{0.69}}$

(c) $n = -\ln(1.00 - 0.90) = \underline{\underline{2.3}}$

(d) $n = -\ln(1.00 - 0.99) = \underline{\underline{4.6}}$

(e) $n = -\ln(1.000 - 0.999) = \underline{\underline{6.9}}$

(f) $n = -\ln(1.0000 - 0.9999) = \underline{\underline{9.2}}$

27-9 (a)
$$m = \text{no. half-lives} = \frac{t}{t_{1/2}} = \frac{-\frac{1}{k}\ln\frac{[A]}{[A]_0}}{-\frac{1}{k}\ln\frac{[A]_0/2}{[A]_0}}$$

$$m = \frac{\ln[A]/[A]_0}{\ln 1/2} = -1.4427 \ln[A]/[A]_0$$

$$m = -1.4427 \ln 0.90 = \underline{\underline{0.152}}$$

(b) $m = -1.4427 \ln 0.50 = \underline{\underline{1.00}}$

(c) $m = -1.4427 \ln 0.10 = \underline{\underline{3.3}}$

(d) $m = -1.4427 \ln 0.01 = \underline{\underline{6.6}}$

(e) $m = -1.4427 \ln 0.001 = 9.96 = \underline{\underline{10.0}}$

(f) $m = -1.4427 \ln 0.0001 = 13.28 = \underline{\underline{13}}$

27-10 (a) From Example 27-3, $[R]_0 = 5.00[A]_0$ where 5.00 is the ratio of the initial reagent concentration to the initial concentration of the analyte.

At 1.00% reaction, $[A] = 0.99[A]_0$

$$[R]_{1\%} = [R]_0 - 0.01[A]_0 = 5.00[A]_0 - 0.010[A]_0 = 4.99[A]_0$$

$$\text{Rate}_{\text{assumed}} = k[R][A] = k(5.00[A]_0 \times 0.99[A]_0)$$

$$\text{Rate}_{\text{true}} = k(4.99[A]_0 \times 0.99[A]_0)$$

$$\text{relative error} = \frac{k(5.00[A]_0 \times 0.99[A]_0) - k(4.99[A]_0 \times 0.99[A]_0)}{k(4.99[A]_0 \times 0.99[A]_0)}$$

$$= \frac{5.00(0.99) - 4.99(0.99)}{4.99(0.99)} = 0.00200$$

$$\text{relative error} \times 100\% = \underline{\underline{0.2\%}}$$

(b) $\text{relative error} = \dfrac{10.00(0.99) - 9.99(0.99)}{9.99(0.99)} = \underline{\underline{0.00100}}$

$\text{relative error} \times 100\% = \underline{\underline{0.1\%}}$

(c) $(50.00 - 49.99)/49.99 = 0.000200$ or $\underline{\underline{0.02\%}}$

(d) $0.01/99.99 = 0.0001$ or $\underline{\underline{0.01\%}}$

(e) $(5.00 - 4.95)/4.95 = 0.0101$ or $\underline{\underline{1.0\%}}$

(f) $(10.00 - 9.95)/9.95 = 0.00502$ or $\underline{\underline{0.05\%}}$

(g) $(100.00 - 99.95)/99.95 = 0.0005002$ or $\underline{\underline{0.05\%}}$

(h) $(5.000 - 4.368)/4.368 = 0.145$ or $\underline{\underline{14\%}}$

(i) $(10.000 - 9.368)/9.368 = 0.06746$ or $\underline{\underline{6.7\%}}$

(j) $(50.000 - 49.368)/49.368 = 0.0128$ or $\underline{\underline{1.3\%}}$

(k) $(100.00 - 99.368)/99.368 = 00636$ or $\underline{\underline{0.64\%}}$

27-11 $\dfrac{d[P]}{dt} = \dfrac{k_2[E]_0[S]}{[S] + K_m}$ (Equation 27–22)

At $v_{\max}$, $\dfrac{dP}{dt} = k_2[E]_0$. Thus, at $v_{\max}/2$, we can write

Chapter 27

$$\frac{dP}{dt} = \frac{k_2[E]_0}{2} = \frac{k_2[E]_0[S]}{[S]+K_m}$$

$$[S] + K_m = 2[S]$$

$$K_m = [S]$$

27-12 (a) Plot 1/Rate versus 1/[S] for known [S] to give a linear working curve. Measure rate for unknown [S], calculate 1/Rate and find 1/[S]$_{unknown}$ from the working curve, and calculate [S]$_{unknown}$.

(b) The intercept of the working curve is $1/v_{max}$ and the slope is K_m/v_{max}. Use the intercept to calculate $K_m = $ slope/intercept, and $V_{max} = 1/$intercept.

27-13 Least-squares analysis of the data (Section 4E) yields

$$\text{slope} = 1.130 \times 10^{-3} \quad \text{and} \quad \text{intercept} = 1.999 \times 10^{-5}$$

$$R = 1.130 \times 10^{-3} c_{Cu^{2+}} + 1.999 \times 10^{-5}$$

$$c_{Cu^{2+}} = \frac{R - 1.999 \times 10^{-5}}{1.130 \times 10^{-3}} = \frac{6.2 \times 10^{-3} - 1.999 \times 10^{-5}}{1.130 \times 10^{-3}} = 5.47 = \underline{\underline{5.5 \text{ ppm}}}$$

27-14 $R = 1.74 c_{Al^-} - 0.225 \qquad c_{Al} = (R+0.225)/1.74$

$$c_{Al} = (0.76 + 0.225)/1.74 = 0.566 = \underline{\underline{0.57 \text{ μM}}}$$

27-15
$$\text{Rate} = R = \frac{k_2[E]_0[\text{tryp}]_t}{K_m + [\text{tryp}]_t}$$

Assume $K_m \gg [\text{tryp}]_t$

$$R = \frac{v_{max}[\text{tryp}]_t}{K_m} \quad \text{and} \quad [\text{tryp}]_t = \frac{RK_m}{v_{max}}$$

$$[\text{tryp}]_t = \frac{(0.18 \text{ μM/min})(4.0 \times 10^{-4} \text{ M})}{1.6 \times 10^{-3} \text{ μM/min}} = \underline{\underline{4.5 \times 10^{-2} \text{ M}}}$$

CHAPTER 28

28-1 (a) *Elution* is a process in which species are washed through a chromatographic column by additions of fresh mobile phase.

(b) The *mobile phase* in chromatography is the one that moves over or through an immobilized phase that is fixed in place in a column or on the surface of a flat plate.

(c) The *stationary phase* in a chromatographic column is a solid or liquid that is fixed in place. A mobile phase then passes over or through the stationary phase.

(d) The *partition ratio* in chromatography is the ratio of the concentration of the analyte in the stationary phase to its concentration in the mobile phase.

(e) The *retention time* for an analyte is the time interval between its injection onto a column and the appearance of its peak at the other end of the column.

(f) The *capacity factor k'* is defined by the equation

$$k' = K_A V_S / V_M$$

where K_A is the partition ratio for the species A and V_S and V_M are the volumes of the stationary and mobile phases respectively.

(g) The *selectivity factor* α of a column toward two species is given by the equation $\alpha = K_B / K_A$ where K_B is the partition ratio of the more strongly held species B and K_A is the corresponding ratio for the less strongly held solute A.

(h) The *plate height H* of a chromatographic column is defined by the relationship

$$H = \sigma^2 / L$$

where σ^2 is the variance obtained from the Gaussian shaped chromatographic peak and L is the length of the column packing in centimeters.

28-2 The variables that lead to *zone broadening* include: (1) large particle diameters for stationary phases; (2) large column diameters; (3) high temperatures (important only in gas chromatography); for liquid stationary phases, thick layers of the immobilized liquid; and (4) very rapid or very slow flow rates.

28-3 In gas-liquid chromatography, the mobile phase is a gas, whereas in liquid-liquid chromatography it is a liquid.

Chapter 28

28-4 In liquid-liquid chromatography, the stationary phase is a liquid which is immobilized by adsorption or chemical bonding to a solid surface. The equilibria that cause separation are distribution equilibria between two immiscible liquid phases. In liquid-solid chromatography, the stationary phase is a solid surface and the equilibria involved are adsorption equilibria.

28-5 The number of plates in a column can be determined by measuring the retention time t_R and width of a peak at its base W. The number of plates N is then given by the equation $N = 16 (t_R/W)^2$.

28-6 Two general methods for improving the resolution of a column are to increase the column length and to reduce the plate height.

28-7 $N = 16 (t_R/W)^2$ (Equation 28 – 14)

(a)

		N	N^2
A	$16 (5.4/0.41)^2 =$	2775	7.7033×10^6
B	$16 (13.3/1.07)^2 =$	2472	6.1110×10^6
C	$16 (14.1/1.16)^2 =$	2364	5.5884×10^6
D	$16 (21.6/1.72)^2 =$	2523	6.3671×10^6
		$\Sigma N = 10135$	$\Sigma N^2 = 25.7654 \times 10^6$

(b) $\overline{N} = 10135/4 = 2534 = \underline{2.5 \times 10^3}$

$s = \sqrt{\dfrac{25.7654 \times 10^6 - (10135)^2/4}{4-1}} = 169 = 0.2 \times 10^3$ (Equation 3 – 4)

$\overline{N} = \underline{2.5 (\pm 0.2) \times 10^3}$

(c) $H = 24.7 \text{ cm}/2534 \text{ plates} = 9.747 \times 10^{-3} = \underline{0.0097 \text{ cm}}$

28-8 (a) $k' = (t_R - t_M)/t_M$ (Equation 28 – 8)

A $k'_A = (5.4 - 3.1)/3.1 = 0.742 = \underline{0.74}$

Chapter 28

B k'_B = (13.3 − 3.1)/3.1 = 3.29 = __3.3__

C k'_C = (14.1 − 3.1)/3.1 = 3.55 = __3.5__

D k'_D = (21.6 − 3.1)/3.1 = 5.97 = __6.0__

(b) Rearranging Equation 28-5 yields

$$K = k'V_M/V_S$$

Substituting the values for k' from part (a) and the numerical data for V_M and V_S gives

$$K = [(t_R - t_M)/t_M]\,1.37/0.164 = k' \times 8.35$$

and

$$K_A = 0.742 \times 8.35 = \underline{\underline{6.2}}$$

$$K_B = 3.29 \times 8.35 = \underline{\underline{27}}$$

$$K_C = 3.55 \times 8.35 = \underline{\underline{30}}$$

$$K_D = 5.97 \times 8.35 = \underline{\underline{50}}$$

28-9 $R_S = 2[(t_R)_C - (t_R)_B]/(W_B + W_C)$ (Equation 28 – 18)

(a) $R_S = 2(14.1 - 13.3)/(1.07 + 1.16) = 0.717 = \underline{\underline{0.72}}$

(b)
$$\alpha_{C,B} = \frac{(t_R)_C - t_M}{(t_R)_B - t_M} = \frac{14.1 - 3.1}{13.3 - 3.1} = 1.08 = \underline{\underline{1.1}} \quad (\text{Equation } 28-11)$$

(c) Proceeding as in Example 28-1(d) in Section 28D-2, we write

$$\frac{(R_S)_1}{(R_S)_2} = \frac{\sqrt{N_1}}{\sqrt{N_2}} = \frac{0.717}{1.5} = \frac{\sqrt{2534}}{\sqrt{N_2}}$$

$$N_2 = 2534 \times (1.5)^2/(0.717)^2 = 11090 = 1.11 \times 10^4$$

From Solution 28-12(c), $H = 9.75 \times 10^{-3}$ cm/plate

349

Chapter 28

$$L = 11090 \times 9.75 \times 10^{-3} = \underline{\underline{108 \text{ cm}}}$$

(d) Proceeding as in Example 28-1(e),

$$\frac{(t_R)_1}{(t_R)_2} = \frac{(R_S)_1^2}{(R_S)_2^2} = \frac{14.1}{(t_R)_2} = \frac{(0.717)^2}{(1.5)^2}$$

$$(t_R)_2 = 14.1(1.5)^2/(0.717)^2 = 61.7 = \underline{\underline{62 \text{ min}}}$$

28-10 (a) From Equation 28-18

$$R_S = 2[(t_R)_D - (t_R)_C]/(W_D + W_C)$$

$$= 2(21.6 - 14.1)/(1.72 + 1.16) = 5.21 = \underline{\underline{5.2}}$$

(b) Proceeding as in part (d) of Example 28-1 we write

$$N_1 = N_2 \frac{(R_S)_1^2}{(R_S)_2^2} = 2534 \times \frac{(1.5)^2}{(5.21)^2} = 210 \text{ plates}$$

$$L = 210 \text{ plates} \times 9.75 \times 10^{-3} \text{ cm/plate} = \underline{\underline{2.0 \text{ cm}}}$$

28-11 $N = 16(t_R/W)^2$ Equation 28-14, text

Letting methylcyclohexane be A, methylcyclohexene be B, and toluene be C,

	N_i	N_i^2
$N_A = 16(10.0/0.76)^2 =$	2770	7.67336×10^6
$N_B = 16(10.9/0.82)^2 =$	2827	7.99265×10^6
$N_C = 16(13.4/1.06)^2 =$	2557	6.53786×10^6
$\Sigma =$	8154	$\Sigma x_i^2 = 22.20387 \times 10^6$

(a) $\overline{N} = 8154/3 = 2718 = \underline{\underline{2.7 \times 10^3 \text{ plates}}}$

(b) Employing Equation 3-4, we obtain

$$s = \sqrt{\frac{22.20387 \times 10^6 - (8154)^2/3}{3-1}} = \underline{\underline{144}} = 0.14 \times 10^3 \text{ plates}$$

$$\overline{N} = 2.7 \times 10^3 \, (\pm 0.1 \times 10^3) \text{ plates}$$

(c) $H = 40/2718 = 1.47 \times 10^{-2} = 0.015 \text{ cm/plate}$

28-12 (a) Substituting into Equation 28-18

$$R_S = 2(10.9 - 10.0)/(0.76 + 0.82) = 1.14 = \underline{\underline{1.1}}$$

(b) $R_S = 2(13.4 - 10.9)/(0.82 + 1.06) = 2.66 = \underline{\underline{2.7}}$

(c) $R_S = 2(13.4 - 10.0)/(1.06 + 0.76) = 3.74 = \underline{\underline{3.7}}$

28-13 (a) Proceeding as in part (d) of Example 28-1 we write

$$\frac{N_1}{N_2} = \frac{(R_S)_1^2}{(R_S)_2^2} = \frac{2718}{N_2} = \frac{(1.14)^2}{(1.5)^2}$$

$$N_2 = 4706 = \underline{\underline{4.7 \times 10^3 \text{ plates}}}$$

(b) From Solution 28-16(c), $H = 1.47 \times 10^{-2}$ cm/plate

$$L = 4706 \times 1.47 \times 10^{-2} = 69.1 \text{ cm} = \underline{\underline{69 \text{ cm}}}$$

(c) Proceeding as in part (e) of Example 28-1, we write

$$\frac{(t_R)_1}{(t_R)_2} = \frac{(R_S)_1^2}{(R_S)_2^2} = \frac{(1.14)^2}{(1.5)^2} = \frac{10.9}{(t_R)_2}$$

$$(t_R)_2 = 18.9 = \underline{\underline{19 \text{ min}}}$$

28-14 (a) $k_1' = (10.0 - 1.9)/1.9 = 4.26 = \underline{\underline{4.3}}$ (Equation 28 – 8)

$k_2' = (10.9 - 1.9)/1.9 = 4.74 = \underline{\underline{4.7}}$

$k_3' = (13.4 - 1.9)/1.9 = 6.05 = \underline{\underline{6.1}}$

Chapter 28

(b) Rearranging Equation 28-5 and substituting numerical values for V_M and V_S gives

$$K_1 = 4.26 \times 62.6/19.6 = 13.60 = \underline{\underline{14}}$$

$$K_2 = 4.74 \times 62.6/19.6 = 15.1 = \underline{\underline{15}}$$

$$K_3 = 6.05 \times 62.6/19.6 = 19.3 = \underline{\underline{19}}$$

(c) $\alpha_{2,1} = (10.9 - 1.9)/(10.0 - 1.9) = \underline{\underline{1.11}}$ (Equation 28 – 11)

28-15 (a) $k'_M = K_S V_S / V_M = 6.01 \times 0.422 = \underline{\underline{2.54}}$ (Equation 28 – 5)

$k'_N = 6.20 \times 0.422 = \underline{\underline{2.62}}$

(b) $\alpha = 6.20/6.01 = \underline{\underline{1.03}}$

(c) Substituting into Equation 28-20 gives

$$N = 16(1.5)^2 \left(\frac{1.03}{1.03 - 1.00}\right)^2 \left(\frac{1.00 + 2.62}{2.62}\right)^2 = \underline{\underline{8.1 \times 10^4 \text{ plates}}}$$

(d) $L = 8.1 \times 10^4 \times 2.2 \times 10^{-3} = 178 = \underline{\underline{1.8 \times 10^2 \text{ cm}}}$

(e) Substituting into Equation 28-21

$$(t_R)_N = \frac{16(1.5)^2 \times 2.2 \times 10^{-3}}{7.10} \left(\frac{1.03}{1.03 - 1.00}\right)^2 \frac{(3.62)^3}{(2.62)^2} = \underline{\underline{91 \text{ min}}}$$

28-16 (a) $k'_M = 5.81 \times 0.422 = \underline{\underline{2.45}}$ $k'_N = 6.20 \times 0.422 = \underline{\underline{2.62}}$

(b) $\alpha = 6.20/5.81 = 1.07$

(c) $$N = 16(1.5)^2 \left(\frac{1.07}{1.07 - 1.00}\right)^2 \left(\frac{1.00 + 2.62}{2.62}\right)^2 = \underline{\underline{1.6 \times 10^4 \text{ plates}}}$$

(d) $L = 1.6 \times 10^4 \times 2.2 \times 10^{-3} = \underline{\underline{35 \text{ cm}}}$

(e) $$(t_R)_N = \frac{16(1.5)^2 \times 2.2 \times 10^{-3}}{7.10} \left(\frac{1.07}{1.07 - 1.00}\right)^2 \frac{(3.62)^3}{(2.62)^2} = \underline{\underline{18 \text{ min}}}$$

CHAPTER 29

29-1 In *gas-liquid chromatography*, the stationary phase is a liquid that is immobilized on a solid. Retention of sample constituents involves equilibria between a gaseous and a liquid phase. In gas-solid chromatography, the stationary phase is a solid surface that retains analytes by physical adsorption. Here separations involve adsorption equilibria.

29-2 Gas-solid chromatography is used primarily for separating low-molecular-weight gaseous species, such as carbon dioxide, carbon monoxide, and oxides of nitrogen.

29-3 Gas-solid chromatography has limited application because active or polar compounds are retained more or less permanently on the packings. In addition severe tailing is often observed owing to the nonlinear character of the physical adsorption process.

29-4 In a soap bubble meter, a soap film is formed in a gas buret through which the effluent from a gas-chromatographic column is flowing. The flow rate is then determined from the time required for the film to travel between two of the graduations in the buret.

29-5 *Temperature programming* involves increasing the temperature of a gas-chromatographic column as a function of time. This technique is particularly useful for samples that contain constituents whose boiling points differ significantly. Low boiling point constituents are separated initially at temperatures that provide good resolution. As the separation proceeds, the column temperature is increased so that the higher boiling constituents come off the column with good resolution and at reasonable lengths of time.

29-6 (a) The thermal conductivity detector is based upon the decrease in thermal conductivity of the helium or hydrogen carrier gas brought about by the presence of analyte molecules.

(b) The flame ionization detector is based upon the electrical conductivity of the gas produced when the column effluent is combusted in a hydrogen/oxygen flame.

(c) The thermionic detector is based upon the ion currents produced when the mobile phase is combusted in a hydrogen flame and then passed over a heated rubidium silicate bead. It is used primarily for detecting analytes that contain phosphorus or nitrogen.

Chapter 29

(d) The flame photometric detector is based upon the radiation emitted by sulfur and phosphorus containing molecules when the eluent is passed into a low-temperature hydrogen/air flame.

(e) The photoionization detector is based upon the ion currents that develop when analyte molecules are irradiated with an intense beam of far-ultraviolet radiation.

29-7 (a) Advantages, thermal conductivity: general applicability, large linear range simplicity, nondestructive. Disadvantage: low sensitivity.

(b) Advantages, flame ionization: high sensitivity, large linear response range, low noise, rugged, ease of use, and response that is largely independent of flow rate. Disadvantage: destructive of sample.

(c) Advantage, thermionic: high sensitivity for compounds containing nitrogen and phosphorus, good linear range. Disadvantages: destructive, not applicable for many types of analytes.

(d) Advantages, flame photometric: selective toward sulfur and phosphorus containing analytes, good sensitivity. Disadvantages: destructive, limited applicability.

(e) Advantages, photoionization: versatility, nondestructive, large linear range.

29-8 A hyphenated gas chromatographic method is a method in which the analytes exiting from a column are identified by one of the selective techniques such as mass spectrometry, absorption or emission spectroscopy, or voltammetry.

29-9 The typical column packing is made up of diatomaceous earth particles having diameters ranging from 250 to 170 μm or 170 to 149 μm.

29-10 (a) A PLOT column is a porous layer open tubular column, which is also called a support coated open-tubular (SCOT) column. The inner surface of a PLOT column is lined with a thin film of a support material, such as a diatomaceous earth. This type of column holds several times as much stationary phase as does a wall-coated column.

(b) A WCOT column is simply a capillary tubing fashioned from fused silica stainless steel, aluminum, copper, plastic, or glass. Its inner walls are coated with a thin layer of the mobile phase.

(c) The SCOT column is described in the answer to part (a) of this question.

29-11 Megapore columns are open-tubular columns that have a greater inside diameter typical open-tubular columns, which range in diameter from 150 to 320 μm. The typical megabore column has a diameter of 530 μm.

29-12 Fused silica columns have greater physical strength and flexibility than glass open-tubular columns and are less reactive toward analytes than either glass or metal columns.

29-13 Desirable properties of a stationary phase for GLC include: low volatility, thermal stability, chemical inertness, and solvent characteristics that provide suitable k' and α values for the various analytes to be separated.

29-14 Currently, liquid stationary phases are generally bonded and/or cross-linked in order to provide thermal stability and a more permanent stationary phase that will not leach off the column. Bonding involves attaching a monomolecular layer of the stationary phase to the packing surface by means of chemical bonds. Cross linking involves treating the stationary phase while it is in the column with a chemical reagent that creates cross links between the molecules making up the stationary phase.

29-15 Film thickness of stationary phases influences the rate at which analytes are carried through the column with the rate increasing as the thickness is decreased. Less band broadening is encountered with thin films.

29-16 (a) Band broadening arises from very high or very low flow rates, large particles making up packing, thick layers of stationary phase, low temperatures, and slow injection rates.

(b) Band separation is enhanced by maintaining conditions so that k' lies in the range of 1 to 10, using small particles for packing, limiting the amount of stationary phase so that particle coatings are thin, and injecting the sample rapidly.

Chapter 29

29-17   16.4/0.60 = 27.333
  45.2/0.78 = 57.949
  30.2/0.88 = 34.318
      Σ   = 119.600

% methyl acetate = (27.333/119.600) × 100% = 22.9%

% methyl propionate = (57.949/119.600) × 100% = 48.5%

% methyl *n*-butyrate = (34.318/119.600) × 100% = 28.7%

29-18   A 32.5/0.70 = 46.429
  B 20.7/0.72 = 28.750
  C 60.1/0.75 = 80.133
  D 30.2/0.73 = 41.370
  E 18.3/0.78 = 23.462
        Σ = 220.143

% A = (46.429/220.143) × 100% = 21.1%

% B = (28.750/220.143) × 100% = 13.1%

% C = (80.133/220.143) × 100% = 36.4%

% D = (41.370/220.143) × 100% = 18.8%

% E = (23.462/220.143) × 100% = 10.7%

CHAPTER 30

30-1 (a) Substances that are somewhat volatile and are thermally stable.

(b) Molecular species that are nonvolatile or thermally unstable.

(c) Substances that are ionic.

(d) Nonpolar, water insoluble compounds and isomeric mixtures.

(e) High-molecular-weight compounds that are soluble in nonpolar solvents.

(f) High-molecular-weight hydrophilic compounds.

(g) Nonvolatile or thermally unstable species that contain no chromophoric groups.

30-2 (a) In an *isocratic elution*, the solvent composition is held constant throughout the elution.

(b) In a *gradient elution*, two or more solvents are employed and the composition of the eluent is changed continuously or in steps as the separation proceeds.

(c) In a *stop-flow injection*, the flow of solvent is stopped, a fitting at the head of the column is removed, and the sample is injected directly onto the head of the column. The fitting is then replaced and pumping is resumed.

(d) A *reversed-phase packing* is a nonpolar packing that is used in partition chromatography with a relatively polar mobile phase.

(e) In a normal-phase packing, the stationary phase is quite polar and the mobile phase is relatively nonpolar.

(f) In *ion chromatography*, the stationary phase is an ion-exchange resin, and detection is ordinarily accomplished by a conductivity detector.

(g) An *eluent-suppressor column* is located after the ion-exchange column in ion chromatography. It converts the ionized species used to elute analyte ions to largely undissociated molecules that do not interfere with conductometric detection.

(h) *Gel filtration* is a type of size-exclusion chromatography in which the packings are hydrophilic, and eluents are aqueous. It is used for separating high-molecular-weight polar compounds.

Chapter 30

(i) *Gel permeation chromatography* is a type of size-exclusion chromatography in which the packings are hydrophobic, and the eluents are nonaqueous. It is used for separating high-molecular-weight nonpolar species.

(j) The *critical temperature* is the temperature above which a gas cannot be condensed to a liquid no matter how high the pressure.

(k) *FSOT columns* are fused silica open-tubular columns used in gas chromatography.

(l) In *two-dimensional planar chromatography*, development is carried with two solvents that are applied successively at right angles to one another.

(m) A *supercritical fluid* is a substance that is maintained above its critical temperature so that it cannot be condensed into a liquid no matter how great the pressure.

30-3 In *gas chromatography*, the gaseous mobile phase serves merely to carry the sample through the column and does not interact in any way with the components of the sample. In contrast, in liquid chromatography, interactions between the liquid mobile phase and the sample components contribute to the separation process. Interactions between the mobile phase and sample components also occur to some extent when the mobile phase is a critical fluid, and separation efficiency is somewhat dependent on the choice of mobile phase.

Because of their low viscosity, diffusion rates of analytes are greater in gas chromatography than in liquids. More rapid separation but greater band broadening are the consequence. Also columns can be much longer thus giving many more plates per column. Critical fluid viscosities are intermediate between those of a gas and a liquid; consequently the behavior of solutes in this type of solvent are intermediate.

30-4 (a) *n*-hexane, benzene, *n*-hexanol.

(b) diethyl ether, nitrobutane, ethyl acetate.

30-5 (a) *n*-hexanol, benzene, *n*-hexane.

(b) ethyl acetate, nitrobutane, diethyl ether.

30-6 In *open-tubular columns*, the stationary phase is held on the inner surface of a capillary tubing, whereas in packed columns, the stationary phase is supported on particles that are contained in a glass or metal tube. Open-tubular columns, which are only applicable in gas and supercritical fluid chromatography, contain an enormous number of plates that permit rapid separations of closely related species. Their disadvantage is their small sample capacities.

30-7 In *adsorption chromatography*, the sample components are selectively retained on the surface of a solid stationary phase by adsorption. In partition chromatography, selective retention occurs in a liquid or liquid-like stationary phase.

30-8 In *size-exclusion chromatography*, separations are based upon the size, and to some extent the shape, of molecules with little interaction between the stationary phase and the sample components occurring. In ion-exchange chromatography, in contrast, separations are based upon ion-exchange reactions between the stationary phase and the components of the sample in the mobile phase.

30-9 Nonvolatile and thermally unstable compounds.

30-10 Supercritical fluid chromatography is particularly applicable to nonvolatile or thermally unstable species that are difficult to detect because they have no chromophoric groups.

30-11 The simplest type of pump for liquid chromatography is a *pneumatic pump*, which consists of a collapsible solvent container housed in a vessel that can be pressurized by a compressed gas. This type of pump is simple, inexpensive, and pulse-free. It has limited capacity and pressure output, it is not adaptable to gradient elution, and its pumping rate depends upon the viscosity of the solvent.

A *screw-driven syringe pump* consists of a large syringe in which the piston is moved in or out by means of a motor-driven screw. It also is pulse-free and the rate of delivery is easily varied. It suffers from lack of capacity and is inconvenient to use when solvents must be changed.

The most versatile and widely used pump is the *reciprocating pump* that usually consists of a small cylindrical chamber that is filled and then emptied by the back-and-forth motion of a piston. Advantages of the reciprocating pump include small internal volume, high output pressures, adaptability to gradient elution, and flow rates that are constant and independent of viscosity and back pressure. The main disadvantage is pulsed output, which must be damped.

Chapter 30

30-12 In *suppressor-based ion chromatography* the chromatographic column is followed by a suppressor column whose purpose is to convert the ions used for elution to molecular species that are largely nonionic and thus do not interfere with conductometric detection of the analyte species. In *single-column ion chromatography*, low capacity ion exchangers are used so that the concentration ions in the eluting solution can be kept low. Detection then is based on the small differences in conductivity caused by the presence of eluted sample components.

30-13 (a) The advantages over HPLC are (1) that it can be used with general detectors, such as the flame ionization detector, and (2) that it is inherently faster or gives better resolution for the same speed.

(b) The advantage over GLC is that it can be used for nonvolatile and thermally unstable samples.

30-14 The difference in the degree of band broadening in the three types of chromatography is largely due to difference in viscosity of the stationary phases, which determines the rate of diffusion of components and thus the extent of longitudinal diffusion. Longitudinal diffusion is greatest in gas chromatography because of the low viscosity of gases; consequently zone broadening due to this phenomenon is a maximum in this medium. It is minimal in liquid solvents because the viscosity of liquids is roughly two orders of magnitude greater than that of gases. Supercritical fluids are intermediate in viscosity as is broadening in this medium.

CHAPTER 32

32-1 The steps in sampling are: (1) identification of the population from which the sample is to be drawn, (2) collection of a gross sample, and (3) reduction of the gross sample to a small quantity of homogeneous material for analysis.

32-2 The object of sampling is to produce a homogeneous laboratory sample of a few hundred grams or less having a composition that is identical to the average composition of the bulk of the material being sampled.

32-3 (a) *Sorbed water* is that held as a condensed liquid phase in the capillaries of a colloid. *Adsorbed water* is that retained by adsorption on the surface of a finely ground solid. *Occluded water* is that held in cavities distributed irregularly throughout a crystalline solid.

(b) *Water of crystallization* is the water contained in a stable solid hydrate, such as $BaCl_2 \cdot 2H_2O$. *Water of constitution* is water that is released from certain compounds when they are heated or otherwise dehydrated. For example

$$Ba(OH)_2(s) \rightleftarrows BaO(s) + H_2O$$

(c) *Essential water* is chemically bound water that occurs as an integral part of the molecular or crystalline structure of a compound in its solid state. *Nonessential water* is that retained by a solid as a consequence of physical forces.

(d) A *gross sample* is a replica of the entire mass of material to be analyzed. It corresponds to the entire sample in both average chemical composition but also in particle size distribution. The *laboratory sample* is a homogeneous material whose chemical composition mimics that of the gross sample and, therefore, the population.

32-4 The weight of a gross sample depends upon how heterogeneous it is, the particle size at which heterogeneity begins, and the uncertainty in the composition that can be tolerated.

32-5 $s_o^2 = s_s^2 + s_m^2$ (Equation 32 – 1)

$\sum x_i = 50.38 + 50.20 + 50.31 + 50.22 + 50.41 = 251.52$

$\sum x_i^2 = 2538.1444 + 2520.04 + 2531.0961 + 2522.0484 + 2541.1681 = 12652.497$

Chapter 32

$$s_m^2 = \frac{12652.497 - (251.52)^2/5}{5-1} = 0.00873$$

$$\Sigma x_i' = 49.53 + 50.12 + 49.60 + 49.87 + 50.49 = 249.61$$

$$\Sigma (x_i')^2 = 2453.2209 + 2512.0144 + 2460.16 + 2487.0169 + 2549.2401$$

$$= 12461.6523$$

$$s_o^2 = \frac{12461.6523 - (249.6)^2/5}{5-1} = 0.15547$$

$$s_s^2 = 0.15547 - 0.00873 = 0.147$$

$$s_s = \sqrt{0.147} = 0.38$$

$$\overline{x}_i' = 249.61/5 = 49.92$$

$$(s_s/\overline{x}_i') \times 100\% = (0.38/49.92) \times 100\% = \underline{\underline{0.76\%}}$$

32-6 (a) $n = (1-p)/\sigma_r^2 p$ (Equation 32–2)

$$\sigma_r = \sqrt{(1-p)/np} \quad \text{where} \quad p = 14/250 = 0.0560$$

$$= \sqrt{(1-0.056)/(250 \times 0.056)} = 0.260 \quad \text{or} \quad \underline{\underline{26\%}}$$

$$\sigma_{abs} = 14 \times 0.26 = 3.6 = 4 \text{ tablets}$$

(b) $90\% = \text{CI} = 14 \pm z\sigma/\sqrt{N} = 14 \pm 1.64 \times 3.6/\sqrt{1} = \underline{\underline{14 \pm 6}}$

(where $z = 1.64$ was obtained from Table 4-1)

(c) $n = (1 - 0.056)/[(0.10)^2 \times 0.056] = 1686 = \underline{\underline{1.69 \times 10^3}}$

32-7 (a) $n = \dfrac{(1-p)}{p\,\sigma_r^2} = \dfrac{(1-0.02)}{0.02(0.25)^2} = \dfrac{49.0}{(0.25)^2} = \underline{\underline{784}}$

(b) $n = 49.0/(0.1)^2 = \underline{\underline{4.9 \times 10^3}}$

Chapter 32

(c) $n = 49.0/(0.05)^2 = \underline{\underline{2.0 \times 10^4}}$

(d) $n = 49.0/(0.01)^2 = \underline{\underline{4.9 \times 10^5}}$

32-8 (a) $250 = \dfrac{(1 - 52/250)}{(52/250) \times \sigma_r^2}$

$\sigma_r = 0.12 = 12\%$

(b) Here, the absolute standard deviation of the estimate is sought.

$\sigma_{abs} = 750 \times 12 (52/250) 0.12 = 224.6 = \underline{\underline{2.2 \times 10^2}}$ broken bottles

(c) 90% CI $= 750 \times 12 \times 52/250 \pm zs/\sqrt{N} = 1872 \pm 1.64 \times 224.6/\sqrt{1}$

$= 1872 \pm 368 = \underline{\underline{1.9 (\pm 0.4) \times 10^2}}$ broken bottles

(d) $n = \dfrac{(1 - 0.21)}{0.21(0.05)^2} = \underline{\underline{1.5 \times 10^3}}$ bottles

32-9 Here, we employ Equation 32-3

$$n = p(1-p)\left(\dfrac{d_1 d_2}{d^2}\right)^2 \left(\dfrac{P_1 - P}{\sigma_r P}\right)^2$$

(a) $d = 7.3 \times 0.15 + 2.6 \times 0.85 = 3.3$

$P = 0.15 \times 7.3 \times 0.87 \times 100/3.3 = 29\%$

$n = 0.15(1 - 0.15)\left(\dfrac{7.3 \times 2.6}{(3.3)^2}\right)^2 \left(\dfrac{87 - 0}{0.010 \times 29}\right)^2 = \underline{\underline{3.5 \times 10^4}}$ particles

(b) mass $= (4/3)\pi(0.20 \text{ cm})^3 \times 3.3 \text{ (g/cm}^3) \times 3.5 \times 10^4$

$= \underline{\underline{3.9 \times 10^3 \text{ g}}}$ or $\underline{\underline{8.5 \text{ lb}}}$

(c) $0.600 = (4/3)\pi(r)^3 \times 3.3 \times 3.5 \times 10^4$

$r = 1.1 \times 10^{-2}$ cm and diameter $= \underline{\underline{0.22 \text{ mm}}}$

Chapter 32

32-10 Proceeding as in Solution 32-9

(a) $d = 0.05 \times 6 + 0.95 \times 4 = 4.1 \text{ g/cm}^3$

$P = (0.05 \times 6 \times 0.80 + 0.85 \times 4 \times 0.03) \times 100 / 4.1 = \underline{\underline{8.3\% \text{ Cu}}}$

$n = 0.05(1 - 0.05)\left(\dfrac{6 \times 4}{4.1 \times 4.1}\right)^2 \left(\dfrac{80 - 3}{0.04 \times 8.3}\right)^2 = \underline{\underline{5.2 \times 10^3 \text{ particles}}}$

(b) mass $= (4/3)\pi(0.25)^3 \times 4.1 \times 5.2 \times 10^3 = \underline{\underline{1.4 \times 10^4 \text{ g} \quad \text{or} \quad 31 \text{ lb}}}$

(c) $0.50 = (4/2)\pi(r)^3 \times 4.1 \times 5.2 \times 10^3$

$r = 0.018$ and diameter $= \underline{\underline{0.36 \text{ cm}}}$

32-11 Proceeding as in Solutions 32-9 and 32-10

(a) $d = 0.02 \times 4.5 + 0.98 \times 3.0 = 30 \text{ g/cm}^3$

$P = (0.02 \times 4.5 \times 0.72 + 0.98 \times 3.0 \times 0.01) \times 100 / 3.0 = \underline{\underline{3.1\%}}$

$n = 0.02(1 - 0.02)\left(\dfrac{4.5 \times 3.0}{3.0 \times 3.0}\right)^2 \left(\dfrac{71 - 2}{0.01 \times 3.1}\right)^2 = \underline{\underline{2.2 \times 10^5 \text{ particles}}}$

(b) mass $= (4/3)\pi(0.10)^3 \times 3.0 \times 2.2 \times 10^5 = \underline{\underline{2.8 \times 10^3 \text{ g} \quad \text{or} \quad 6.1 \text{ lb}}}$

(c) $0.750 = (4/3)\pi(0.10)^3 \times 3.0 \times 2.2 \times 10^5$

$r = 6.4 \times 10^{-3} \text{ cm}$ and diameter $= \underline{\underline{0.13 \text{ cm}}}$

32-12 The required sample size can be calculated as in the three previous solutions.

$d = 7.3 \times 0.01 + 2.6 \times 0.99 = 2.6 \text{ g/cm}^3$

$P = 0.01 \times 7.3 \times 0.87 \times 100 / 2.6 = 2.4\%$

$n = 0.01(1 - 0.01)\left(\dfrac{7.3 \times 2.6}{2.6 \times 2.6}\right)^2 \left(\dfrac{87 - 0}{0.05 \times 2.4}\right)^2 = \underline{\underline{4.1 \times 10^4 \text{ particles}}}$

mass $= (4/3)\pi(0.25)^3 \times 2.6 \times 4.1 \times 10^4 / 454 = \underline{\underline{15 \text{ lb}; \quad \text{insufficient sample}}}$

CHAPTER 33

33-1 *Dry ashing* is carried out by igniting the sample in air or sometimes oxygen. *Wet ashing* is carried out by heating the sample in an aqueous medium containing such oxidizing agents as H_2SO_4, $HClO_4$, HNO_3, H_2O_2, or some combination of these.

33-2 A *flux* is a solid salt, usually containing an alkali metal cation, that liquifies upon heating to a red heat. It is used to decompose refractory materials and render them soluble in aqueous solution.

33-3 B_2O_3 or $CaCO_3/NH_4Cl$.

33-4 Refractory metal oxides are generally decomposed by potassium pyrosulfate, $K_2S_2O_7$.

33-5 When hot concentrated $HClO_4$ comes in contact with organic materials or other oxidizable species, explosions are highly probable.

33-6 (a) Samples for halogen determination may be decomposed in a Schöniger combustion flask, combusted in a tube furnace in a stream of oxygen, or fused in a peroxide bomb.

(b) Samples for the determination of sulfur are treated in the same way as described in part (a).

(c) Samples for nitrogen determination are decomposed in hot concentrated H_2SO_4 in a Kjeldahl flask or oxidized by CuO in the tube furnace in the Dumas method.

(d) Samples for heavy metal determination may be dry ashed by ignition in a hot flame or wet ashed with an oxidizing solution such as nitric acid, sulfuric acid, perchloric acid, or some combination thereof.

CHAPTER 34

34-1 A *masking agent* is a complexing agent that reacts selectively with one or more components of a solution to prevent them from interfering in an analysis.

34-2 (a) In an *exhaustive extraction*, the sample, in solution or as a solid, is extracted repeatedly with fresh portions of a solvent. In a *countercurrent extraction* the sample is distributed repeatedly between fresh portions of both immiscible liquids.

(b) For the equilibrium

$$x A_y (aq) \rightleftarrows y A_x (org)$$

the *distribution coefficient* K_d is given by the equation

$$K_d = [A_x(org)]^y / [A_y(aq)]^x$$

where the bracketed terms are *species* molar concentrations. The *distribution ratio D* is given by

$$D = c_A^y(org) / c_A^x(aq)$$

where $c_{A(org)}$ and $c_{A(aq)}$ are total or *analytical* concentrations of the various A-containing species.

34-3 $K_d = c_{org}/c_{aq} = 9.6 = D$

Applying Equation 34-9 gives

(a)
$$(c_{aq})_n = \left(\frac{V_{aq}}{V_{org} D + V_{aq}} \right)^n (c_{aq})_0$$

$$(c_{aq})_1 = \left(\frac{50.0}{40.0 \times 9.6 + 50.0} \right)^1 0.150 = \underline{\underline{1.73 \times 10^{-2} \text{ M}}}$$

(b)
$$(c_{aq})_2 = \left(\frac{50.0}{20.0 \times 9.6 + 50.0} \right)^2 0.150 = \underline{\underline{6.40 \times 10^{-3} \text{ M}}}$$

(c)
$$(c_{aq})_4 = \left(\frac{50.0}{10.0 \times 9.6 + 50.0} \right)^4 0.150 = \underline{\underline{2.06 \times 10^{-3} \text{ M}}}$$

(d) $$(c_{aq})_8 = \left(\frac{50.0}{5.0 \times 9.6 + 50.0}\right)^8 0.150 = \underline{\underline{6.89 \times 10^{-4} \text{ M}}}$$

34-4 Proceeding as in Solution 34-3, we obtain

(a) $$(c_{aq})_1 = \left(\frac{25.0}{25.0 \times 6.25 + 25.0}\right)^1 0.0600 = 8.28 \times 10^{-3}$$

$$\%Z = 8.28 \times 10^{-3} \times 100\% / 0.0600 = \underline{\underline{13.8\%}}$$

Proceeding in the same way

(b) 5.88% (c) 1.73% (d) 0.78%

34-5 (a) Substituting into Equation 34-9

$$1.00 \times 10^{-4} = \left(\frac{25.0}{25.0 \times 9.6 + 25.0}\right)^n 0.0500 = (0.0943)^n \, 0.0500$$

$$(0.0943)^n = 1.00 \times 10^{-4} / 0.0500 = 2.00 \times 10^{-3}$$

Taking the log of both sides of this equation gives

$$n \log 0.0943 = \log 2.00 \times 10^{-3}$$

$$n = \frac{\log 2.00 \times 10^{-3}}{\log 0.0943} = 2.63 = 3$$

total volume = 3 × 25.0 = $\underline{\underline{75 \text{ mL}}}$

(b) $$1.00 \times 10^{-4} = \left(\frac{25.0}{10.0 \times 9.6 + 25.0}\right)^n 0.0500 = (0.2066)^n \, 0.0500$$

$$n \log 0.2066 = \log 2.00 \times 10^{-3}$$

n = 3.94 = 4 extractions

vol = 4 × 10.0 = $\underline{\underline{40 \text{ mL}}}$

(c) n = 10.9 = 11 and vol = $\underline{\underline{22 \text{ mL}}}$

34-6 Proceeding as in Solution 34-5, we obtain

(a) n = 3.49 = 4 and vol = $\underline{\underline{200 \text{ mL}}}$

Chapter 34

  (b) $n = 4.78 = 5$ and vol = <u>125 mL</u>

  (c) $n = 8.08 = 9$ and vol = <u>90 mL</u>

34-7 (a) $(c_{aq})_2 = \left(\dfrac{50.0}{25.0D + 50.0}\right)^2 (c_{aq})_0$

$(c_{aq})_2/(c_{aq})_0 = 1.000 - 0.990 = 0.0100$

$0.0100 = \left(\dfrac{50.0}{25.0D + 50.0}\right)^2$

$\sqrt{0.0100} = 0.100 = 50.0/(25.0D + 50.0)$

$D = K_d = (50.0 - 50.0 \times 0.100)/(25.0 \times 0.100) = \underline{\underline{18.0}}$

 (b) In the same way, $K_d = \underline{\underline{7.56}}$

34-8 (a) $(c_{aq})_4 = 0.0500 \times 1.0 \times 10^{-6} = 5.00 \times 10^{-8}$

$5.0 \times 10^{-8} = \left(\dfrac{30.0}{10.0D + 30.0}\right)^4 \times 0.0500$

$(1.00 \times 10^{-6})^{1/4} = 30.0/(10.0D + 30.0) = 3.16 \times 10^{-2}$

$K_d = D = (30.0 - 3.16 \times 10^{-2} \times 30.0)/(0.316) = \underline{\underline{91.9}}$

 (b) In the same way, $K_d = \underline{\underline{27.0}}$

 (c) In the same way, $K_d = \underline{\underline{50.3}}$

34-9 (a) Assume that in the presence of HClO$_4$, HA is not dissociated to an appreciable extent, whereas in the presence of NaOH, dissociation is complete. From the data for Solution 1, we write

$K_d = [HA]_{org}/[HA]_{aq} = 0.0454/\left(0.150 \times \dfrac{50.00}{100.0} - 0.0454\right) = \underline{\underline{1.53}}$

 (b) $[HA]_{org} = 0.0225$

368

$[HA]_{aq}$ = 0.0225/1.53 = <u>0.0147</u>

$[HA]_{org} + [HA]_{aq} + [A^-]_{aq}$ = $50.0 \times 0.150/100.0$ = 0.0750

$0.0225 + 0.0147 + [A^-]_{aq}$ = 0.0750

$[A^-]_{aq}$ = <u>0.0378</u>

(c) $[H_3O^+]$ = $[A^-]_{aq}$ = 0.0378

K_a = $0.0378 \times 0.0378/0.147$ = <u>9.7×10^{-2}</u>

34-10 (a) K_d = $[I_2]_{CCl_4}/[I_2]_{aq}$ and $[I_2]_{aq}$ = 1.12×10^{-4}

$[I_2]_{CCl_4}$ = $\dfrac{(25.0 \times 0.0100 - 25.0 \times 1.12 \times 10^{-4}) \text{ mmol } I_2}{10.0 \text{ mL CCl}_4}$ = 0.0247

K_d = $0.0247/(1.12 \times 10^{-4})$ = <u>221</u>

(b) After the second extraction

(25.0×0.0100) mmol I_2 = $25.0\,[I_2]_{aq} + 25.0\,[I(SCN)_2^-]_{aq} + [I_2]_{CCl_4}$

$[I_2]_{aq}$ = $[I_2]_{CCl_4}/K_d$ = $1.02 \times 10^{-3}/221$ = 4.62×10^{-6}

Substituting into the previous equation gives

0.250 = $4.62 \times 10^{-6} \times 25.0 + [I(SCN)_2^-] + 10.0 \times 1.02 \times 10^{-3}$

$[I(SCN)_2^-]$ = $(0.250 - 1.16 \times 10^{-4} - 1.02 \times 10^{-2})/25.0$ = 9.59×10^{-3} = $[I^-]$

K_f = $\dfrac{[I(SCN)_2^-][I^-]}{[I_2][SCN^-]^2}$ = $\dfrac{9.59 \times 10^{-3} \times 9.59 \times 10^{-3}}{4.62 \times 10^{-6} \times (8.08 \times 10^{-2})^2}$ = <u>3.05×10^3</u>

$[SCN^-]$ = $0.100 - 2 \times 9.59 \times 10^{-3}$ = 8.08×10^{-2}

Chapter 34

34-11 1 mol Ba^{2+} ≡ 2 mol H^+ ≡ 2 mol NaOH ≡ 2 eq acid

(a)
$$18.1 \text{ mL NaOH} \times \frac{0.1006 \text{ mmol NaOH}}{\text{mL NaOH}} \times \frac{10^{-3} \text{ mol}}{\text{mmol}} \times \frac{1 \text{ eq acid}}{\text{mol NaOH}} =$$

$$1.821 \times 10^{-3} \text{ eq acid (as the Ba salt)}$$

$$\frac{0.393 \text{ g Ba salt}}{1.821 \times 10^{-3} \text{ eq acid}} = \frac{215.8 \text{ g Ba salt}}{\text{eq acid}}$$

To obtain the mass of acid per equivalent of acid we subtract the weight of 1/2 mol Ba and add the weight of 1 mol H. Thus,

$$\text{eq wt acid} = \frac{215.8 \text{ g Ba salt}}{\text{eq acid}} - \frac{137.3 \text{ g Ba}}{2 \text{ eq acid}} + \frac{1.008 \text{ g H}}{\text{eq acid}} = \underline{\underline{148 \text{ g/eq acid}}}$$

(b) The acid has two titratable protons. Thus its formula weight is twice its equivalent weight. That is,

$$\text{fw } H_2A = \frac{148 \text{ g acid}}{\text{eq acid}} \times \frac{2 \text{ eq acid}}{\text{fw acid}} = \underline{\underline{296}}$$

34-12 (a) HL(aq) ⇌ HL(org)

HL(aq) + H_2O ⇌ H_3O^+ + L^-(aq)

Cu^{2+}(aq) + 2L^-(aq) ⇌ CuL_2(aq)

CuL_2(aq) ⇌ CuL_2(org)

(b) This system is analogous to the one discussed in Section 34D-1 and Equation 34-18 therefore applies. That is,

$$D = \frac{c_{org}}{c_{aq}} = \frac{K_{ex} c_L^2}{[H_3O^+]_{aq}^2}$$

At pH = 5.65 or $[H_3O^+]_{aq}$ = 2.239 × 10^{-6}, it is given that

$$c_{aq} = c_{org}$$

370

Therefore,

$$D = \frac{c_{org}}{c_{aq}} = 1.00 = \frac{K_{ex}(0.0100)^2}{(2.239 \times 10^{-6})^2}$$

$$K_{ex} = 1.00 \, (2.239 \times 10^{-6})^2 / (0.0100)^2 = \underline{5.01 \times 10^{-8}}$$

(c) Applying Equation 34-9, where $(c_{aq})_0 = 5.00 \times 10^{-5}$ and

$$(c_{aq})_n = 5.00 \times 10^{-5} (1.000 - 0.990) = 5.00 \times 10^{-7}$$

$$5.00 \times 10^{-7} = \left(\frac{50.0}{25.0 \times 5.01 + 50.0}\right)^n 5.00 \times 10^{-5}$$

$$0.01000 = (0.2853)^n$$

$$n = \log 0.100 / \log 0.2853 = 3.7 \text{ or } \underline{4 \text{ extractions}}$$

(d) Proceeding as in part (d), $n = 5.51 = \underline{6 \text{ extractions}}$

34-13 (a) Here the predominant equilibrium is

$$2Ag^+(aq) + H_2Z(org) \rightleftarrows Ag_2Z(org) + 2H^+(aq)$$

Because two moles of Ag^+ are transferred to the organic phase per mole of ligand,

$$D = \frac{c_{org}}{c_{aq}^2} \approx \frac{[Ag_2A]}{[Ag^+]^2} \quad \text{(compare with Equation 34-14)}$$

$$c_{aq} = 5.41 \times 10^{-6} \text{ M}$$

$$c_{org} = \frac{[\text{original no. mmol } Ag^+(aq) - \text{final no. mmol } Ag^+(aq)]/2}{10.0 \text{ mL}}$$

$$= \frac{25.0 \, (2.24 \times 10^{-4} - 5.41 \times 10^{-6})/2}{10.0} = 2.732 \times 10^{-4}$$

$$D = c_{org}/c_{aq}^2 = 2.732 \times 10^{-4} / (5.41 \times 10^{-6})^2 = \underline{\underline{9.34 \times 10^6}}$$

Chapter 34

(b) Proceeding in the same way as the derivation in Section 34D-1 we obtain an equation analogous to Equation 34-18. That is,

$$D = \frac{c_{org}}{c_{aq}^2} = \frac{K_f K_{d2} K_1 K_2}{K_{d1}} \times \frac{c_L}{[H_3O^+]_{aq}^2} = \frac{K_{ex} c_L}{[H_3O^+]^2}$$

where K_1 and K_2 are the first and second acid dissociation constants of H_2Z in water.

$$[H_3O^+] = \text{antilog}(-4.30) = 5.012 \times 10^{-5}$$

$$K_{ex} = D[H_3O^+]^2/c_L = 9.34 \times 10^6 (5.012 \times 10^{-5})^2/0.025 = \underline{0.938}$$

(c) When 99.5% of Ag^+ is extracted, the number of millimoles of Ag^+ remaining a_n is given by

$$a_n = a_0 \times 0.005 = 2.24 \times 10^{-4} \times 0.005 = 1.12 \times 10^{-6} \text{ mmol } Ag^+$$

At equilibrium after one extraction of solution, a_1 mmol of Ag^+ remain in the aqueous solution and

$$c_{aq1} = a_1/V_{aq} \quad \text{and} \quad c_{org1} = (a_0 - a_1)/2V_{org}$$

Substituting into the expression for D

$$D = \frac{(a_0 - a_1)/2V_{org}}{a_1^2/V_{aq}^2} = \frac{(a_0 - a_1) V_{aq}^2}{a_1^2 \times 2V_{org}}$$

$$2a_1^2 D V_{org} = V_{aq}^2 a_0 - V_{aq}^2 a_1$$

$$a_1^2 + \frac{V_{aq}^2}{2D V_{org}} a_1 - \frac{V_{aq}^2 a_0}{2D V_{org}} = 0$$

$$a_1^2 + ka_1 - ka_0 = 0$$

Where

$$k = V^2/(2D V_{org}) = (25.0)^2/(2 \times 9.34 \times 10^6 \times 10) = 3.34 \times 10^{-6}$$

$$a_1^2 + 3.346 \times 10^{-6} a_1 - 3.346 \times 10^{-6} a_1 = 0$$

Chapter 34

In a similar way it can be shown for the second extraction

$$a_2^2 + 3.346 \times 10^{-6} a_2 - 3.346 \times 10^{-6} a_1 = 0$$

and so forth.

Before the first extraction

$$a_0 = 2.24 \times 10^{-4} \times 25.00 = 5.60 \times 10^{-3} \text{ mmol Ag}^+$$

$$a_1^2 + 3.346 \times 10^{-6} a_1 - 3.346 \times 10^{-6} \times 5.60 \times 10^{-3} = 0$$

$$a_1 = 1.352 \times 10^{-4}$$

$$a_2^2 + 3.346 \times 10^{-6} a_2 - 3.346 \times 10^{-6} \times 1.352 \times 10^{-4} = 0$$

$$a_2 = 1.966 \times 10^{-5}$$

$$a_3^2 + 3.346 \times 10^{-6} a_3 - 3.346 \times 10^{-6} \times 1.966 \times 10^{-5} = 0$$

$$a_3 = 6.609 \times 10^{-6}$$

After

4 extractions, $a_4 = 3.318 \times 10^{-6}$

5 extractions, $a_5 = 2.056 \times 10^{-6}$

6 extractions, $a_6 = 1.438 \times 10^{-6}$

7 extractions, $a_7 = 1.08 \times 10^{-6}$

Thus, <u>7 extractions</u> are required to reduce the number of millimoles of Ag$^+$ to less than 1.12×10^{-6}.

(d) At $[H_3O^+] = 1.00 \times 10^{-3}$

$$D = K_{ex} c_L / [H_3O^+]^2 = 0.938 \times 0.025 / 1.00 \times 10^{-6} = 2.345 \times 10^4$$

$$k = V_{aq}^2 / 2D V_{org} = (25.00)^2 / (2 \times 2.345 \times 10^4 \times 10) = 1.333 \times 10^{-3}$$

373

Chapter 34

$$a_1^2 + 1.333 \times 10^{-3} a_1 - 1.333 \times 10^{-3} \times 5.15 \times 10^{-4} = 0$$

When 99.5% of Ag^+ has been extracted

$$a_n = 5.15 \times 10^{-4} \times 0.005 = 2.58 \times 10^{-6}$$

Solving the series of quadratics

$$a_1 = 3.968 \times 10^{-4}$$

$$a_2 = 3.20 \times 10^{-4}$$

$$a_3 = 2.66 \times 10^{-4}$$

$$a_4 = 2.78 \times 10^{-4}$$

$$a_5 = 1.98 \times 10^{-4}$$

Obviously, a separation will require a very large number of extractios (~ 520) and therefore is not feasible at this pH.